Women and Science

Forthcoming titles in
ABC-CLIO's

Science and Society
Series

The Environment and Science, Christian C. Young

Exploration and Science, Michael S. Reidy, Gary Kroll, and Erik M. Conway

Imperialism and Science, George N. Vlahakis, Isabel Maria Malaquias,
Nathan M. Brooks, François Regourd, Feza Gunergun, and David Wright

Literature and Science, John H. Cartwright and Brian Baker

Race, Racism, and Science, John P. Jackson, Jr., and Nadine M. Weidman

Advisory Editors
Paul Lawrence Farber and Sally Gregory Kohlstedt

Women and Science

Social Impact and Interaction

Suzanne Le-May Sheffield

ABC-CLIO

Santa Barbara, California • Denver, Colorado • Oxford, England

12-04

Library of Congress Cataloging-in-Publication Data

Sheffield, Suzanne Le-May, 1967–
     Women and science : social impact and interaction / Suzanne Le-May Sheffield.
         p. cm. — (Science and society)
     Includes bibliographical references and index.
     ISBN 1-85109-460-1 (hardcover : alk. paper) ISBN 1-85109-465-2 (e-book)
1. Women in science—Social aspects. 2. Feminism and science. 3. Women scientists.
I. Title. II. Series: Science and society (Santa Barbara, Calif.)

     Q130.S44 2004
     305.43'5—dc22   2004015402

This book is also available on the World Wide Web as an e-book.
Visit abc-clio.com for details.

ABC-CLIO, Inc.
130 Cremona Drive, P.O. Box 1911
Santa Barbara, California 93116-1911

This book is printed on acid-free paper.
Manufactured in the United States of America

To Scott and Aidan for all your love and support

Contents

Series Editor's Preface

T he discipline of the history of science emerged from the natural sciences with the founding of the journal *Isis* by George Sarton in 1912. Two and a half decades later in a lecture at Harvard Sarton explained, "We shall not be able to understand our own science of to-day (I do not say to use it, but to understand it) if we do not succeed in penetrating its genesis and evolution." Historians of science, many of the first trained by Sarton and then by his students, study how science developed during the sixteenth and seventeenth centuries and how the evolution of the physical, biological, and social sciences over the past 350 years has been powerfully influenced by various social and intellectual contexts. Throughout the twentieth century the new field of the history of science grew with the establishment of dozens of new journals, graduate programs, and eventually the emergence of undergraduate majors in the history, philosophy, and sociology of science, technology, and medicine. Sarton's call to understand the origins and development of modern science has been answered by the development of not simply one discipline, but several.

Despite their successes in training scholars and professionalizing the field, historians of science have not been particularly successful in getting their work, especially their depictions of the interactions between science and society, into history textbooks. Pick up any U.S. history textbook and examine some of the topics that have been well explored by historians of science, such as scientific racism, the *Scopes* trial, nuclear weapons, eugenics, industrialization, or the relationship between science and technology. The depictions of these topics offered by the average history textbook have remained unchanged over the last fifty years, while the professional literature related to them that historians of science produce has made considerable revision to basic assumptions about each of these subjects.

The large and growing gap between what historians of science say about certain scientific and technological subjects and the portrayal of these subjects in most survey courses led us to organize the Science and Society series. Obviously, the rich body of literature that historians of science have amassed is not regularly consulted in the production of history texts or lectures. The authors and editors of this series seek to overcome this disparity by offering a synthetic, readable, and chronological history of the physical, social, and biological sciences as they developed within particular social, political, institutional, intellectual, and economic contexts over the past 350 years. Each volume stresses the reciprocal relationship between science and context; that is, while various circumstances and perspectives have influenced the evolution of the sciences, scientific disciplines have conversely influenced the contexts within which they developed. Volumes within this series each begin with a chronological narrative of the evolution of the natural and social sciences that focuses on the particular ways in which contexts influenced and were influenced by the development of scientific explanations and institutions. Spread throughout the narrative readers will encounter short biographies of significant and iconic individuals whose work demonstrates the ways in which the scientific enterprise has been pursued by men and women throughout the last three centuries. Each chapter includes a bibliographic essay that discusses significant primary documents and secondary literature, describes competing historical narratives, and explains the historiographical development in the field. Following the historical narratives, each book contains a glossary, chronology, and most importantly a bibliography of primary source materials to encourage readers to come into direct contact with the people, the problems, and the claims that demonstrate how science and society influence one another. Our hope is that students and instructors will use the series to introduce themselves to the large and growing field of the history of science and begin the work of integrating the history of science into history classrooms and literature.

—*Mark A. Largent*

Introduction: Marie Curie, an Icon for Women Scientists

What properer object can there be of womans emulation than the deeds of other famous women?
—Thomas Heywood, *The Generall History of Women* (1657)

One woman immediately comes to mind when someone asks, Were there any women scientists in history? Manya Sklodowska-Curie, otherwise known as Marie Curie (1867–1934), is held up as an icon, a representation of what it means to be a successful woman in science. She was praised and rewarded by contemporaries for her discovery of the elements radium and polonium and for her research work on radioactivity, work that contributed to further discoveries about the atom and leading to the development of a treatment for cancer. She is the subject of numerous biographies and children's books. Writers hold up Marie Curie to women scientists and to girls aspiring to be scientists as someone for them to emulate in a profession that appears at first glance to have very few female role models.

Was Curie an exceptional woman who "made it" in a man's world, an aberration impossible to imitate? Clearly, there have been fewer female scientists in history than there have been male scientists, and certainly Marie Curie is exceptional, not only among female scientists but among scientists in general. Curie was a two-time winner of the Nobel Prize for science for her important work on radioactivity, and her achievements would be difficult for any man or woman to emulate. Moreover, few scientists garner the popular fame and renown that Marie Curie achieved. But she was not the only woman in history to achieve success in science. There have been many such women; indeed, *The Biographical Dictionary of Women in Science* (2000) lists 2,500 outstanding women scientists of the western world from ancient times to the present, and thousands more have pursued scientific work.

Marie Curie (1867–1934), noted physical chemist, poses in her Paris laboratory. (Bettmann/CORBIS)

Gaining access to science, long a solely male province, has not been easy for women. The life and work of Marie Curie demonstrates what can be achieved by women scientists, and it is this triumphal story that is most often told. However, another story is interwoven with the tale of Curie's success: the story of her struggle to practice science when the odds were against her. Both stories are equally valid and equally important. Marie Curie does—and should—stand as an icon for women in science. She is an icon not simply because of her exceptional achievements but also for her sacrifices and struggles against barriers that had been placed in her way simply *because* she was a woman. Patriarchal society in general, and male scientists specifically, have worked, both consciously and unconsciously, to keep women out of science. Often women were barred by their inability to gain access to knowledge and learning, or by rules of social behavior that dictated that science was not for girls, roles that funneled women toward socially acceptable and respectable positions as wives and mothers in western society. At other times, male scientists have used scientific knowledge itself to "prove" that women's minds were incapable of practicing science, and that their bodies were unable to stand the intense intellectual work and physical labor required. Marie Curie's life reflects both the success women can have as scientists and the struggles with which most women scientists must contend. Though her talent was exceptional, her life story reflects many of the conundrums faced by women scientists.

Marie Curie practiced physics and chemistry in the first half of the twentieth century, when these disciplines were still very much male enclaves. This introduction will follow her life achievements in science to outline the struggles, constraints, and barriers that she faced and to explain how and why she became the iconic woman scientist. Marie Curie's life story epitomizes the main themes of this book. Women in the western world, from the seventeenth century to the present, were ostracized from science by individual men and the strictures of a patriarchal society. However, women often coped with, negotiated, and overcame these restrictions to participate in and contribute to science.

Marie Curie: The Heroic Story of a Woman Scientist

In a 1930s children's book series entitled "Health Heroes," Grace T. Hallock authored a volume on Marie Curie. She began the tale of Curie's life as follows:

> On November 7, 1867, a little girl whose life was to be like a fairy tale was born in the ancient Polish city of Warsaw. Her parents were of noble birth, but poor, and she herself was beautiful and gifted. As a young woman she endured terrible hardships in order to study science in the city of Paris. There she met a man whose genius and ideals matched her own. The two fell in love with each other and were happily married. After years of almost incredible labors, they discovered a wonderful substance called radium, which is now used by physicians everywhere to relieve the distress and often to save the lives of people suffering from a dreadful disease. The name

given to this little girl by her parents was Marya Sklodowska. The whole world remembers her today as Marie Curie.

Hallock wrote of a fairy-tale hero for children to admire. Her telling of Curie's story was based on the first biography ever written about Marie Curie. This biography was written by Curie's younger daughter, Eve Curie, who was intent on portraying her mother in a heroic light. Eve Curie's biography of Marie begins by informing the reader that "The life of Marie Curie contains prodigies in such number that one would like to tell her story like a legend." Eve Curie was amazed by her mother's achievements, despite her daily, personal relationship with her.

Other commentators have been equally amazed by Curie's achievements. In 1906, Andrew Carnegie, a wealthy American businessman and patron of the arts and sciences, set up a scholarship fund to endow Curie's students' research. Carnegie refused to have the scholarship named after himself. Instead, he suggested it be named the Curie Foundation. Even in China, in the temple of Confucius at Taiyuan-fu, there was a portrait of Curie hanging, among other "benefactors of humanity." Her portrait was in the company of Descartes, Newton, the Buddhas, and the emperors of China. Public and scientific awe continued through to 1967, when a special symposium was held in Warsaw to celebrate the centenary of Marie Curie's birth. Many of the papers presented were scientific in nature, but several paid historical tribute to Curie and her work. In his "Introductory Address," the deputy prime minister of Poland referred to Curie's "immortal achievements" and "self-sacrificing labor" that resulted in the "great discoveries . . . by our great countrywoman." One of the conference papers referred to Curie as the "Copernicus of the world of the small." Curie continues to be referred to in these glowing, larger-than-life terms to the present day. Recent scientific journal articles refer to her as a "scientific entrepreneur," a "scientific pioneer," and "a giant connecting two centuries." In some cases, Curie's life takes on religious overtones, baptizing this scientist a saint. Her fascinating life story, argues science journal writer Deborah Noble in *Analytical Chemistry*, makes it difficult to get "any realistic picture of Curie as a working chemist" (1993, 215). The story of the life of Marie Curie has certainly reached mythic proportions and, to a certain extent, the celebration and veneration of Marie Curie's life is well justified.

As is the case in all tales of famous intellects, Curie's family recollected childhood moments that give a glimpse into the promise and abilities to be nurtured within this little girl. For example, at the tender age of four she sat with her family while her elder sister, Bronia, struggled to read a passage from a book aloud. Marie, growing frustrated, looked over her shoulder and, to her parents'

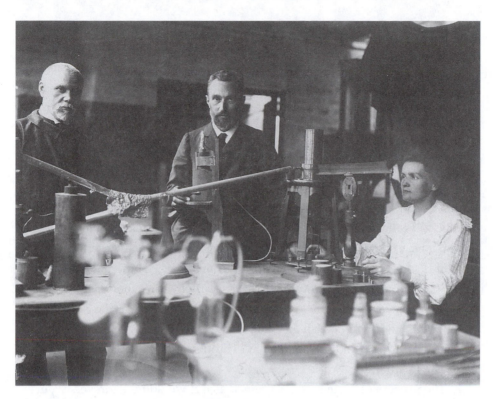

French physicists Marie Curie and Pierre Curie work together in their laboratory. The Polish-born scientist and her husband shared the 1903 Nobel Prize for physics with Henri Becquerel for their discovery of radioactivity. Marie Curie was also the sole winner of the 1911 Nobel Prize in chemistry. (Hulton Archive/Getty Images)

and sister's stunned amazement, read the entire sentence without hesitation. No one even knew that she could read. Marie's facility with languages was to stand her in good stead in later life, but even in childhood it served its purpose. Living in Russian-occupied Warsaw, Poland, her primary school teachers were expected to teach the Russian syllabus. In defiance of the authorities, they also taught the Polish language and Polish history. When school inspectors came to Marie's school to test the children, Marie was most frequently called upon to answer the questions posed, not only because was she an intelligent child but because her Russian was perfect. Other children, including her sisters, envied her abilities. Yet, Marie Curie was not the only person in her family with academic successes. Her parents, Vladislav and Bronislawa Sklodowska, were impoverished lesser nobles who clung to their intellectual curiosity and Polish nationalism. Both teachers, they instilled in their children a passion for learning and raised boys and girls who consistently won awards and pursued degrees with ardor. In addition to her facility for book learning, Marie would later write in her autobiography that "All my life through, the new sights of Nature made me rejoice like a child." Her father encouraged this love of nature through his teachings and by providing Marie opportunities to travel the countryside. For all their intellectual rigor, the Curie family was a close one that shared their joys and sorrows, comforting and supporting one another throughout their lives.

As a result of her familial support, particularly support from her sister Bronia and her father, Marie Curie was a successful young student in Poland. Graduating from the Russian gymnasium at age sixteen, she left with a gold medal as top student and won awards in mathematics, history, literature, German, English, and French. Time and again Marie Curie's love of learning wrecked her health as she ignored food and rest in the pursuit of knowledge. Despite her childhood achievements at school, Marie Curie, as a woman, could not attend university in Poland. As well, her father had lost money in an investment and had the funds to send only her brother, Jozef, to university. Marie and Bronia thus began the arduous work of tutoring to save for their own schooling. During this time Marie joined a secret, underground organization known as the Floating University, which educated Polish men and women while promoting nationalist fervor and nurturing positivist philosophy. Through this experience she had her first taste of laboratory work.

The process of saving money seemed as if it would take forever. So, eventually, Marie suggested to Bronia that she would commit herself to work as a governess full-time and pay for Bronia's medical education. Then, when Bronia had become a doctor, she could in turn assist the younger Marie to fund her schooling. As a result, Marie spent seven years, 1884 to 1891, as a governess. During this time she sat up every night to read or to practice mathematics. While feeling that governess work was beneath her dignity—socially and intellectually—she persevered, and Bronia not only finished her education and became a doctor but had also married a doctor and set up a home in Paris. Bronia wasted no time in inviting Marie to Paris to live with them and pursue her education at the famous Sorbonne, the Paris university. Thanks to her family's persuasiveness, Marie Curie overcame her trepidations about traveling and studying at the university after so many years of provincial life as a tutor. Eve Curie wrote that, at this time, her mother "was far, very far, from thinking that when she entered this train she had at last chosen between obscurity and a blazing light, between the pettiness of equal days and an immense life." However, a whole new world and life was about to open up to her.

Marie wrote in her autobiography that her time at the Sorbonne was one of the happiest in her life: "I shall always consider one of the best memories of my life that period of solitary years exclusively devoted to the studies, finally within my reach, for which I had waited so long." After living with Bronia and Kazimierz Dluski for a short time, she rented an attic that allowed her more space for her studies. Here, at the beginning of her scientific work, she lived in a mean abode, unheated, with very few comforts of home. Marie hardly ate or slept. She was engrossed in her studies and knew she had to catch up with the other students in mathematics, attend lectures, and study for exams. Eve Curie wrote, "For more than three solid years she was to lead a life devoted to study alone: a life in conformity with her dreams, a 'perfect' life in the sense in which that of the monk or the missionary is perfect." Between 1881 and 1894, beginning at age twenty-four, she earned two degrees, both equivalent to a baccalaureate, one in physics and one in mathematics. In 1894 she won the Polish Alexandrovitch Scholarship, which allowed Polish students to pursue their education outside of Poland. The rarity of a woman attending university, living independently, and studying mathematics and science at this time can not be stressed enough. But what could a woman *do* with university degrees? Marie Curie had originally planned to return to Poland, live with her father, and teach. However, the opportunity arose for her to work for the Society for the Encouragement of National Industry in Paris on the variations in the magnetic properties of certain hardened steels. She accepted the offer. While trying to find laboratory space for the project, she was introduced to Pierre Curie, who, it was believed by friends, would be able to give her the room she needed. Pierre was already an authority on magnetism.

Meeting Pierre changed the course of Marie's life. With no plans to marry, she found herself ardently courted by Pierre, who, also planning never to marry, changed his mind upon meeting Marie. Pierre marveled at Marie because she not only understood his own scientific endeavors but also held similar interests. He had written in his diary at the age of twenty-two, "Women of genius are rare." Now he believed he had found one, and he did not rest until she agreed to marry him. She hesitated on the grounds that Pierre was established in Paris and that she had intended to return to Poland to live with and care for her aging father. She left Paris in 1894 for the summer, planning never to return, but Pierre kept up a correspondence urging her to return to Paris and suggesting that, possibly, they could live together if she preferred. She did, in the end, return to Paris, claiming her desire to continue her work in science. Pierre was soon a constant visitor at her door and, in July 1895, she agreed to marry him. They had a modest wedding ceremony, with only close family and friends in attendance, and their honeymoon consisted of a bicycle tour through the French countryside. They both wanted to get back to their work as quickly as possible.

Marie completed and published her work on the magnetic properties of steel in the 1897 *Bulletin of the Society for Encouragement of National Industry* and decided to pursue a teaching certificate, which would allow her to become a professor and to teach young girls. In 1896 she qualified in the examination with the highest marks in her class. But by 1897, a year after the birth of her first daughter, Irene, she had resolved to earn her doctorate degree and began to search for a subject that interested her. Fascinated by the recent experiments with X-rays conducted by the French scientist Henri Becquerel, she wanted to investigate the source of energy that caused uranium salts to give off penetrating rays. This radiation from uranium and, as she was to discover, also thorium and pitchblende, became the subject of her thesis. In 1903 she would become the first woman in France to receive a Ph.D. in physics for her thesis "Research on Radioactive Substances." From her research she postulated that these rays were a property of the atoms themselves, and not caused by any environmental factors. She named this property "radioactivity" and went on to study hundreds of substances. In doing so, she discovered two elements as yet undiscovered in pitchblende, both of which proved to be sources of radioactivity. She named these new elements polonium (after Poland) and radium. Curie's work on radioactivity proved to be revolutionary. Classical physics stated that the atom is immutable, meaning no force could break it down nor could it be changed into something else. Marie Curie noted that radioactivity was a characteristic of some atoms, not the result of interaction between molecules. Radioactive atoms emit particles, implying that an atom is made up of small particles and is therefore mutable; thus Curie's discovery set the groundwork for the creation of the new physics. Twentieth-century physicists would uncover the atomic structure and eventually establish that the atom's structure was not fixed but could be transformed. This was such an earth-shattering proposition that even the Curies hesitated to believe the implications of their own work. These discoveries would lead to important practical applications including new medical treatments for cancer. It would also lead, as Pierre Curie had feared, to the creation of the atomic bomb.

Despite Marie's research, many scientists demanded more evidence that polonium and radium actually existed. They challenged her to produce these elements in their pure form. Pierre was working during this time at the University of Paris, and Marie was teaching at the Ecole Normale Superieure de Sevres, a government boarding school for girls. This enabled them to earn just enough money to support their family and to allow Marie and Pierre to work on research that would provide convincing evidence for the existence of polonium and radium. Realizing the importance of Marie's work, Pierre set aside his own work temporarily to assist Marie. They worked together beginning in 1898 and for the next eight years. Their "laboratory," provided by the School of Physics where Pierre was teaching, was nothing more than an unheated shed with a leaking roof, containing virtually no scientific apparatus. Often the delicate work of attempting to chemically extract the elements from tons of pitchblende was ruined or interfered with due to these impossible working conditions. But the

Curies persevered, spending every possible moment in their laboratory shed and constantly exposing themselves to radioactivity. Their health suffered as a result. They were always tired and found themselves losing weight. Finally, though, in 1902, they were successful, and Marie was able to provide the atomic weight of radium, having extracted the pure form from 200 tons of sandstone and rock.

Awards and honors quickly recognized their work on radioactivity. Marie won the Prix Gegner three times—in 1898, 1900, and 1902. In 1903, Pierre won the 10,000-franc biannual Prix La Caze and was invited by the Royal Society of London to give one of its prestigious Friday lectures. Together, Pierre and Marie were awarded the Davy Medal by the Royal Society several months later. In November 1903, Marie Curie became the first woman to be awarded a Nobel Prize for physics, won jointly with Pierre Curie and Henri Becquerel, for their research on radiation. Marie Curie became one of only two people to win the Nobel Prize twice, when she won it a second time in 1911, this time for chemistry. On that occasion she received it for her work alone, suppressing the notion that she was merely an assistant, or a secondary player, in her husband's work. The 1911 Nobel Prize was awarded to her for her work on radium and polonium. Later, in 1923, she would be awarded a substantial lifetime pension by the French government for her discovery of radium. According to Eve Curie, "The idea of choosing between family life and the scientific career did not even cross Marie's mind. She was resolved to face love, maternity and science all three, and to cheat none of them. By passion and will, she was to succeed."

In the midst of such great public triumph, great personal tragedy struck Marie's life. In 1906 Pierre was hit by a horse-drawn vehicle and died instantly, his head crushed by the wheel of the wagon. The loss of Pierre was unbearably painful for Marie; she had lost not only her husband but also her working partner in science. In her journal she wrote to her husband, "life is atrocious without you, it is anguish without a name, a desolation without limits." She wanted to continue her work, but how could she now that Pierre was gone? Pierre had held the chair in physics at the Sorbonne at the time of his death, and Marie was chief of the school's laboratory. After some discussion, the Sorbonne offered Marie her husband's chair; she accepted, and in doing so she became the first woman to teach at the Sorbonne and the first woman to hold a full science professorship at a university. In November 1906, the hall was packed with students, colleagues, and press who came to see the spectacle of a woman lecturing on physics. Always suffering from stage fright before class, Marie, controlled and unmoved, entered the lecture hall and began her lecture exactly where Pierre had left off, without any special speech or foreword. The audience was mesmerized by a lecture the majority did not understand and by the sight of the petite widow carrying on the work that she and her lost husband had begun together. By 1910 she had written and published her 1,000-word *Treatise on Radioactivity*, summarizing the progress of research in radioactivity since her first observations in 1897.

Curie's work was interrupted by World War I. At first she remained stoically behind in her empty laboratory in Paris, her students having been drafted. She did not wish to appear to be fleeing from her adopted country. Besides, as Eve Curie recorded in her biography, Marie Curie argued, "If I am there . . . perhaps the Germans will not dare plunder them [her new laboratories]: but if I go away everything will disappear" (292). Curie found herself wanting to help the Allied war effort in some way, and she quickly came up with the idea of forming a mobile X-ray service for doctors at the front lines. Though relatively unfamiliar with this equipment, she trained herself and other men and women to work the X-ray machines. Using her fame, she cut through government and military red tape and managed to convince well-to-do friends to loan her their cars. With the help of eldest daughter Irene and many other trained X-ray technicians, more than a million X-rays were conducted in the field. Men could now be operated on with much greater efficiency, as X-rays clearly showed the projectiles embedded in the wounds, allowing for their quick removal. Marie wrote of her work and experiences in her book *Radiologie et la guerre* (1921). Marie also converted her prize money from her second Nobel Prize into war bonds. When the French government needed gold, Marie Curie attempted to donate her scientific medals. The official refused to melt down the medals, much to Marie's disgust. Marie had voluntarily taken on Pierre's disregard for material possessions and awards for their work and believed the reward lay in the work itself.

After the war, the death of Pierre Curie set Marie Curie on a new life mission to uphold the memory of her beloved husband. As the first woman to head a large laboratory, she used her status and fame as a Nobel Prize winner to garner money, largely from the United States, to fund the purchases of radium for her laboratory, establish student scholarships, and create two new Institutes of Radium in Paris and Warsaw. Much of Curie's fame came from the fact that her discoveries had prompted medical research into the use of radioactivity as a cure for cancer. The Curie family was constantly short of money, largely because Marie and Pierre both refused to apply for any patents on their discoveries. They believed that scientific knowledge should be open to all—scientists, industrialists, and the public. Marie upheld this tradition after Pierre's death. She absolutely abhorred crowds, dinners, and speeches, but quickly realized their potential and suffered through them to fund her increasingly expensive scientific research. Marie Meloney, a fund-raiser par excellence, saw Marie Curie as her hero and lauded Curie in the press as a poverty-stricken woman scientist too retiring in manner to ask for funds herself but who desperately needed and deserved the financial assistance of the American people. Susan Quinn has argued in *Marie Curie: A Life* (1995) that this "fictional version" of Marie Curie "produced a brilliantly successful fund-raising campaign" (385). Marie Curie was actually well established in her institute with funds to provide for research, but the United States could provide much more. After her first U.S. visit, in 1921, to receive a gift of radium from the women of America, she was dubbed "the radium woman." While continuing to prefer to ignore fame, she steeled herself to face the adulating crowds so that she might receive the fortune that came with it. She was always careful, however, to make sure donations of radium and monetary funds were made out to her institute and not to her personally. She nevertheless did receive several honorary degrees from U.S. universities, including doctor of sciences from the University of Chicago, Columbia University, and Wellesley College. In 1922, Marie Curie was made a member of the Academy of Medicine of Paris for the part she played in discovering radium, a discovery that led to the new treatment for cancer. The president of the academy, M. Chauchard, called her "a great scientist, a great-hearted woman who has lived through devotion to work and scientific abnegation, a patriot who, in war as in peace, has always done more than her duty." She was the first woman admitted to an academy in France.

Marie Curie worked incessantly up until a few months before her death in 1934, continuing in later life to relish her laboratory work despite increasing ill health due to her lifelong exposure to radiation, and cataracts that, before her operation, slowed her work considerably. She died only a year before her daughter Irene and Irene's husband, Frederic Joliot, received the Nobel Prize in chemistry for discovering artificial radioactivity. Not only a famous scientist in her own right, Marie Curie had succeeded in raising a daughter in her own image.

Curie's life was a life of firsts, and her work is certainly a testament to what women scientists can achieve. Her struggle for scientific education, her scientific work under terrible conditions, her perseverance in the face of adversity, and her scientific discoveries and subsequent rewards and honors mark Curie as an exceptional woman and scientist worthy of heroic and iconic status. But as many of her own students found, such a person could be an intimidating presence in the laboratory, despite her treatment of them as members of her family to be carefully nurtured and trained. Even her son-in-law, whom she treated fondly and respected as a scientist, found himself tongue-tied in her presence. Marie Curie had been Joliot's childhood hero, whose picture he had cut from a newspaper clipping and hung over his makeshift laboratory in his parents' bathroom along with pictures of Pierre Curie and Louis Pasteur. Despite her small physical size, she was a towering presence and continues to be so even today, more than half a century after her death.

Struggles to be Heard: Marie Curie in the Shadow of her Gender

Francoise Giroud noted in her 1986 biography of Marie Curie that "we want Marie to be both saint and martyr," but when we force her into this mold, "we not only falsify her but we take away another of her dimensions, her experience of guilt and the drama into which this reticent and modest woman was plunged when her private life was exhibited publicly" (99). Some biographers and historians of science, while recognizing Curie's exceptional achievements, nevertheless explore the life and work of Marie Curie more fully, taking us beyond the simplified, iconic status that Curie holds as a woman scientist. They want to know Curie the human being, not just Curie the icon. They have tried to understand her life and work within its social, political, and cultural context and to give value to her life—not just her intellectual achievements and scientific discoveries, but her struggles to achieve success in a man's world; her life as a wife, mother, and widow; and her collaborative experiences in science. Through such examinations, Curie can still remain an icon for women in science, but a more human and more attainable figure to emulate.

While making important strides in science and receiving the accolades that go along with such achievements, Marie Curie was also a woman, and throughout her life she faced the gender inequities embedded in European society. Frustrated by the rule that forbid women to enter the University of Warsaw, Marie and her sisters could only enjoy the stories of their brother's experiences at university. But the sisters had received enough education to enable them to become teachers, and, besides, marriage and motherhood were considered to be their ultimate goal in life. Finding a husband to take care of them financially was not only socially acceptable, it was a social expectation.

Even if there had been money, admission to a Polish university would have been impossible for Vladislav Sklodowska's daughters. In Poland, universities required that students fulfill a language requirement, and girls were not educated in the necessary languages. By default they were unable to obtain admission to the universities. Their only option was to attend a foreign university. Eventually both Bronia and Marie attained their degrees in France, but not before Marie suffered the humiliation of serving as a governess to two wealthy families. Governesses filled a socially problematic position in European families. On the one hand they were hired help, servants of a kind. On the other hand, most governesses had some pretensions, educational or familial, to the same social status as the families for whom they worked. Governesses lived in a netherworld between upstairs and downstairs, not belonging fully to either class. It was often a lonely and degrading experience for the women who filled these positions.

President Warren G. Harding escorts Marie Curie down steps to the south grounds of the White House in May 1921. (Library of Congress)

Marie certainly felt that she was the social equal, not to mention the intellectual and moral superior, of her employers. Yet, when she fell in love with Casimir Zorawski, the elder son of the family for whom she was working, and he requested his parents' blessing for their marriage, Marie suffered their horrified rebuffs. She was not the sort of woman suitable for their son, as she had no money or social standing in the community. To add insult to injury, Marie had no choice but to swallow her pride and stay in her position as governess for fifteen more months, when Bronia finally invited Marie to join her in Paris.

Marie depicted her student life in Paris as ideal. Her upbringing and personality and the relatively open nature of Parisian society in the late nineteenth century allowed her, for this brief period, to largely ignore the usual constraining social mores. Certainly women could enter the universities in France, but women students were still in the very small minority. When Marie entered the Sorbonne in 1892, there were 210 women enrolled out of a student population of nine thousand. Few of these women were actually pursuing degrees; most were taking courses for pleasure. Moreover, Marie chose to study first science and then mathematics, both subjects not generally considered suitable for women. Though she at first lived with her sister and brother-in-law, she quickly moved herself to her own quarters, living in abysmally poor conditions to attain independence and a solitary, uninterrupted working environment. She traveled to school alone, lived alone, and even worked and studied largely by herself.

Her family had long supported her intellectual attainments regardless of the fact that she was a woman, so her desire to succeed in her education is unsurprising. But her years in confinement as a governess, tightly constrained within the circumscribed world of middle-class women, had made her long for freedom—physical, intellectual, and social. While still a governess, she had written to her cousin, Henrietta, in 1886, with a certain amount of sarcasm, "If you could only see my exemplary conduct! I go to church every Sunday and holiday, without ever pleading a headache or a cold to get out of it. I hardly ever speak of higher education for women. In a general way I observe, in my talk, the decorum suitable to my position." Marie had been an agnostic for some time, longed for her chance at higher education, and was hardly ever, for the rest of her life, to be concerned with decorum. Her life as a governess forced her into a role that deeply depressed her and caused her to long all the more to escape.

When she reached Paris she was determined to live every moment of her freedom to its utmost and to work hard at her studies to the point of exhaustion. Her brother-in-law wrote to her father, with a tone of alarm and concern, "She is a very independent young person, and in spite of the formal power of attorney by which you placed her under my protection, she not only shows me no respect or obedience, but does not care about my authority and my seriousness at all. I hope to reduce her to reason, but up to now my pedagogical talents have not proved efficacious." Eve Curie wrote of the years in which Marie ignored the advances of the young men with whom she studied and "drew nearer to men who did not pay court to her and with whom she could talk about her work. . . . Marie had no time to give to friendship or to love. She loved mathematics and physics." She did not want to be distracted from her goal to achieve an education. At the end of her university studies she had achieved an education equal to that of a man's. In 1893 she obtained the equivalent of a master's degree in physics, placing first in her class at the Sorbonne, and in 1894 she obtained a similar degree in mathematics. With two degrees to her name, she had become an exceptional scholar and longed to be a scientist in her own right. But there was a price. For many years she felt guilty about not returning to Poland to live with her father (her mother had since died), as she knew a dutiful daughter ought to have done. Instead, she had moved into another world, one neither *of* women nor *for* women. She nevertheless desired to be a part of the scientific community despite the difficulties she knew would confront her. For the rest of her life Curie was a scientist, but in the eyes of many of her contemporaries she was a *woman* scientist. Her gender would, to different degrees, disadvantage her throughout her life, and she would continually have to struggle to overcome the barriers placed in her way.

Marrying Pierre Curie was both advantageous and disadvantageous for this ambitious woman scientist. Her marriage brought her love and children, both of which she cherished, as well as a man who was perhaps an ideal scientific partner for Marie. Pierre seemed indifferent toward the idea of women scientists. Moreover, his reticent and modest personality, together with his strict sense of social justice and commitment to the purity of scientific research, meant that Marie practiced science without censure from her husband. Marie described Pierre in her biography of him as "liv[ing] on a plane so rare and so elevated that he sometimes seemed to me a being unique in his freedom from all vanity and from the littlenesses that one discovers in oneself and in others, and which one judges with indulgence although aspiring to a more perfect ideal." Although he believed it was rare for a woman to choose what he referred to as "the anti-natural path" in life—that is, a life in which one denied oneself the pleasures that human beings (especially women) "naturally" desire—he clearly recognized Marie's commitment to her scientific endeavors, or "obsession," as several biographers refer to it.

Pierre and Marie became collaborators in science, complementary partners who joined forces to produce groundbreaking research. Pierre saw Marie as his scientific equal, and they painstakingly itemized who had done what in their experiments and discoveries when they published their work. Marie also published some of her own work independently. When it appeared that Pierre was going to be nominated for the Nobel Prize without Marie, he protested to the committee and asked that they be nominated together. When he finally received the chair in physics at the Sorbonne, he made sure that Marie was given the paid position of chief of work in his laboratory. Clearly Pierre was a supportive career partner who respected Marie's intellect and helped to advance her career. As he himself noted, he would not have chosen the path of isolating radium. It was Marie's insight, fortitude, and determination that kept them on this course.

Unfortunately, to many outsiders, Marie was perceived as succeeding by riding the coattails of her husband. For many men and women of the time there could be no other explanation for her startling scientific achievements—they must belong to her modest husband. The Curies foresaw this problem and, as noted earlier, carefully recorded the roles each played in their work. But this care did not seem to hinder either the press or other male scientists, such as Bertram Borden Boltwood and Ernest Rutherford, from believing that it was Pierre who was the scientific genius and Marie who was the helpful, hard-working assistant. In 1903 the *New York Herald* reported that "Mrs. Curie is a devoted fellow laborer in her husband's researches and has associated her name with his discoveries." The same year London's *Vanity Fair* noted that Marie had been responsible for "[fanning] the sacred fire in him whenever she saw it dying out."

Hertha Ayrton, a physicist in her own right, found herself constantly reading in the British newspapers that Pierre Curie was the discoverer of radium. Exasperated, she wrote in March 1909 to the *Westminster Gazette*, "Errors are notoriously hard to kill, but an error that ascribes to a man what was actually the work of a woman has more lives than a cat." Despite the Curies' best efforts, the myth of the male genius and his faithful female assistant persisted.

At the same time, Marie suddenly found herself confronted with household tasks she had ignored as a student and that now needed to be addressed. Uneducated in the arts of cleaning, cooking, and raising babies, she scrambled to learn the basics. She found herself preparing meals in the early hours of the morning before going off to the lab, shopping on the way home, keeping the family's finances in order, and comforting sleepless children at night. Pierre noted that sometimes Marie "finds her double task beyond her powers." Nevertheless, Marie found time to continue her work, producing her most innovative insights while she was pregnant with her daughters. Despite continued illnesses suffered during both pregnancies and a miscarriage in between the two, she refused to rest or to take time off from her work. As with most career women, she found herself living a dual life. Her attention to household concerns, however, was tempered by hired nursemaids and the constant care and affection that the children's grandfather, Pierre's father, lavished upon them as the fifth resident in their home. Without outside help and family assistance, Marie would have been unable to continue her work. Such was the lot of a woman at the beginning of the twentieth century. No matter their profession, women could rarely fulfill both their personal and professional obligations at the same time. Women usually had to make a choice between family and career.

Curie, though, never had to make such a choice. Yet, despite every effort to afford her daughters the best care, her own included, some biographers have noted that Curie's work and fame had an adverse affect on them. Despite her glowing biography of her mother, Eve Curie noted that "in spite of the help my mother tried to give me, my young years were not happy ones." She could not join the conversations of her mother and older sister Irene as they engaged in spirited scientific discussion over the dinner table, and Eve consequently felt left out. Even Irene felt the pull of her mother toward her work, and on one occasion, when Marie's female students from her Sevres class visited for tea, Irene appeared disturbed by their intrusion on her time with her mother and kept repeating, "You must take notice of me." Both daughters knew that science was their mother's third child and feared that possibly it was her favorite child. At the time of the awarding of Marie's first Nobel Prize, there had been much concern in the press about whether a woman could retain her femininity and fulfill her family duties as a wife and mother. Such discussion evolved into commentary upon the larger issues of the validity of feminism and equality between the sexes. Most commentators at the time believed that the differences between the sexes needed to be retained for the good of civilization, though most noted that a minority of women would not fit the mold. The press played a delicate balancing game between Marie's scientific accomplishments and her ability to maintain her femininity.

Marie Curie would not receive equal professional recognition for her work until after the death of Pierre in 1906. Initially, the Sorbonne's first reaction was to offer Pierre's widow and his two children a pension for their financial support. Marie refused what she considered to be charity. University administrators then suggested that Pierre's chair in physics be left empty and that Marie be named director of the laboratory. In other words, Marie would do Pierre's job without officially assuming the chair. It was only after much encouragement from her friends and colleagues that the Sorbonne offered her Pierre's chair, which Marie immediately accepted. Objectively speaking, Marie deserved the chair as much as Pierre, yet it is unlikely that she would have ever received such a position if it were not for the untimely death of her husband. She was the first woman to hold such a position in the university and the first woman to win a Nobel Prize, but she continued to face opposition to her inclusion in the top scientific ranks. For example, when she visited the United States in 1921, she found herself lauded by the American people, but some universities, such as Yale, hesitated to give her an honorary degree, and Harvard University refused entirely. The retired president of Harvard, Charles Eliot, believed the reasons for this rejection were that "the credit for the discovery of radium did not belong entirely to her and that, furthermore, she had done nothing of great importance since her husband died in 1906." Eliot was insinuating, of course, that the true master scientist was Pierre, not Marie. More important, though, Marie never gained membership in the French Academy of Sciences. A member of the academy, M. Amagat, had stated with "virtuous indignation" that "women cannot be part of the Institute of France." Marie Curie was refused membership by one vote.

Some historians have argued that Curie's failure to gain admission to the French Academy of Sciences went deeper still. At about the same time, a scandal involving Marie had surfaced in the French press. Her intimate relationship with a long-time family friend and colleague was revealed publicly shortly after her failure to gain membership in the academy. Although Marie was a widow, Paul Langevin, the man in question, was married and had children. Curie was accused of being a foreigner intent on stealing a French woman's husband. According to friends of Curie and Langevin, Langevin's marriage was not a happy one. He had begun as a student of Pierre's and quickly became a close friend of Pierre and Marie. After Pierre's death he and Marie became closer. Although

Marie never publicly declared that the relationship truly existed, friends believed that it did, and surviving letters certainly indicate that they were deeply affectionate toward one another. It is also known that Langevin's wife, Jeanne, was so jealous of Marie that on several occasions she threatened to kill Marie. Jeanne Langevin eventually sued for divorce from Paul, using letters she had stolen from her husband's desk as proof of the affair with Marie. Curie, always terrified of public incursion upon her life, was made physically ill by the attack upon her in the press and by the feeling that she had besmirched the name Curie and dishonored her dead husband. The iconic, saintlike Marie Curie that the press had created had fallen from grace.

Though such an affair would have been unlikely to affect a male scientist's candidacy, Marie's perceived moral and ethical conduct played a significant role in the question of her admittance to such an austere body as the French Academy of Sciences. Not only was she a woman, but, it would seem, she was an immoral woman. She was completely beyond the pale of acceptable, respectable womanhood and as such could not expect to be elected. When the affair became public, there was even talk at the Sorbonne about her losing her chair as a result. Furthermore, biographer Robert Reid has suggested in *Marie Curie* (1974) that Curie's second Nobel Prize was "concocted for the best of sympathetic and humanitarian motives" but that "it is difficult to avoid the embarrassing realization that Marie Curie was awarded the Nobel Prize twice for the same work, once for physics and once for chemistry" (212–213). According to Reid, the second Nobel Prize was a sympathy vote by many foreign scientists in an attempt to console Curie after her rejection by the academy and for the scathing attacks by the press as a result of the Langevin affair. The scandal eventually subsided, but Marie never forgot the backlash she received for living as a woman outside of conventional gender norms. For the rest of her life she presented to the world and to all but her very closest friends and family an emotionless facade, carefully guarding her privacy and always fearing some sort of retribution. As a result of the scandal, her life became solitary and to some degree lonely. As a woman scientist of great achievements she set herself apart. In doing so she was also forced to play into the Parisian tabloid press's depiction of herself as "the vestal virgin of radium" who had consecrated her life to science and lived for nothing but her research.

Ironically, as Sharon Bertsch McGrayne has pointed out in *Nobel Prize Women in Science* (1992), "by creating an almost impossible standard for women scientists to live up to, Marie Curie may have made their professional progress more difficult. Although universities did not expect every male scientist to be an Albert Einstein, women scientists were continually measured against Marie Curie—and naturally found wanting" (34). Margaret Rossiter has argued in

Women Scientists in America (1982) that women scientists in attempting to emulate Marie Curie "adopted a new, more conservative, and less confrontational strategy of deliberate overqualification and personal stoicism" (129). Susan Quinn, however, posits in *Marie Curie: A Life* (1995) that Curie's visit possibly "inspired" women, as more women chose to pursue a science degree in 1932, three years after Curie's second visit to the United States, than had done so before (397). Marie Curie herself had approved a large number of women for the period to work in her lab, but according to Eve Curie she did not like to be an example for other women, noting to her daughters that "it isn't necessary to lead such an antinatural existence as mine, I have given a great deal of time to science because I wanted to, because I loved research. . . . What I want for women and young girls is a simple family life and some work that will interest them."

Historians' attempts to reshape our image of Marie Curie do not represent an effort to knock her off her pedestal, but rather to bring her into closer focus. Marie Curie was exceptional, but she was ordinary as well. She faced all the trials and tribulations that many women of her time faced; furthermore, she had to contend with the barriers against women's wider contribution in educational institutions, in scientific practice, and with regard to professional attainments and awards. Her successes—as well as the successes of other women scientists—made it possible for other women to succeed. For instance, Curie's daughter Irene sat on the stage side-by-side with her husband and spoke first upon receiving their Nobel Prize in 1935. In contrast, on receiving her first Nobel Prize, Marie Curie had sat in the audience while her husband gave the speech. But the retelling of Marie Curie's struggles and failures allows us to recognize her battles as a reflection of the cultural norms and values that have, and often still do, inhibit women's ability to achieve scientific renown in a patriarchal society and a male-dominated profession.

Conclusion

Despite the ever-increasing numbers of women in science during the twentieth century, the image of the male-only scientist has persisted. Recent studies of children's perceptions of scientists and studies of depictions of scientists in popular American magazines between 1910–1955 make it clear that the general reader in the twentieth century still had difficulty imagining a typical scientist as female. Although exceptions exist, Marcel LaFollette has argued that the image of women scientists as anomalies—superscientists who give up their life for the cause of science—persists. The image of the white male scientist in his pristine lab coat, intently working away in his laboratory, is the current icon of the sci-

Marie Curie with two men at the Standard Chemical Company radium extraction plant in Canonsburg, Pennsylvania (Library of Congress)

entist, but this perception may change in part due to works such as this one that illuminate the history of women scientists. Photographs of Marie Curie at work in her lab show her working intently, apparently unaware of the camera, engrossed in her research and surrounded by her equipment. However, unlike the stereotypical male scientist, she often looks tired and unkempt, and her working conditions are often not the best. These images thus remind us that Marie Curie's scientific achievements did not come easily to her in part because she struggled with the prejudices against women scientists.

The surviving photographs of her also remind us that she lived outside of the lab as well as within it. She fell in love with a man who shared and appreciated her scientific talents, and she gave birth to two daughters, one of whom followed in her mother's scientific footsteps. Photographs show her and her daughters leaning upon one another, belying the notion that Marie was always cold and distant. She had a close family circle with whom she shared her time, and a circle of like-minded colleagues and students who appreciated her worth as a scientist and as a teacher. Marie Curie was not only a famous Nobel Prize scientist of great achievements but also a woman scientist who struggled to be educated,

Nobel Prize–winning chemist Marie Curie poses with her daughters Irene (left) and Eve (right). (Bettmann/CORBIS)

heard, and recognized. Furthermore, she was also a human being who had foibles, but who nevertheless had a fulfilling family life, supportive friendship network, and successful scientific collaboration with her husband. Marie Curie was more than a superscientist. Her life also represents the difficulties for women in science, yet her life story reflects a possibility for more than the stereotypical, lonely career path for women. It is in this complete story that we find a truly useful icon for woman scientists, attesting to what women can do, reminding us of the social barriers she had to face and overcome, and showing us that despite her icon status, she was a woman of deep feeling and varied experience. Her life story still can and should serve as a role model for girls and women working in science, but as a real-life woman scientist instead of as a distant, unattainable iconic figure.

Marie Curie's life story exemplifies the struggles women faced to gain access to and participate in science in the late nineteenth and early twentieth centuries. Many women throughout modern western history have faced similar barriers, but like Curie they also managed to find a path to science. This volume will examine chronologically the various ways in which women were ostracized from the scientific community in various countries and at different times in history, but it will also examine how women managed to participate in science. Very few women at any time in history have achieved Curie's scientific success in the male world of science. Yet women have managed to negotiate niches for themselves on the peripheries of established science, to engage in informal networking to gain access to scientific knowledge, and to break into the formal organizations and institutions of male science. Each woman was driven, like Curie, to study the natural world. This volume will examine how they made this possible.

Bibliographic Essay

There are numerous short biographical descriptions of Marie Curie's life. The most useful are the entry under "Curie, Marie" in Marilyn Ogilvie and Joy Harvey's *The Biographical Dictionary of Women in Science* (2000), pp. 311–317; Marlene and Geoffrey Rayner-Canham, *Women in Chemistry: Their Changing Roles from Alchemical Times to the Mid-Twentieth Century* (1998), pp. 97–107; and Sharon Bertsch McGrayne, *Nobel Prize Women in Science: Their Lives, Struggles, and Momentous Discoveries* (1993), pp. 11–36.

Curie's life has, of course, been the subject of many full-length biographies. The earliest work, Eve Curie's biography of her mother entitled *Madame Curie* (1938), focused on telling the often-selective, triumphant, heroic story of Curie's life. Eve Curie wrote her biography with admiration for all her mother's struggles and disappointments, stressing how Marie was always able to overcome and succeed no matter what the obstacles. Here, Marie Curie is the hero of her own life and a model for others, as she overcomes all odds to achieve her dreams to study and work as a scientist. Yet, despite her great successes, she remains personally unaffected by fame, rising above it and utilizing it only to further the cause of science rather than for her own personal or financial gain. Eve Curie's biography does provide voluminous quotations from private letters and Marie Curie's diary, which are used effectively to reveal the thoughts and concerns of Marie Curie. For those who wish an even closer glimpse into the mind of Marie Curie, her biography *Pierre Curie* (1923), and her own short autobiography appended to it, are very useful but still selective. Marie Curie praises Pierre for his scientific achievements and high moral character while remaining modest about her own scientific achievements and making only brief comments about her personal life and relationships.

More recent biographies, such as those by Robert Reid, Francoise Giroud, and Rosalynd Pflaum, draw heavily on archival and other biographical materials and attempt to modify the heroic portrait given by Eve Curie's biography. Reid's work, *Marie Curie* (1974), is tantamount to a psychological biography, following Curie's intellectual growth from an insecure, neurotic schoolgirl to an ambitious and stoic woman scientist of great renown. Giroud's volume, *Marie Curie: A Life* (1986), originally published in French, relies heavily on other biographical works, although it does reference archival material as well. Giroud contests the notion that Marie rode the coattails of Pierre to scientific success and stresses instead their joint work. She also pays close attention to Marie's personal life and reveals in detail the Curie-Langevin affair, quoting extensively from contemporary journals that published articles about the affair. Nevertheless, Giroud's work is still firmly situated within the heroic biography tradition. Pflaum's work, *Grand Obsession: Madame Curie and Her World* (1989), is based on extensive interviews with some of the main figures in Curie's life as well as other archival and published material. Almost half the book is dedicated to a biography of Irene Curie, paying close critical attention to the effect that Marie's scientific work and successes had on her daughters' lives and psyches. Moreover, Pflaum notes that much is made of Marie Curie's sufferings as a student but points out that such suffering was largely self-inflicted, as Marie could have quite easily continued to live with her sister and brother-in-law. Pflaum also paints a less rosy picture of Marie's personality than did Eve Curie, indicating that Marie could be stubborn and generally very difficult to get along with. Pflaum clearly establishes the importance of Marie's work but seems to want to put her second to Pierre, whom she views as the true scientific genius, never receiving the full credit he deserved. While stressing the difficulties that Curie faced as a woman, Pflaum and Reid establish Marie's general disinterest in the feminist cause; this is in contrast to both Giroud and Eve Curie, who interpreted Marie's interest in teaching young girls as being based on feminist inclinations.

The most recent biography to date is an in-depth, revisionist work by Susan Quinn titled *Marie Curie: A Life* (1995). Quinn's aim is to try to reveal not just

Marie Curie the hero but also the human being, noting that Curie has been characterized as a timid, unemotional figure, largely in control of her personal and professional life. Quinn argues that Curie was in fact an intensely emotional woman with a fiery and stubborn personality. Revealing this other side of Curie does not detract from her scientific achievements, says Quinn, but rather refocuses our veneration on the human figure rather than the mythic one. Quinn's work provides an important and extensive biographical background to the work being done in the specific area of the history of women and science. Much of this work pays close attention to the relationship between Marie and Pierre on both personal and professional levels and places their lives and work in the context of the ability of, and ways for, women to enter into science at the turn of the century. Such works aggressively attack the notion that Marie Curie was simply an assistant to Pierre Curie, instead stressing the importance of their successful collaboration to their scientific success and Marie's entry into the professional scientific world. Helena M. Pycior's important work can be situated within the framework that Quinn established in her biography. Over a period of a decade, Pycior has explored, in several articles, the collaboration between Marie and Pierre and their approach to balancing personal and professional life. These articles include "Marie Curie's 'Anti-natural Path': Time Only for Science and Family," in Pnina G. Abir-Am and Dorina Outram (eds.), *Uneasy Careers and Intimate Lives: Women in Science, 1789–1979* (1987):191–214; "Reaping the Benefits of Collaboration while Avoiding Its Pitfalls: Marie Curie's Rise to Scientific Prominence," *Social Studies of Science* 23 (1993): 301–323; and most recently, "Pierre Curie and 'His Eminent Collaborator Mme. Curie': Complementary Partners," in Helena M. Pycior, Nancy G. Slack, and Pnina G. Abir-Am (eds.), *Creative Couples in the Sciences* (1996):39–56.

Other useful scholarly articles deal with specific aspects of Marie Curie's life and work rarely explored in any depth in the biographies. Stanley Pycior's work on the relationship between Marie Curie and Albert Einstein specifically investigates their involvement in the League of Nations. (See Pycior, Stanley. "Marie Curie and Einstein: A Professional and Personal Relationship." *The Polish Review* 44 [1999]: 131–142). He notes that while the press made much of Einstein's involvement in the league, in fact the part he played was minimal in comparison with that of Curie. Einstein and Curie's close friendship led Einstein to follow Curie's lead in all things relating to their committee work for the league's International Committee on Intellectual Cooperation. Xavier Roque's article titled "Marie Curie and the Radium Industry: A Preliminary Sketch," in *History and Technology* 13 (1997):267–291, focuses on Curie's involvement with the industrial sector and explores another understudied area of her life and work, arguing that the characterization of Marie Curie as a pure scientist is misleading because it belies Curie's "integrated vision" of industry and scientific research, which she upheld throughout her life. Finally, analysis of the image/myth of Curie has also been only minimally explored, first in an unpublished paper on the portrayal of Curie in children's literature by Susan Lindee, titled "The Scientific Romance: Purity, Self-sacrifice, and Passion in Popular Biographies of Marie Curie" (February 1998), read at the American Association for the Advancement of Science in Philadelphia; and second in a short article by Alberto Elena that explores the depiction of Curie's life in the motion picture by MGM Studios called *Marie Curie* (1943), titled "Skirts in the Lab: *Madame Curie* and the Image of the Woman Scientist in the Feature Film," *Public Understanding of Science* 6 (1997):269–278. Lindee and Elena argue that both mediums have played a role in constructing not only the specific, heroic image of Curie but also the image of women scientists more generally in contemporary culture. Such representations, they point out, often offer the "woman scientist" as a surprising, if sometimes important, aberration in the history of science, promoting the notion that a successful woman scientist is an exception to the norm as well as a good candidate for sainthood, sacrificing her life for the scientific calling.

Constructing a New Science:
The Masculine Tradition

Whilst sharing in her Toil, she shar'd the Fame,
And with the *Heroes* mixt her interwoven Name.
No longer, *Females* to such Praise aspire,
And serfdom now We rightly do admire.
So much, All Arts are by *Men* engross'd,
And Our few Talents unimprov'd or cross'd."

> —Anne Finch, Countess of Winchilsea, from
> "A Description of One of the Pieces of Tapestry at Long-Leat" (1713)

During the sixteenth century and the early seventeenth century, the image of the goddess of Science often graced the frontispieces of scientific works. Such images represented a muse for both male and female authors, a source of inspiration and a guide to knowledge. By the eighteenth century, readers were more likely to find a portrait of the author engraved on the frontispiece of a work of science when the author was a man. When the author was a woman, the image of the goddess of Science persisted. The change of emphasis from outside guidance to self-reliance for male scientists symbolized the profound rethinking of the conceptualization of the natural world and how it should be studied. Few women experienced this shift to self-reliance. It was more acceptable for female scientists to continue to see themselves as guided rather than as self-reliant. The changes taking place in the emphasis of male scientists and the conceptualization of the natural world reflected the negative assertions about and attitudes toward both the intellectual and scientific abilities of women and toward nature as a whole.

The story of scientific women in history is not one of a simple progression over time. During the Middle Ages and the Renaissance, women had numerous opportunities to study nature, although what we think of as science today did not exist. Attitudes toward nature, ideas about how the natural world functioned, and the place of men and women in the study of that world were quite different from what they would become in the seventeenth century. In general, medieval

Engraving from Gregor Reisch's Margarita Philosophica *(1503) in which a female fig-ure representing arithmetic watches over Pythagoras using the arithmetic of the aba-cus and Boethius using the arithmetic of the decimal number system. (The Art Archive)*

scholars perceived the world as a living organism, nature as female, and male and female as two complementary parts of a greater whole. The spiritual realm was as equally available to study as the physical world and the two were insepa-rable from one another. Although medieval Europe did not support ideas of social or political equality between men and women, the religious and occult nature of the medieval world, the conceptualization of nature, and the empowerment of the reproductive woman did mean that women had access to some forms of learning and that their knowledge about nature was socially acceptable. As a result, women were respected in society for their knowledge of medicine, human reproduction, food production, and the life cycles and uses of animals and plants—all of which gave women an important economic role.

During the seventeenth century, a period now referred to as the Scientific Revolution, the approach to the study of the natural world changed. Seventeenth-century natural philosophers reconstructed science as a method of investigating the physical world through experimentation, mathematics, and correct reasoning, all reported in an impersonal manner. They believed that this was an objective approach to studying the natural world. Although historians have subsequently argued about whether or not this change was really revolutionary, it is clear that these scholars created a substantially new approach to the study of the natural world, even if it took a while for practitioners to solidify its strict practice. The new science they developed contrasted markedly with the old science. Instead of viewing the world from a holistic perspective as a living organism, the new philosophy imagined the world as a machine, the investigation and understanding of its physical parts being of the utmost importance. Scientists increasingly separated the study of the physical world from theology in this period, although they often saw nature as the realm of divine law. They sought to find the universal laws in nature. In their attempt to separate the new science from the old, scientists, then referred to as natural philosophers, formed societies and academies that would foster and protect their new approach to acquiring natural knowledge. They wanted to completely separate themselves from those men and women who continued to insist on practicing old methods of science, but in doing so they also isolated women from the new practice. Including women, they felt, would undermine their new study, in part because women tended to be followers of old practices, but also because of the "natural" character of women, which they believed was irrational, emotional, spiritual, and lacking intellectual rigor. Legitimizing their new objective study meant keeping out those who either through their low social status, inappropriate "natural" character, or wrong-headed approach to natural knowledge would undermine the prestige of the new science.

The story of the Scientific Revolution is, then, partly one about the ostracization of women from investigations into the workings of nature. Historian David Noble has called this new science "a world without women." Natural philosophers, like the men of the church in the same period, believed they had to oust women from their midst to solidify their authority and prestige within the larger society. Yet Noble's bold assertion does not reveal the whole story. Certainly natural philosophers generally argued that women should not, and could not, practice the new science, barring them from entry into their inner sanctum of societies and academies. Nevertheless, some women did continue to practice science, often through discussions with men and the reading of their published works. Although the nature of science and science practice was changing during this period, there were still many spaces in which women could practice science.

Some women were concerned about the implications of the new science, not just for themselves and other women but for the natural world, the spiritual realm, and civilized society. As a result they spoke and wrote against the new science and in some cases provided alternative worldviews. Equally, some women were fascinated by the new science, supporting it through discussion and published writings. In either case, the lack of well-established institutions of science during this period meant that women could still gain access to scientific knowledge, despite the new philosophy and the efforts of the natural philosophers to keep them out.

Women, Nature, and Knowledge in the Medieval World

Prior to the introduction of a new kind of science in the seventeenth century, European society's attitudes toward the study of the natural world incorporated three different influences: First, religious ideas and beliefs about nature outlined by the Bible and Christian theology suggested that God had created the world and placed humans in it as his supreme creation, giving human beings dominion over nature. Second, alongside the Christian tradition there still existed a mystical, magical, or occult tradition that believed the natural world was powerful in its own right and must be respected by humans. And third, the texts of the Greek scholars expounded philosophies about the natural world that incorporated both the material and the spiritual in their study of nature. These three different branches of knowledge coalesced with one another to create a medieval vision of the natural world that was harmonious, organic, and fluid. Nature was alive, both in an organic and in a spiritual sense. But though the medieval world was certainly not a world of social or political equality for women, this did not mean that women were not respected within their appropriate sphere or that they did not have any power at all within European society. In fact, while the Christian, mystical, and Greek traditions all supported the subordinate position of women within society, at the same time these bodies of thought created space for women to think and learn about nature, and to have influence in the realm of natural knowledge.

The use of the term *Mother Nature* throughout history has led women to be given power in society and to have it taken away from them. The representation of nature as female had two major implications for women. First, it implied that women had a natural insight into and knowledge of the natural world around them that men did not have. Second, the term implied that women were closer to the natural world and therefore further away from the civilized world of mind and intellect, which men tended to represent. Although these assumptions were not true of all women, prior to the Scientific Revolution this association of women with the knowledge of the natural world was, to some degree, a powerful position for women in society. Generally speaking, European societies believed that nature had to be treated with kindness and respect if human beings wished to harvest and mine the land successfully, and so too did women have to be treated with similar respect if men and women were to live in harmony with one another. They respected nature as a powerful, creative force and as a dangerous, destructive force, and, by association, women were seen in the same light. Women were producers—of life, of food, and of healing remedies. They had these abilities because of their close ties with the natural world, which in turn provided them with the knowledge to bear forth the means of subsistence, survival and reproduction. As such, they were integral members of the community. Medieval men who studied natural knowledge or who worshiped the natural world as a manifestation of God could certainly also come to know nature. But in order to do so, they had to wait for nature to guide them to her secrets. Ideally, those who wanted to know the natural world in medieval Europe believed they had to work with and within nature, not attempt to control, manipulate, or dominate over nature. Not everyone held such attitudes and beliefs, but those who did believed that creating a harmonious relationship between human beings, the land, and God was of paramount importance.

These ideas about nature, and about the characteristics of men and women, were bound up in Greek, Christian, and mystical doctrines. The ancient Greek philosopher Plato believed that the whole world was an organism brought to life by a female soul. Aristotle, Plato's student, also posited the organic nature of the world through the interconnectedness of matter, form, and purpose. Centuries later, an early form of Christianity, referred to as Gnosticism, although arguing for a separation between spirit and matter, good and evil, and God and nature, nevertheless believed that God had created a female generative principle, which in turn generated angels and the physical world. God was, for them, androgynous, encompassing attributes of both the male and the female. As a result, many women were attracted to this early form of Christianity. The gnostic tradition influenced the mystical, alchemical tradition, in which practitioners believed it was possible to turn base metals into gold and to discover the elixir of long life. They upheld the gnostic belief in an equality of male and female generative principles, and as a result, women wrote many of the early treatises on alchemy. Twelfth-century Christians, who worked Platonic and Aristotelian ideas into their philosophy, believed that nature was subservient to God but nevertheless recreated female nature as a goddess who, while answerable to God, was nevertheless more powerful than human beings.

As women were the bearers and nurturers of new life within human society, so was the earth bearer and nurturer of all life on earth. The association of women's biological and social roles with those of the earth resulted in a symbolic and literal relationship between the two. This general philosophical belief that the earth was female resulted in respect for the land. During the sixteenth-century Renaissance, European societies still strongly held these same beliefs. Although they certainly cultivated and mined the land prior to the seventeenth century, to some extent the representation of nature as a mother led to a degree of restraint upon the harvesting of plants and minerals from the earth. There was plenty for all without having to ravage the bowels of the earth. Pliny of Rome (A.D. 23–79) warned in his *Natural History* against overmining, noting, "We penetrate into her entrails, and seek for treasures . . . as though each spot we tread upon were not sufficiently bounteous and fertile for us!" He argued that earthquakes were Mother Nature's way of protesting against the defiling of her body. It was obvious to many, even in this early period, that the avarice of human beings could cause pollution of the air, water, and land. In 1530, Henry Cornelius Agrippa in his *The Vanity of Arts and Sciences* warned that mining had "made the very ground more hurtful and pestiferous." Such voices struggled against the increasingly capitalist, not to mention Protestant, notion that the earth was God's gift to human beings to exploit at their will.

The feminization of nature, according to the historian Carolyn Merchant, resulted not only in the respectful treatment of the natural world but in giving women an important, if not equal, position in society. The association of women and nature with reproduction meant that medieval society took for granted that women would have more knowledge about the human body than would men. As a result, women oversaw reproductive and healing processes, fulfilling roles as midwives and healers in their communities. These women had usually apprenticed and were thus well trained and well paid for their work. Yet despite this separation of men and women's spheres in regard to natural reproduction, men and women could nevertheless be considered partners in economic production. Women born into peasant farming families or into urban artisan families certainly did their fair share of the work. Male family members clearly recognized the paramount importance of women to the economic survival of the family, and these women's contributions to the family economy were also more formally recognized by the craft guilds, organizations that regulated the workers of specific skilled trades. The recognition of women by such institutions and in law, as either their husband's partners or, after their husband's death, the rightful heir to the family land or trade, could result in a strong economic position for women within society. Occasionally, women became guild members of crafts as individuals in their own right, and some women organized their own guilds, such as the women silk producers in London. Similarly, aristocratic men considered their wives capable of overseeing their business interests when they were away from home and of having the ability to take over the running of a business after their husbands' deaths. Such positions gave women some limited civic rights as well. Londa Schiebinger has argued that although sciences such as astronomy or entomology, for example, never constructed guilds, these craft traditions that recognized women's rights as workers facilitated women's entry into the sciences.

Another way for women to enter into the realm of knowledge was through the convents. Women were often founders and leaders of monastic communities in the early medieval period. While still subject to the Catholic hierarchy, they found independence in such a life from the constraints of a patriarchal society and freedom to pursue their own interests without the encroachment or the demands and concerns of the outside world. Prior to the Protestant Reformation in the sixteenth century, monasteries not only were places of worship where men and women consecrated their lives to God but also places of learning. Particularly in the life of double monasteries (which housed both men and women), a female presence was often at the center of learning. The study of natural philosophy was an obvious corollary to the study of theology, for the ultimate purpose of both was to celebrate the glory of God.

Despite the decline of the monasteries after the Protestant Reformation that swept through much of Europe, the Renaissance was nevertheless a period of increased learning. During this period, the focus of learning shifted from religious institutions to the courts, where humanist scholars encouraged a new sort of study. Humanists were scholars who were rediscovering an interest in human culture and who promoted serious study of the arts and sciences within a growing court life. Developing urban communities, centered on the courts, drew together a significant number of aristocrats, who were encouraged to share and explore their newfound interests. Aristocratic men and women alike were involved in this reinvigoration of learning. In fact, the humanist approach to learning, which posited the importance of individual study of original texts, fostered the idea that women were just as capable as men of reading, writing, and thinking about the works they studied. Humanists considered intellectual ability to be a Christian virtue and thus one as amenable to women as to men.

Perhaps the most explicit practice of scientific learning by women took place in the household. The medieval and Renaissance tradition of the aristocratic housewife, familiar with the domestic sciences including a knowledge of medicine, pharmacy, and chemical science, meant that women could be as equally well-versed in science as men. Technical books for women proliferated in the sixteenth century, attesting not only to their participation and skill in such activities but also to their increasing literacy, their familiarity with Greek mythology and Latin pharmacological terms, the importance society placed in women's education and training in these fields, and, in essence, as Elizabeth Tebeaux and Mary Lay have argued in "Images of Women in Technical Books from the English Renaissance" (1992), their "freedom to enter fully into the world" (206). Women such as Henrietta Maria, wife to King Charles I of England, and the sisters Elizabeth Grey and Alethea Talbot, ladies of the bedchamber to Queen Henrietta Maria, published their knowledge and shared their learned information within circles of like-minded women. Many other now-forgotten women would have had the same knowledge, even though they did not publish. They would have tended not only to their family's needs but to those of the community. The mistress of the house had to know and understand recipes and instructions for everything used in the house, including for cooking, planting and harvesting of crops, dyeing, medical treatments, midwifery, cleaning products, making of ink, brewing and distillation, wine preparation, bread-making, beekeeping, animal husbandry, horse training, silkworm production, dairy work, drying, storing, pickling, brining, and conserving. A knowledge of gardening did not go amiss, either. John Gerard's well-known *The Herbal or Generall Historie of Plantes* (1597) acknowledged that in many instances women were the source of the knowledge he conveyed in recipes attached to specific plants for medical or food preparations. Knowledge of how to grow medicinal ingredients was of the utmost importance for women.

Though sixteenth-century scientific men would have preferred to keep women out of their scientific practice, the lack of institutional support of any kind for science at this time meant that women were often involved in one way or another. John Dee, a sixteenth-century English natural philosopher renowned for his alchemy, astrology, and conversations with angels, and Ulisse Aldrovandi, a Bolognese naturalist, both shared the same situation. They had to practice their science in the domestic sphere. Although both men were relatively wealthy in the context of the times, they, like their other counterparts in science, had no official place of work. John Dee, employing several apprentice-assistants who lived and worked in his home, found that it was nearly impossible to keep his scientific work separate from the domestic work of the household performed and overseen by his wife, Jane Dee. Although John Dee had many outbuildings on his property that accommodated his work, Jane Dee often found him and his work underfoot. Though John notes with apprehension his wife's anger at the poverty and disreputable reputation his practices brought upon their household, the fact of the matter was that he needed his wife's assistance to be able to practice science.

Ulisse Aldrovandi, fearing similar difficulties, made strenuous attempts to physically separate the domestic household from his scholarly endeavors. At his palace in the city he not only had a separate study room but also built a separate museum building for his collections. He believed that the life of the mind was a masculine activity, separate from the daily workings of the household, which his wife, Francesca Fontana, oversaw. So insistent was he upon this separation that when he built a villa in the country to better accommodate his study and writings on nature, he constructed separate apartments for his wife. Just in case there was any doubt about her position, he had inscribed above the entrance to her dressing room, "It is proper that women be clever not in civic but in domestic affairs." Yet despite these efforts Aldrovandi was no more successful than Dee in keeping his wife secluded from his work in the domestic realm.

Jane Dee's role in her husband's science involved the difficult and onerous task of "managing an experimental household," including budgeting their finances, overseeing the apprentices as well as up to twenty household servants, and maintaining the respectability of the household in the eyes of the community despite the mysterious studies her husband conducted. Jane even became the subject of her husband's work as he tracked their sexual activity and Jane's menstrual cycles, going so far as to examine the fetus of a miscarriage Jane suffered, and wondering over its shapelessness. While John was certainly the philosopher, Jane, to a large extent, controlled the physical and social settings of his practice.

Francesca Fontana was even more deeply connected to her husband's scientific work. While relying upon Francesca, as John relied on Jane, to oversee the household, Ulisse Aldrovandi also needed Francesca's assistance when women visitors came to view his museum. Moreover, it was Francesca's dowry that allowed Aldrovandi to build his country villa. But most important, she was a learned lady in her own right. As such she helped Aldrovandi assemble his books by editing his writing and finding useful passages for his books. For the posthumous 1606 work, *On the Remains of Bloodless Animals*, which was their shared work, she was responsible for writing the Latin preface. For ten years after his death she controlled access to his museum. Unlike David Noble's "world without women," Deborah Harkness (1997) has argued that sixteenth-century natural philosophy was "a world *among* women, for natural philosophy was the guest of the household during this period" (251).

The New Philosophy: Keeping Women Out

Whereas the medieval philosophies contained holistic, organic, and vitalist components, within these systems existed the seeds of a different science waiting to be exploited. Some scientists, such as Philippus Aureolus Paracelsus (1493–1541), a Swiss alchemist and medical practitioner, posited the equal part played by men and women in reproduction. But the dominance of Aristotelian theories about human generation and, consequently, their implications for men and women held firm sway in the medieval and Renaissance periods. Aristotle firmly upheld the notion that there were significant biological differences between men and women based on their reproductive functions and roles that in turn dictated separate and different social roles for each of them. Male seed was the creative force in reproduction; the female womb was simply a carrying receptacle that housed and nourished the baby until birth. Such a theory posited, by extension, that men were naturally active beings and women were naturally passive beings. Men were made for the world, women for the home. Men were producers, women reproducers. This theory was made analogous to the universe, where the male heavens impregnated the passive female earth.

Within Judeo-Christian traditions, religious men and women believed that women were subordinate to men and that the natural world ought to submit to the will and demands of human beings. According to Genesis, man's dominion on earth was a God-given right. Woman's place in Christian society was as the subservient role of mother, wife, and daughter. Though woman was made from man, she was different from him. She had brought disorder into the world by defying God's law and eating fruit from the tree of knowledge in the Garden of Eden. As a result, human beings had been punished by God. Banished from paradise, men were forced to work the soil and women to bring forth children in pain. Medieval Christians considered it was man's role, as the ruler on earth, to regain control over nature, knowledge, and women. In this way paradise might be regained. Moreover, changes were taking place in the Catholic Church that increasingly encouraged a male-only community, whereby church authority ruled over individual, prophetic authority. The Church increasingly placed men at the head of religious households. This new ideal, contrasting with that of the earlier mixing of men and women in religious groups, was an attempt to solidify Christian ideals. The clergy turned their backs on women leaders, or at least insisted upon the separation of men and women within the religious community, as they increasingly associated women with heretical and radical ideas and with sexual temptation. In doing so, the clergy also raised themselves institutionally above women as they kept women apart from church rituals, no longer permitting women to perform them, or to hold high positions within the church.

In this context, philosophers posed a new approach to the study of nature. The two most famous proponents of this new philosophy were the Englishman Francis Bacon (1561–1626) and the Frenchman René Descartes (1596–1650). Bacon, a Cambridge-educated lawyer, was frustrated by the men of the old

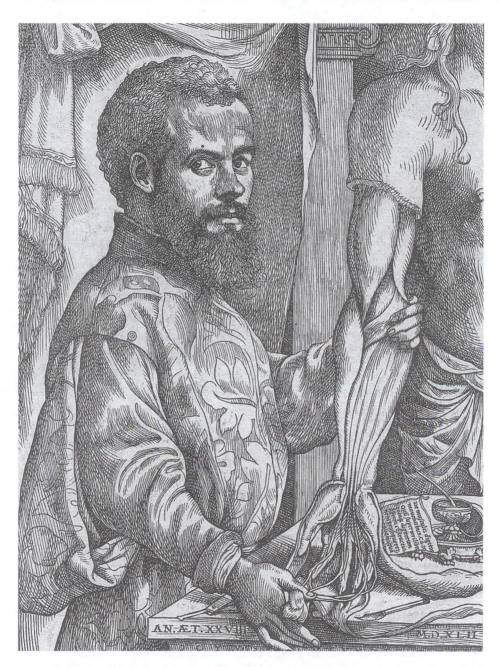

The only known likeness of Andreas Vesalius (1514–1564) shows the formally attired Belgian anatomist grasping the dissected arm of a cadaver. (Courtesy of the National Library of Medicine)

schools of thought that paid more attention to talking about things and reading what others had written than in actively seeking knowledge for themselves. In his works *The New Organon* (1620) and *New Atlantis* (1627) Bacon made clear his belief that if one wanted to discover answers to questions about the natural world, one had to look directly to nature. He wanted to get away from abstractions and instead wanted men of learning to consider objective phenomena. Bacon emphasized fact collection and experimentation, but he also stressed that experiments ought to pose specific questions capable of being answered experimentally. Such assertions encouraged the investigation of the material world and the setting aside of the spiritual world as unamenable to direct observation or experimentation. Collections of experimental facts would lead, eventually, to valid generalizations. This method of acquiring knowledge, moving from the specific data acquired through experimentation to general theories derived from the results of numerous experiments, is called *induction*. Bacon also posited a new world order based on his new method, wherein scientists would work communally in institutions established for the purpose and would be paid for their work. He also realized the important connection between advances in scientific knowledge and advances in technology, which, in turn, he argued, would improve human life.

In contrast to Bacon, René Descartes was a proponent of the *deductive method*, which involved using the mind to reason and thus theorize about the natural world. He was not as interested as was Bacon in experimentation, but he did accept the usefulness of experiments and observations to a point. Instead, Descartes posited the notion of mind/body dualism, which meant that he believed that the mind could exist separately from the body and was not dependent on the body. The individual mind was pure and was only tainted by outside influences; thus he advocated a reliance on the right reasoning of the individual mind to explore knowledge for itself without any prejudgments, prejudices, or outside interferences. The body was merely a machine that allowed human beings to move around and did not impinge upon the mind's ability to reason. Although Descartes firmly underpinned his philosophy with God, some of his religious contemporaries complained that God was peripheral in Descartes's methodology. His major work that explored these themes was *Discourse on the Method of Rightly Conducting the Reason and Seeking Truth in the Sciences* (1637). Eventually, different aspects of Bacon's and Descartes's ideals about the search for natural knowledge joined to formulate a modern scientific method that included experimentation, direct observation of nature, mathematics, and theoretical reasoning.

Sir Isaac Newton, famous for articulating the theory of gravity, has often been upheld as an icon of the new philosophy in the seventeenth century. Unlike others before him, Newton had described what he called "The System of the World," establishing universal laws for the motion of bodies on earth, tides, and planets through the application of mathematical principles. Yet the new philosophy did not come into being immediately. Newton himself was still very much intrigued by the occult and by alchemical researches. At the same time he found himself wrestling with the "impertinently litigious Lady" Philosophy, who time and again seemed to mar his mathematical and experimental systems. Newton argued that gravity was a "force." Other natural philosophers viewed Newton's idea of "force" as containing occult qualities. On the one hand, Newton believed the world could be reduced to mathematical equations, but on the other hand, the world was also a mystical place, a world infused with feminine qualities, and a world created and influenced by God.

The new men of science had to learn the new scientific methods introduced by Bacon and Descartes and subsequently had to protect these ideas from thinkers who preferred traditional ways of seeing and knowing the natural world. Men of science constructed new metaphors to describe the natural world and the way they believed it ought to be investigated, in an attempt to change the old mind frame of the unconverted. Bacon's work suggested a new attitude toward the investigation of nature. Rather than waiting for nature to unfold her secrets, the natural philosopher had to force nature to reveal them. In *The Masculine Birth of Time* (1603) he wrote, "For you have but to follow and as it were hound nature in her wanderings, and you will be able when you like to lead and drive her afterward to the same place again. . . . Neither ought a man to make scruple of entering and penetrating into these holes and corners, when the inquisiting of truth is his whole object." Moreover, this systematic approach to the study of nature that separated God, feeling, and spirituality from science left a dead, inanimate world for investigation. The focus of study was no longer to understand the interconnections and workings of the cosmos as a whole but to understand the separate parts and how they functioned. This mechanical philosophy removed many of the characteristics associated with women in society from their application to the study of nature.

Historians have argued that such descriptions of and theories about nature resulted in a backlash against women and the natural world. They suggest that such language reinforced the notion that women were incapable of becoming part of the new scientific community and that nature and women were objects to be studied, controlled, and utilized for man's selfish purposes. Historians have interpreted the metaphors that Bacon used to describe the natural world as condoning and encouraging men's rape of the natural world. The violence of this image replaced the more reverential attitude toward nature that had existed beforehand and, argue historians like Carolyn Merchant, reflected the loss of status and respect for women too. Other historians have cautioned against such blanket statements about Bacon and by extension all of science and its practitioners, arguing that the characterization of them as brutal rapists is unfair. Bacon also used metaphors that could suggest he saw a role for women in science; for instance, in his *The Advancement of Learning* (1605) he referred to "knowledge that is delivered as a thread to be spun on," spinning being the work of women. Moreover, he clearly remained anxious for investigators to treat nature with care and respect, writing in his *New Organon* (1620) that science is "the servant and interpreter of Nature" and that "Nature to be commanded must be obeyed." As Evelyn Fox Keller (1985, 42) has pointed out, "mind" for Bacon "was both phallus and womb"; alternatively male and female, he confounded the gender of both nature and knowledge. Equally, Descartes, writing in French rather than in Latin and arguing for the power of the individual mind regardless of the biological body to which it was attached, certainly provided space for women to read and put into practice the method of investigation he outlined. Nevertheless, in the end, the ideas they presented in their works did serve to elevate the value of stereotypical male characteristics and to demote those of women.

This ideological shift in the conceptualization of nature also resulted in a shift from the view of a harmonious natural world to a chaotic one. The role of the scientist was to establish order and control over the natural world and, by extension, over the lives of women. One of the most disturbing aspects of the relationship between women and nature for early modern scholars was the power that such a relationship engendered for women. Though women's relationship to, and knowledge of, reproduction and healing had long been established, this state of affairs began to change. Women's knowledge in these areas was increasingly seen as incompetent during the sixteenth century. Natural philosophers argued that women's ideas and beliefs were grounded in mystical and superstitious practices rather than in correct reasoning or knowledge about nature.

At the same time, villagers began to often accuse their wise-women of being responsible for the ills that beset their community. The death of animals, destruction of property, or an unexplained illness or death of a family member could provoke neighbors to accuse others of witchcraft against them. These women were accused of having sexual relations with the devil to gain illicit knowledge and to satisfy their lustful passions. Like Eve in the Garden of Eden, her knowledge could bring the downfall of humankind. Such women were considered out of control and a danger to their community, for they had the power of the devil on their side. The witch trials that took place all over Europe from the end of the fifteenth century to the end of the seventeenth century resulted in the torture and brutal murder of thousands of women condemned to death for witchcraft. In the years 1645–1647, an English lawyer named Matthew Hopkins, also known as "the witch-finder general," was responsible for convicting two hundred women under the Acts of Parliament that made witchcraft a statutory offense punishable by death. Accused women often confessed under extreme physical duress and mental anguish. The witchcraft trials of this period were perhaps the most terrifying result of the attempt by both religious and scientific men to gain control over nature, women, and society.

The nature of women's education also separated them from science. During the seventeenth and eighteenth centuries many educational treatises were being published that argued for the importance of a good education. In this context, writers outlined the nature of the most desirable form of education for children. But it quickly became clear that the term *children* actually meant boys, and that the discussion of girls' education was seen as separate and different from that of children more generally. Boys were educated in Latin, long the language of scholars and necessary for admittance to colleges. Girls were not. Educational commentators stipulated that boys needed an education in order to fulfill the duties of their occupation in life whereas girls needed an entirely different kind of education that was sex-specific, automatically disallowing their entry into any sort of scholarly institution or occupation. Girls' education usually consisted of learning some French, music, needlework, and basic reading. In more strongly religious homes, training in the domestic arena, as well as inculcating a sense of piety and duty, was considered important. Marriage and motherhood together were considered women's sole purpose in life.

Seventeenth- and eighteenth-century writers who believed in the importance of serious women's education argued that men's progressive education led to their ability to become responsible, independent individuals whereas women's education encouraged dependency, ignorance, and boredom. Bathsua Makin, a seventeenth-century educator, wrote in her *An Essay to Revive the Antient Education of Gentlewomen* (1673): "The barbarous custom to breed women low is grown general amongst us and hath prevailed so far, that it is verily believed . . . that women are not enbued with such reason as men, nor capable of improvement by education, as they are. . . . A learned woman is thought to be a comet, that bodes mischief whenever it appears." Though commentators made suggestions for the improvement of girls' education, they were rarely put into action. Kept outside of the system of liberal education that was developing for boys, women's education took place at home. Their education, based upon their sex, denied them access to the public schools and universities.

As seventeenth-century men set up scientific societies alongside these educational institutions, women were absented from them. These men founded the Royal Society in 1660 in England, the French Academy of Sciences in 1666, the Prussian Academy of Sciences in 1700, the Academy of Sciences of the Institute of Bologna in 1711, and the Imperial Academy of Sciences in Russia in 1725. These societies oversaw the production, evaluation, and reporting of new scientific knowledge both to their members and the public at large, establishing library and museum collections in the process. They also held themselves responsible for defining and controlling scientific standards and for providing a protected, institutional space for scientific practice and discussion, free from theological interferences. Henry Oldenburg, secretary of the Royal Society, indicated that the Society's purpose would be to "raise a Masculine Philosophy . . . whereby the Mind of Man may be ennobled with the knowledge of Solid Truths" (quoted in Boyle 1664). Women, it appeared, were nowhere to be found; indeed, the academies refused membership to women almost entirely until much later. The new science had triumphed as a controlled body of knowledge that only a circumscribed group of male individuals could possibly hope to acquire. The new philosophy had laid down the rules that outlined the path to this knowledge and, over time, the new men of science marginalized and ostracized those who did not seek natural knowledge in this specified way, increasingly silencing their objections.

Women's Science Practice in the Seventeenth Century

Anna Maria van Schurman (1607–1678) was a famed and learned Dutch scholar who supported education for women and who rejected Cartesian philosophy, defending Aristotelian philosophy and the Christian tradition. She concluded that "the results of women's thoughts and deeds left almost as little trace in history as a ship leaves on the sea" (Rang 1996). Of course women did think, write, and act on their beliefs and ideas, but historians have had to work hard to retrace women's participation in intellectual activities in the seventeenth century, including uncovering their attitudes toward and participation in the new science. The resurrection of Anna Maria van Schurman's work has been due to the kind of painstaking research that has revealed many women who practiced science in this period. The new scientists did not firmly solidify their male-only enclaves until the late seventeenth and early eighteenth centuries, so women found themselves able to push through into the unsealed domain of male learning in various ways. Moreover, although fascinated by the new philosophy, women did not receive it uncritically, and they at times provided insightful objections and alternative philosophies to counter the new trends arising in science.

In the seventeenth century, aristocratic women could move beyond their domestic science studies via male family members' interest in science. Through familial contacts, women supported men in science, became their patrons, and sometimes participated in scientific activity. For example, Elisabetha Hevelius (1647–?), wife of renowned astronomer Johannes Hevelius of Danzig (now Gdansk, Poland), assisted her husband with his astronomical observations. She

Johannes Hevelius with second wife Elisabetha, using a sextant. Six feet in radius, the sextant was modeled after that of Tycho Brahe. (Bettmann/Corbis)

became so adept that after his death she completed and published his unfinished works. In England, Katherine, Lady Ranelagh (c. 1614–1691), the older sister of Robert Boyle, who was one of the Founding Fellows of the Royal Society and who epitomized the experimental method through his work in chemistry, was part of an intellectual circle of women who produced books on household science, including medicine and pharmacy. Her works exhibited a practice of chemistry and a knowledge of herbals. Boyle had his sister to thank for his early introduction into the inner circles of London literary and philosophical society. As well as sharing the same social circle, the brother and sister also shared laboratory facilities, built by Boyle onto the back of Katherine's house in Pall Mall. The similarity between chemical and kitchen technologies may also have meant that Katherine performed her scientific work in her kitchen. Though Katherine was still ostensibly working in the older traditions of a home-based, domestic science, she was nevertheless using new experimental techniques. Obversely, while reaching into new and uncharted territory in chemistry, Robert Boyle nonetheless recognized the usefulness of the kind of knowledge his sister produced to his own work. Only in the 1660s, the first decade of the existence of the Royal Society, did evidence arise of the disparaging of "Ladies Chemistry" to separate it from the more gentlemanly and aristocratic endeavors of chemists working within the new science.

Not all women needed to have familial connections to a man to practice science. Queen Christina of Sweden (1629–1689) studied theology, philosophy, and astrology and was "an adept like Newton" (Cook 1997, 3) in alchemy. She gave her patronage to the Accademia Reale in Rome and supported another academy in Valicella, allowing scientists to study natural philosophy despite the hostility of the religious community. Clearly, women were not completely cloistered from science upon the introduction of the new philosophy, nor were they relegated to any particular disciplines of natural philosophy; their interests and their practice ran the full gamut of possibilities.

Women were also in contact, through writing, with men who could assist their scientific learning. Descartes, considered one of the most important philosophers of the new science, had a number of female patrons and students among his correspondents. One of the ironies of Descartes's new method of philosophy, considering its role in ousting women from the realm of natural philosophy, was that he had written it in straightforward language hoping to reach a broader audience and as a result thought that even women who lacked formal training would be able to understand and follow it. Aristocratic women were intrigued to learn about this new philosophy. They read Descartes's *Discourse on Method* and wrote to him with their questions and philosophical concerns. Descartes took this correspondence seriously, finding that in some instances it assisted him in thinking through the philosophical difficulties and problems he faced. One such correspondent was Princess Elisabeth of Bohemia (1618–1680), with whom he carried on a six-year correspondence. Elisabeth numbered among her other correspondents the philosophers Gottfried Leibniz and Nicolas de Malebranche and the Dutch scholar Anna Maria van Schurman. She stood out in the Bohemian court for her learning. Contemporaries considered her "incomparable," as she was equally adept at mathematics and metaphysics. Taking up Descartes's proposed separation between body and spirit, she found herself confused about how the immaterial soul could influence the material body. Her question led to a lengthy debate between herself and Descartes, and eventually resulted in the latter's writing of his *Passions of the Soul* in 1649. Earlier, in 1644, he had dedicated his *Principles of Philosophy* to her. Elisabeth was fascinated by Descartes's methodical doubt, sex-neutral reason, universal moral rules, and scientific method. She believed his principles allowed anyone who wished to do so to participate in science. Nevertheless, she could not reconcile his objective, abstract stance to the study of the world with his need for a subjective individual searching for truth. Elisabeth's own debilitating physical ailments led her to doubt the feasibility of the mind to overcome the onslaught of disease and the passions, and thus to subsume subjectivity under objectivity, body under mind. Her personal experience told her otherwise. In addition, her religious faith made it difficult for her to see how she could reconcile Descartes's idea of an infinite universe with the special place of human beings on earth destined by God. Although giving her faith in her own rational faculties, Descartes could not persuade Elisabeth that human objectivity was possible.

Discussion of these new philosophies, and science more generally, also took place in a more public forum. Women were often responsible for running what are now referred to as philosophical salons. They held salons in the domestic sphere, the receiving rooms of their homes. Yet they were public in the sense that nonfamily members were invited. Moreover, discussion in salon conversations could assist the more public display of men's and women's philosophical disputations both in lectures and in published works. The emphasis on women's association with nature and the passions in this period makes it easy to overlook the fact that many seventeenth-century women were consciously looking for ways to assert themselves as rational beings capable of high-level reasoning. As science moved slowly out of the domestic realm, the salon became an intermediary space in which both sexes could meet and discuss the issues of the day, including politics, religion, literature, and science. Women who followed Descartes's philosophy became known in the French salons as *Cartesiennes*. These women, like Princess Elisabeth of Bohemia, debated and responded to the new philosophy.

In England, women also responded to Descartes both positively and negatively. Anne Conway (1631–1679) engaged in a lengthy dialogue with Henry More, one of the foremost Cambridge scholars of Cartesian philosophy in Britain. Though their letters did not contain discussions on philosophy like that of Descartes and Elisabeth, it was clear in their exchange of Descartes's works that they shared an interest in his ideas. Other philosophers, such as Francis Mercury van Helmont and Joseph Glanville, spent extensive amounts of time at her home, Ragley Hall, in Warwickshire, reading, talking with her and with one another, and working in the laboratory set up there. Conway, suffering from debilitating headaches throughout her life that often confined her to darkened rooms, nevertheless found the opportunity to write, in half-illegible scribble, her only book, *The Principles of the Most Ancient and Modern Philosophy* (1690?). Her notes were compiled by her friends More and van Helmont and published on the Continent and in England after her death. Conway's work expressed her religious and philosophical concerns with the new science and proposed instead her own "monastic vitalist" philosophy, rejecting the Cartesian philosophy and opposing her friend and correspondent, Henry More. She argued that nature was a living organism, that there was no such thing as dead matter, and consequently that there was unity, not separateness, between mind and body. As a result, she believed in the reincarnation of spirits in new bodies after death, denying that the body was simply a machine of knowable parts. Furthermore, she retained a distinction between God and his creation. Serious and intense study of Descartes's work led Conway to reject Cartesian philosophy.

In many ways a contrast to Conway, Margaret Cavendish, Duchess of Newcastle (1623–1673), was a self-educated, prolific writer of poetry, literature, and philosophy. Her husband, William Cavendish, supported these endeavors, which was just as well, for she noted in her *Philosophical and Physical Opinions* (1655) that "I cannot for my life be so good a housewife as to quit writing." As a self-proclaimed materialist she began by writing in support of the new philosophy, arguing that all substances were made from atoms. She quickly realized, however, that atomists' philosophy led to a mechanistic worldview with which she did not agree. She feared that such a philosophy would only further the social and political chaos she witnessed during the English Civil War because the theory eliminated God and hierarchy and substituted chance, individual inclination, and democracy in their place. She did, like Bacon, continue throughout her life to separate the study of faith and reason, leading to the accusation that she was an atheist. She was not an atheist. She simply believed that though all men and women were religious, the imposition of religion was a source of discord and cause for war between factions. Cavendish was searching for a philosophy that would resist, if not prevent, social discord that she believed was an integral part of

Anne Finch, later Lady Conway (1631–1679)

Anne Finch was born into an aristocratic London family. Her father recognized her as an intelligent woman and had her home-educated by tutors in Latin and mathematics, among other subjects. She was weak in health from early childhood but nevertheless relished intensive study. She suffered throughout her life with excruciating headaches. Continually attempting to seek cures, she was brought into contact with some of the most renowned doctors, surgeons, and chemists of her time. In 1650, at nineteen years of age, she married Edward Conway (later third Viscount Conway and Kuillutag and first Earl of Conway). Immediately after their marriage she lived at Ragley Hall in Warwickshire, but after the death of her only son in infancy from smallpox, she lived for some years at Portmore, her husband's property in Ireland. Anne's mental and physical health had been badly affected by the loss of her son. Nevertheless, supported in her philosophical studies by her brother, John Finch, and her husband, she continued her tutoring after marriage. Her brother introduced her to Henry More, a Cambridge tutor, whose philosophical interests lay in studying both Plato and Descartes. He quickly came to recognize her as an intellectual colleague rather than a student. He later introduced her to the chemist Francis Mercury van Helmont, and through him she was introduced to the Kabbalah and Quakerism. As a result she became a Quaker late in life. During her lifetime she wrote a philosophical treatise refuting the Cartesian system and promoting her own vitalist philosophy in which she argued that all of nature was a living organism, denying the mind/body dualism of Cartesian philosophy. This work was published after her death as *The Principles of the Most Ancient and Modern Philosophy*, in or about 1690, by More and van Helmont, in Amsterdam. Some scholars believe that her work heavily influenced Leibniz, a German mathematician and philosopher, who found he shared many of Conway's ideas.

human nature. Though she dismissed Bacon's experimental method as ultimately flawed, she followed Descartes's philosophy that posited that right reason came from depending on one's own thoughts and ideas, not on those of others. Her experience of a hostile external world led her to turn inward to seek answers within herself about the world and to invent new worlds rather than discover those existing. She rejected Cartesian dualism, believing, like Princess Elisabeth of Bohemia, that subjectivity could not be overcome and that reason was a part of all matter. Her own philosophy contained many of the same features as those of the men advocating new approaches to the study of nature. Men of the new science ignored and ridiculed her philosophy because she rejected experimentation as a useful method of studying the natural world. Aspects of Cavendish's philosophy, such as her belief that the earth had "Restoring beds or Wombs" that produced new life at its center every springtime, seem easy to ridicule today. Yet many of her contemporaries held to the same or similar beliefs.

To promote her own program for the study of nature and women's involvement therein, Cavendish utilized aspects of the very philosophy that male scientists would use to oust women from scientific study. She adhered to the rational and reasoning aspects of Cartesian philosophy but nevertheless found within it space for women's transgression of their supposed nature. If women could reason as the Cartesian philosophy suggested, why should they not be an equal part of the intellectual and rational world of scientific investigation? Despite her own happy marriage and ability to write and publish, she was well aware of the increasing restrictions placed on women's learning. In her epistle "to all noble and worthy ladies" prefixed to her *Poems and Fancies* (1653) she wrote, "we [women] are like worms that only live in the dull earth of ignorance, winding ourselves sometimes out by the help of some refreshing rain of good education, which is seldom given us." Although on one occasion the Royal Society invited Cavendish as their guest, for the most part many of her contemporaries dismissed her as unlearned, nothing but a social pariah, even arguing that she was probably mad. This was the price she paid not only for working outside of the realms considered suitable for women but for her attempts to undermine the new philosophical basis of masculine authority. Her *Observations upon Experimental Philosophy* and the appended *Blazing World* (1666), dubbed an early work of science fiction, were her vehicles for attacking and satirizing the Royal Society and their experimentalism that posited that the world was objectively knowable.

Whereas Cavendish and Conway remained outside of the new science in different ways, other women found ways to pursue their scientific interests within the male-dominated world of institutions and publications. Two German women, Maria Sibylla Merian (1647–1717) and Maria Winkelmann (1670–1720), are examples of seventeenth-century women who moved into the increasingly masculine world of science. However, both women remain, in Natalie Zemon Davis's term, "on the margins" of the new, male science. This does not, however, detract from their remarkable achievements. Merian, the wife of an artist, was a skilled artist-naturalist, well-versed in watercolors, oils, painting textiles, and copperplate engraving. In 1699, at the age of fifty-two, she boarded a boat to Suriname (Dutch Guinea) with the purpose of studying and drawing nature. In 1705 she published her work as *Metamorphosis of the Insects of Suriname*, which achieved the expectations of natural history of this time. She directly observed and accurately drew the plants and insects that appeared in her book. But she was also inspired by her religious wonder of the natural world and, unlike the still, lifeless depictions of nature she so often found in the works she studied, was determined to give life to the insects she observed, to understand their life cycles and to place them within an ecological context. Moreover, she

Margaret Cavendish, Duchess of Newcastle (1623–1673)

The life of Margaret Cavendish (neé Margaret Lucas) was profoundly affected by the English Civil War. Born into a Royalist landowning family, she received a typical girl's education. As a young woman she entered the service of Queen Henrietta Maria. They fled into exile to France together in 1642. Here, she met and married William Cavendish, an exiled Royalist commander and later Duke of Newcastle. Margaret was William's second wife. In France, through the philosophical salon of her husband that became known as the Newcastle Circle, she met some of the foremost men of the new science: van Helmont, Descartes, and Pierre Gassendi. The experience of the salons and her husband's interest in science led to her own interest and early studies. Despite Margaret's lack of education, she was fascinated by the ideas of her time. During her life she published, with her husband's full support and encouragement, twenty-three volumes ranging from literature and poetry to philosophy and science fiction. She studied the major works of the philosophers of her day and the Greeks, but in the end rejected all their philosophies and instead constructed her own philosophy of the natural world, out-

(Bettmann/Corbis)

lined in her *Observations upon Experimental Philosophy* (1666). In her mind the best course to knowledge was one that sailed a safe distance between the chaotic theories of atomism and the overly mystical theories of the vitalist philosophies. After the restoration of King Charles II to the throne of England in 1660, the couple returned to England. During their time in exile, Margaret's brother had been executed as a Royalist and her mother's and sisters' graves desecrated. Both her family of origin and her husband's family lost property during this period that was never recovered. The duke was disappointed about not being admitted to the king's inner circle on his return to London. At the same time, Margaret found herself ostracized by the Royal Society members for her rejection of the experimental method they promoted. The couple retreated to their country estate, Welbeck Abbey. Although they remained childless, Margaret had several step-children who believed she was intent on stealing their father's property from him, putting their inheritance at risk. In the end, however, Margaret died suddenly before her eighty-year-old husband, at the age of fifty. In her honor the duke built an impressive tomb for her in Westminster Abbey.

peppered her text with tales and knowledge shared by Amerindian and African women with her about the use to which certain plants could be put. In order to accomplish her aims, Merian had to travel far at a time when ocean travel was rarely undertaken, even by men, and to live and work in a foreign land. Although

Maria Sibylla Merian (1647–1717)

Merian was born to German parents. Her father was an etcher, a physician, and a naturalist. Later, her stepfather was a still-life painter, engraver, and art-dealer, and her half-brothers were engravers and painters as well. Thus, Merian was brought up in a family that practiced both art and science. Her stepfather educated Merian alongside his male pupils, and she went on to marry an artist, Andreas Graf, who had come to study with her stepfather. She gave birth to two daughters (Dorothea Maria and Johanna Helena) who, in their turn, were also educated in medicine and art. Living with her husband in Nuremberg in 1675, still a young mother, Merian was included in the painter Joachim Sandrart's German Academy, furthering her art education. In these early years Maria was a skilled flower painter and employed her time painting on tablecloths and making copperplate engravings. After attending Sandrart's school she published a flower book, between 1675 and 1680. Though the work was ostensibly a pattern book for artists and embroiderers, its renditions of flowers, spiders, butterflies, and caterpillars were accurate. In 1679 and 1683 she published her two-volume *Wonderful Transformation and Singular Plant-Food of Caterpillars*, containing one hundred copperplate engravings, which, in contrast to the flower book, contained accompanying text about the life of the insects she represented. However, in 1685, her life took an abrupt change. For the next three years she retreated to the Labadist community in Holland where she could renounce worldly endeavors. Here she had the time to focus her attention on learning natural history and the Latin she needed to study it. Leaving behind her husband, who eventually divorced her, but taking her daughters and mother with her, she relished her life in this community until 1691. She then moved with her daughters to Amsterdam and made a living by selling her paintings. At the age of fifty-two, she managed to obtain financial assistance from the City of Amsterdam to take a trip to Suriname (Dutch Guinea) to study and paint from life the flora and fauna there, accompanied by one of her daughters. These trips resulted in the publication of her most well-known work, *Metamorphosis of the Insects of Suriname* (1705), which expanded her earlier three-volume work, *European Insects* (1679, 1683, and 1687). As a result of her travels and the ensuing publications, she became an important member of the Amsterdam scientific circle. She died of a stroke in 1717.

(Time Life Pictures/Getty Images)

her work was, in some respects, pushing the boundaries of accepted practice in natural history and art, naturalists nevertheless utilized it as an important resource. In some respects Merian studied the natural world with the objectivity demanded of the new science, yet in other respects she formulated a new view of the natural world that did not wrench its subjects for study from their natural context.

Maria Winkelmann was an astronomer who, like Elisabetha Hevelius, assisted her husband, Gottfried Kirch, with his astronomical work. She came to their marriage already well-versed in astronomy. In this context, she conducted observations and, as a result, discovered a previously unknown comet in 1702. She went on to publish three astronomical tracts between 1709 and 1711 in her own name. During this time her husband had been astronomer to the Berlin Academy of Sciences, but when he died in 1710, the academy would not consider giving her the position her husband had held as academy astronomer despite her solid astronomical reputation, and despite the fact that as he had aged she had taken most of his work upon herself. The academy did not doubt her expertise, but its members feared the precedent such an election would set for other women and what an influx of women into the academy would do to their efforts to build prestige for their institution. They did agree, however, to let Winkelmann continue in her role as assistant, along with her two daughters, to her son Christfried, who was an observer for the academy. Winkelmann was furious and frustrated but had to make the best of an imperfect situation. Women certainly managed to practice science in this period. But as science moved out of the home and into public institutions, women's hold on even the limited role they had previously played in science slowly diminished.

Conclusion

By the beginning of the eighteenth century, science was clearly moving out of the domestic sphere. Gottfried Kirch, Maria Winkelmann's husband, had recorded in his diary for 1704 that although he could see through the attic window that the sky was light, he was unable to make observations "since the washing from two households was hanging there. It was a pity because I missed the conjunction of Jupiter and Venus." In contrast, his son was to know only the better-equipped observatory of the academy his father had struggled to create. As a result of this change, women were less and less likely to be found as managers of, or partners in, their husbands' work. The best they could hope for was to be silent and invisible assistants, a far less powerful position within the scientific community, although not necessarily a less important one for the work of science. Yet always there are exceptions, and some women did make the transition from domestic to institutional science. Whereas the Berlin Academy turned down Winkelmann's request for membership and employment in the early eighteenth century, at the same time the aristocratic Laura Bassi (1711–1788) received a doctorate in philosophy at the University of Bologna in 1733 and quickly became a celebrated professor of physics there and a member of the Academy of Sciences in Bologna.

The origins of the sexing of science as a male activity began in the seventeenth century. At this time, male philosophers attempted to bring prestige to their occupation and its institutions and to assert control and order over the natural world. In order to accomplish their aims, they believed they needed to ostracize women from the practice of science. As the eighteenth century progressed, women were increasingly being depicted as subservient and dependent members of society. As they lost their role as producers and were increasingly circumscribed to fulfill domestic roles in the home and as their status as reproducers was diminished in importance, women maintained their association with the natural world. As a result, men no longer believed women capable of making any substantial intellectual contribution to either science or society.

The image of the goddess of Science was disappearing; no longer was Science a woman leading her followers along the path to knowledge. Moreover, Nature was increasingly represented as a willing woman, susceptible to the lure of men who wished to know her, unveiling herself before science. This metaphorical change highlights women's increasing absence from science as researchers, the decline of old traditions in science, and the rise of the new mechanical and experimental philosophy. But while women and their work, scientific or otherwise, were relegated to the shadows, the metaphors associated with women haunted institutions such as the Royal Society and scientists such as Sir Isaac Newton. Attempts to suppress women and the world they had come to represent were difficult. Even so, the struggle for women to regain a place in the new scientific community established in the seventeenth century would not be easy. The Nobel Prize medal, designed in 1902, has engraved on one side the image of Alfred Nobel, the inventor and patron, but on the other side there is an image of the female goddess of Science revealing the goddess Nature with her horn of plenty. As Londa Schiebinger (1988) has observed, even in the twentieth century, "Woman could serve as the image of science, but women were not yet welcome in the fellowship of science" (691). Still, Maria Sybilla Merian's face, not as goddess of Science or Nature, but as a woman scientist in her own right, long appeared on the DM500 banknote, attesting to the wide recognition and status that some women scientists have achieved, however belatedly.

Bibliographic Essay

Any account of gender and women in the history of science in the early modern period must make reference to Carolyn Merchant's work *The Death of Nature: Women, Ecology, and the Scientific Revolution* (1980). Although published over twenty years ago, Merchant's reconstruction of the seventeenth century as a period of loss—for nature, women, and science, and thus for society as a whole—is still considered to be a ground-breaking work. Previously, historians had depicted the science of the seventeenth century as the triumphant success of reason over religion and irrationality. Merchant's work, however, not only brings about a rethinking of this assessment of the seventeenth century but prompts the reader to ask whether a science that studies the world as a lifeless machine is really the best approach to the study of the natural world. Merchant's work asserts that the Baconian and Cartesian approaches to science gave human beings permission to damage the natural world, leading to environmental crisis. Furthermore, she argues, the human psyche has been damaged, as we now live in a world devoid of spirit and respect for nature and, by extension, absent of respect for women and women's ways of knowing the world. Feminist philosophers of science have taken up Merchant's call to reexamine the nature of science, resulting in the reconstruction of Baconian science as a male tool of repression over women and the natural world. Evelyn Fox Keller and Sandra Harding, along with Merchant, have examined the metaphors used by Bacon to argue that his philosophical system was misogynistic. However, it is worth noting that Fox Keller tempers her arguments in this vein by the admission that Bacon's metaphors shift frequently, resulting in a "hermaphroditic mind" (Evelyn Fox Keller, "Baconian Science: The Arts of Mastery and Obedience," *Reflections on Gender and Science* (1985):42; see also Sandra Harding, *Whose Science? Whose Knowledge?* (1991).

More recently, this line of research has been revised by scholars who feel that this feminist interpretation of Bacon makes too much of a few, often misinterpreted or misquoted, phrases in a large body of work. Alan Soble in his "In Defense of Bacon," *Philosophy of the Social Sciences* 25, 2 (June 1995):192–215; Kathleen Okruhlik in "Birth of a New Physics, or Death of Nature?" in Elizabeth D. Harvey and Kathleen Okruhlik (eds.), *Women and Reason*, (1992):63–76; and Sarah Hutton in "The Riddle of the Sphinx: Francis Bacon and the Emblems of Science," Lynette Hunter and Sarah Hutton (eds.), *Women, Science, and Medicine, 1500–1700: Mothers and Sisters of the Royal Society* (1997):7–28, all argue that the feminist attacks on Bacon are overwrought and need to be tempered with an assessment of Bacon within the context of his own time and within the context of his intended meaning. Soble's angry diatribe against Merchant, Harding, and Keller's interpretation of Bacon graphically makes the point that Bacon did not mean for scientists to go about raping and torturing women and that, in such assessments, Bacon's metaphors are taken too literally, "project[ing] into the canon horrors that are not there" (212). Hutton and Okruhlik take a more constructive approach, reassessing Bacon rather than attacking prior perspectives, yet they still clearly wish to moderate the tone of earlier feminist claims. Hutton notes that Bacon uses many different metaphors and that whereas some of his metaphors denigrate women and women's place in science, others suggest that Bacon recognized the power and knowledge of women. Okruhlik argues that instead of focusing on a dichotomy of theories about nature—either masculine *or* feminine—historians and feminists would do better service to science and to society if they recognized the need to make a theory choice that incorporated the best of both worlds. Okruhlik does not necessarily see holism, for example, as essentially feminine and reason as essentially masculine.

Feminists have argued that a female-centered view of the world has been and would be more sensitive to the ecology of both nature and society. The problem with this view is that it implies that women's holistic perspective does not and cannot incorporate any aspects of the new science of reason; that perhaps women are incapable of and/or disinterested in reason. Yet, historians have clearly uncovered much evidence of women's interest in and practice of the new science. An excellent example of such a work is Londa Schiebinger's *The Mind Has No Sex?: Women in the Origins of Modern Science* (1989), which makes the argument that although the new science supported the idea that 'mind' or 'reason' was not connected to one particular sex, social and intellectual characteristics long associated with women were defined as being incompatible with the needs of scientific study and were used by scientific men to provide evidence for the need to keep them out of science. Certainly some women rejected the new science, but even in this context their ability to read, understand, and in the end reject it implies that they are as equally intellectually capable as men of understanding the new science. Schiebinger provides evidence of how women have demonstrated themselves capable of science, via such examples as Maria Winkelmann and Margaret Cavendish, but argues that male contemporaries ignored such examples and ostracized women purely on the basis of their sex.

Other works, building on the claims of both Merchant and Schiebinger that women did practice science in the early modern period, demonstrate further ways in which women participated in science. Some have examined the physical spaces that allowed women to participate, to various degrees, in science. Such works point to the location of science practice in the home, and the domestication of some sciences, as ways that facilitated women's entry into and practice of science. See Deborah Harkness, "Managing an Experimental Household: The

Dees of Mortlake and the Practice of Natural Philosophy," *Isis* 88, 2 (June 1997):247–262; Paula Findlen, "Masculine Prerogatives: Gender, Space, and Knowledge in the Early Modern Museum," in Peter Galison and Emily Thompson (eds.), *The Architecture of Science* (1999):29–57; Lynette Hunter, "Women and Domestic Medicine: Lady Experimenters, 1570–1620" and "Sisters of the Royal Society: The Circle of Katherine Jones, Lady Ranelagh," in Lynette Hunter and Sarah Hutton (eds.), *Women, Science, and Medicine, 1500–1700: Mothers and Sisters of the Royal Society* (1997):89–107 and 178–197. A study of how women could play a role in the new science in the salons can be found in Erica Harth's *Cartesian Women: Versions and Subversions of Rational Discourse in the Old Regime* (1992). Arguing that women were capable of studying Descartes, and did do so through the space of the salon, Harth proposes that both Cartesian women and the space of the salon were, by the end of the eighteenth century, eventually ostracized from the intellectual science community, marginalized within feminine forms of scientific discussion.

The work of Ruth Salvaggio in *Enlightened Absence: Neoclassical Configurations of the Feminine* (1988) has demonstrated that women also fill metaphorical space in science. She notes that though women may have been excluded from literary and scientific discourse, this "discourse is filled with references to woman" (5). Biographical works that look at individual women, such as Anna Maria van Schurman, Anne Conway, and Margaret Cavendish, examine in detail the evidence for women's participation in new and old science traditions. All such works point to the importance of recognizing the role these women played in the new philosophy and in studying their thoughts on the relationship of human beings to the natural world, regardless of whether they eventually embraced or rejected it. On this, see Anna Battigelli, *Margaret Cavendish and the Exiles of the Mind* (1998); Mirjam de Baar, Machteld Lowensteyn, Marit Monteiro, and A. Agnes Sneller, *Choosing the Better Part: Anna Maria van Schurman* (1607–1678) (1996); Sarah Hutton's edition of Marjorie Hope Nicolson, *The Conway Letters* (1992); Rebeca Merrens, "A Nature of 'Infinite Sense and Reason': Margaret Cavendish's Natural Philosophy and the 'Noise' of a Feminized Nature," *Women's Studies* 25 (1996):421–438; Sophia B. Blaydes, "Nature Is a Woman: The Duchess of Newcastle and Seventeenth-Century Philosophy," in Donald C. Mell Jr., et al. (eds.), *Man, God, and Nature in the Enlightenment* (1988):51–64; and Lisa T. Sarasohn, "A Science Turned Upside Down: Feminism and the Natural Philosophy of Margaret Cavendish," *Huntingdon Library Quarterly* 47 (1984):289–307. Such works counteract the influence of David Noble's *A World without Women: The Christian Clerical Culture of Western Science* (1993). Noble argues that men ostracized women from science in this period because the changing nature of the church set the precedent for masculinizing the professions. Nevertheless, he overstates his point by trying to convince the reader that there was a total absence of women from science in this period until the end of the nineteenth century, when attempts of women to gain entry were stymied yet again by the professionalizing men of science. Clearly the afore-mentioned works provide plenty of evidence for women's participation in science during this period despite the best efforts of men to keep them out. In fact, many men welcomed and benefited from women's contribution to science during that time.

2

Women's Bodies, Women's Minds: The Science of Women

There are no women of genius; the women of genius are men.
—Goncourt in Cesare Lombroso, *The Man of Genius* (1881)

In Pedra, Italy, in 1601, soldier Daniel Burghammer gave birth to a baby girl who was duly christened by the church and nursed at his breast, much to the astonishment of his wife and his army captain. Burghammer admitted his body was half male and half female, but was he a man or a woman? Which social, political, and religious roles could such a person fill? The mixed sex of Daniel Burghammer's body resulted in him/her being denied full participation in either gender's social roles by his community, exhibiting just how important a part the sex of an individual played in determining his/her gendered position within society. Daniel Burghammer had produced a child through sexual reproduction born of his/her own body and nursed the newborn, but Daniel was considered neither a fit mother to raise a child nor fit to be a husband and keep a wife. Daniel's wife sought and achieved a divorce, the church finding that Burghammer's ability to give birth was incompatible with the role of husband. The child was put up for adoption. (Recounted in Fausto-Sterling 2000, 35.)

Burghammer's story highlights the dichotomies in western thought, which often seeks to understand one concept in contrast to its opposite. For example, reason can only be understood with reference to emotion, culture with reference to nature, mind to body, and male to female. Social institutions and individuals use a person's biological sex to determine the gendered social role a man or a woman can play within the community. Whereas society has strict rules about gender roles based on the sex of an individual, clearly nature does not always comply, as in the case of Burghammer. Nevertheless, scientists long ago turned to the biology of sex to establish clear guidelines about what it meant to be a man and what it meant to be a woman in western society.

Prior to the advent of the new science in the seventeenth century, scientists were only one voice among many who defined and constructed the world. In the medieval period, men of the church, religious doctrine and texts, ancient philosophies, medical texts, and legal, political, and social bodies all played a role in defining the roles of men and women. Their ideas and opinions on the subject were varied. But as scientists gained in social prestige, their voices became increasingly predominant. Their strict categorization of the natural world resulted in the stricter definitions of *male* and *female* in the human species as two dichotomous groups. Whereas hermaphrodites such as Burghammer (persons born with a mix of male and female sex organs) had always been an acknowledged natural occurrence, if not always easily accepted by society, hermaphrodites in the eighteenth and nineteenth century became monstrous anomalies. In fact, the new, strict definitions of sex by scientists made hermaphrodites almost invisible to science and to society.

With their social ideologies about the role of men and women in reproduction, the family, and the larger society, scientists found themselves "proving" the natural inferiority of the female sex and the superiority of the male sex. Informed by their social ideals, it seemed to scientists that there were no other categories, no space to maneuver beyond the boundaries so obviously set by nature and reflected in culture. Their social and cultural ideals about men and women informed the questions they asked about the sexes and influenced the answers they were willing to accept. They read their social ideals back onto nature, onto the sex of human beings. Though many of their theories were often convoluted and contradictory, scientists nevertheless believed they had succeeded in validating social norms about men's and women's roles in society by using biology to naturalize these norms.

Although scientists' voices dominated and were given significant social cachet from the seventeenth century onward, at every turn there were often contesting voices and competing theoretical explanations. Scientific theories and observational and experimental evidence were questioned and contradicted. Some findings opposed the idea of women's physical, mental, and moral inferiority to men posited by mainstream scientists, putting forth instead explanations for their equality as human beings with men in society. Other findings indicated women's superiority over men, turning the scientists' arguments upside down.

The Weaker Sex: The Science of Woman

Women were understood to be the weaker sex long before the advent of the new science in the seventeenth century. Religion, philosophy, the legal system, and the western social hierarchy all reflected the belief that women were physically weaker than, and intellectually and morally inferior to, men. Scientists have long found supporting evidence in nature to lend legitimacy to the social order: Men's bodies were (and are) on average bigger and stronger than women's bodies. Reproduction was socially empowering for men and did not interfere with their physical and mental capabilities; for women, having children was physically, mentally, and socially disempowering, encompassing all their vital energies. Intellectually, men were believed to have a clear grasp of abstract concepts such as justice whereas women were believed to be only capable of emotion and the lower forms of reasoning. Early modern societies believed that women who moved outside of their natural physical and mental roles risked physical and mental illnesses, even death.

Ancient theories posited that there was only one sex, the male, and that females were merely less developed males. In Aristotle's biology, heat was the central principle of all life. He argued that the more heat an animal generated, the more developed it would be. Working from the physiological and psychological differences he observed between men and women, Aristotle concluded that men were hot and women were cold. Similarly women were considered to be wet, and men were dry. To be hot and dry was superior to being cold and wet because in a circular argument these attributes were connected with the physiological and psychological characteristics of men and women.

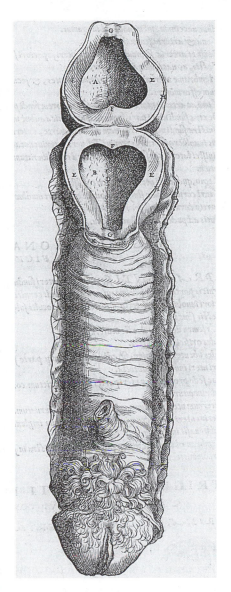

The uterus and vaginal canal. Anatomist Andreas Vesalius provided this visual rendering of Galen's conception in De humani corporis fabrica *(Basel, 1543). (Courtesy of the National Library of Medicine)*

The Greeks also compared men's and women's genitals and observed that men and women had the same organs, the only difference being that because men were hot their genitals were external to their body and women, who were cold, had their genitals on the inside of their bodies. The Greek physician and theorist Galen argued in the second century A.D. that women's genitals were an inferior, interior copy of men's genitals, explaining in his *On the Usefulness of the Parts of the Body* that they "were formed within her when she was still a fetus, but could not because of the defect in the heat emerge and project on the outside." As a result, well into the seventeenth century, anatomical drawings of female genitalia depicted the vagina as an internal penis and the ovaries as internal testicles.

Women appeared central to reproduction, yet their reproductive role still did not accord them superior status to men even in this realm. The Greeks believed that a woman's uterus was central to her well-being and that its tendency to become disconnected and move around her body, causing illness, ought to be prevented by moderate amounts of sexual activity and pregnancy. Women, they believed, were at the mercy of their reproductive organs, not empowered by them. In addition, Aristotle argued that men contributed the superior qualities of intellect and reason to the fetus in reproduction whereas women contributed the nutritive and generative aspects. Moreover, they believed that sperm contributed the soul, and that a woman's body was made solely to nurture the soul placed in her by man. In this way men participated in the divine; women did not. Women were merely the soil, and men provided the seed.

Religious traditions reinforced the biological evidence that men were superior to women. In both ancient and Christian religions, creation stories show man created first, woman second. This secondary chronological status implied a lesser status for women in the human hierarchy imposed by God or the immortals. Accordingly, woman was human, she was above the animals, but she was neither physically nor mentally the equal of man. In Greek creation mythology, woman was brought into being as a punishment to man. Woman was created to be physically beautiful: irresistible to men, yet the bane of their existence. Hesiod in his *Theogony* tells the story of the creation of the world and describes woman in the following manner:

> They live among mortal men as a nagging burden
> and are no good sharers of abject want, but only of wealth.
> Men are like swarms of bees clinging to cave roofs
> to feed drones that contribute only to malicious deeds.
> (lines 592–595)

Plato's creation myth noted that the world was created with only men in the beginning. When a man failed to govern his emotions, he would be reincarnated as a woman. Again, woman was imperfect man and as such had an inferior body, mind, and even soul.

In the Christian tradition the first woman, Eve, is placed in the Garden of

Eden with the first man, Adam. God gives Adam and Eve everything they could possibly need and desire but warns them that they must not eat from the tree of knowledge. A serpent convinces Eve to eat from the tree, and she in turn convinces Adam to eat its fruit. As a result, God banishes Adam and Eve from the garden. Eve is depicted as the cause of the fall of man from paradise, and thus women become the cause of men's suffering on earth. The first-century Jewish philosopher Philo pointed to the fact that Eve was made of Adam's rib and thus made of different matter, which he deemed inferior, causing him to associate women with the senses. Augustine and Thomas Aquinas, two important medieval Christian theologians, agreed with Philo but did point to the fact that there was one thing that women could do that men could not: they could give birth to new life. This explained women's closer connection to the senses. Men were removed from the labor of reproduction and could thus rule their passions with reason. During the Protestant Reformation, Martin Luther posited the equality of man's and woman's creation while acknowledging the differences between them. However, unable to resist the notion that woman's difference makes her inferior, Luther went on to argue that Eve lost equality when she ate the fruit from the tree of knowledge. He wrote in his *Lectures on Genesis* (1535), "For the punishment, that she is now subjected to the man, was imposed on her after sin and because of sin, just as the other hardships and dangers were: travail, pain, and countless other vexations." Biological arguments clearly reinforced theological assertions.

In the moral realm, women were depicted as sensitive and sensual beings ruled by their passions, both emotive and sexual. As such, women represented a danger to men as they might tempt them to undermine their moral judgment and thus their reason for sexual pleasure. Women's lasciviousness was underscored biologically by the assumption that women's orgasm was considered necessary for reproduction, although the problem of conception resulting from rape that assumed absence of female pleasure continued to trouble medieval scholars. Nevertheless, while it was generally considered that men had to induce orgasm in women in order to reproduce, at the same time they had to be sure to control women's sexual passions so order would be maintained in society and civilization maintained and allowed to progress. The Greeks and early Christians believed that because of these passions, women were prone to being deceitful and at the same time could easily be deceived. As a result, women's participation in witchcraft and their tendency to be led into consorting with the devil were two predominant fears in the early modern period. Women were considered less moral than men, whose higher reasoning allowed them to make a clear distinction between good and evil, right and wrong. This fear of women's wild sensuality continued into the eighteenth century. In his book *On Women* (1772), Denis

Diderot clearly expressed his anxiety about the inability of women to control their own sensuality: "The woman dominated by hysteria feels I know not what internal or celestial emotions. At times she makes me shiver. I have seen and heard in her the raging of the ferocious beast which is a part of her." A woman's entire being, according to Diderot, was defined by her body.

The new science of the seventeenth century contributed scientific authority to earlier claims of women's inferiority. Drawing upon these early Greek and Christian theories of the body, scientists began with the already established premise that men were superior to women. So, for example, the theory of preformation was introduced in the late seventeenth century; it argued that the embryo was fully formed in the parent. This theory was presented to counteract the problem of embryonic development out of unorganized matter for which scientists had no mechanical explanation. Ancient and medieval scholars had long debated the individual contributions of the male and female to reproduction, with no clear conclusion on the matter, although they always attempted to couch contribution within the context of appropriate social roles for men and women. In the seventeenth-century debate the question arose as to which parent, the male or the female, carried the preformed embryo. Anton van Leeuwenhoek (1632–1723) presented evidence to the Royal Society in 1677 that he had observed male semen under the microscope and had seen "spermatic animalcules" therein. He argued that these animalcules were the embryos for which preformationists had been searching. The female egg, he argued, contained only matter for sustaining the embryo as it gestated. Although a group known as Ovists argued that the egg contained the embryo, Animaculists won the day, and their theory of men's preeminence in reproduction remained popular until the late eighteenth century, when scientists returned to theories of the equal contribution of males and females to the creation of an embryo. Clearly, seventeenth-century scientists were influenced by earlier biological theories about sex and by gendered social ideals. They, in turn, reinforced these theories and ideals and thus ensured the reproduction of the gendered status quo via their scientific theory.

Despite the fact that men were considered the primary and even the sole contributors in reproduction, women's reproductive functions drew particular attention in the eighteenth century. In fact, during this period the science of women, gynecology and obstetrics, was consciously formulated into a discipline by the male medical profession. In so doing, they ousted women from the knowledge and control over birthing and began the medicalization of the birth process. Menstruation, pregnancy, birth, lactation, and menopause became medical conditions to be treated rather than normal aspects of the female body's life cycle. Although male midwives stirred fears about the violation of women's virtue via their intimate access to the bodies of other men's wives, gradually the male mid-wives established themselves as authorities on birthing over the traditional female midwives.

The male medical community increasingly believed that women's repro-ductive functions overwhelmed every facet of their lives. The onset of menstru-ation meant physical and mental incapacity for women during a portion of every month as they rested while they bled and were considered unable to undertake any physical or mental task. Pregnancy, birth, and lactation were obviously phys-ically and mentally exhausting. Women who attempted to turn their attention and energy away from their reproductive bodies toward intellectual pursuits faced the possibility that they might fall ill, become infertile, or perhaps go mad or die—or so thought the nascent medical establishment. Of course the fact was conveniently overlooked that lower-class women continued to work during men-struation, pregnancy, nursing, and child rearing.

Female midwives and male medical practitioners were still disputing the vulnerability of women's bodies during the eighteenth century. But by the end of the nineteenth century, the apparent permanent fragility and incapacity brought upon women in every stage of their life due to their reproductive role was believed to have been scientifically confirmed. The Scottish psychiatrist T. S. Clouston in his *Clinical Lectures on Mental Diseases* (1883) warned: "The regu-lar normal performance of the reproductive functions is of the highest impor-tance to the mental soundness of the female. Disturbed menstruation is a con-stant danger to the mental stability of some women; nay, the occurrence of normal menstruation is attended with some risk in many unstable brains. The actual outbreak of mental disease, or its worse paroxysms, is coincident with the menstrual period in a very large number of women indeed." Interestingly, although there exist medical conditions and illnesses specific to men, no paral-lel science of men arose to deal with these male sex-specific issues. This new sci-ence of gynecology and obstetrics raised the profile of the men who practiced it. As their studies intensified throughout the eighteenth and nineteenth centuries, so were they able to confirm the constant infirmity, or at least weakness, of women's bodies due to functions associated with reproduction. This was one more way in which men attempted to control women's lives via the knowledge they constructed about women's bodies.

In the nineteenth century new scientific theories continued to reinforce the biological inferiority of women to men. Craniology, the recapitulation the-ory, the theory of evolution with reference to sexual selection, and the physical theory of the conservation of energy were all believed by scientists, directly or indirectly, to support women's lesser status. Craniology and recapitulation the-ory went hand in hand. According to the theory of recapitulation, an individual organism will pass through the stages of development of its ancestors during the embryonic stage of its life. Anthropologists and psychologists picked up on this theory. Noting that women, children, and "the lower races" shared physical, mental, and behavioral qualities with one another, they extrapolated that all three groups were undeveloped forms of the final type: white man. Craniology lent scientific credence to this theory. By measuring the dimensions of the skull from all possible angles and the size and weight of the brain, anthropologists and psychologists believed they had found scientific evidence of the intellectual and physical inferiority of women and blacks to white males. Male skulls and brains were bigger than women's, and the facial plane was less angular. The only conclusion to be drawn from this data was that men's bigger brain size meant superior intellect.

Other prominent scientific theories were utilized to support women's inferiority. In addition to his theory of evolution by natural selection, Charles Darwin posited the idea of sexual selection. Darwin argued that certain individuals have a greater chance of reproducing than others due to advantages drawn from secondary sexual characteristics—among goats and sheep, for example, the large horns that males use for fighting off competitors for a mate; among birds, the beautiful tail feathers that attract a prospective mate to choose a certain male. Darwin considered mental attributes to be another of these sexual characteristics. Although men and women inherited intellectual traits, Darwin argued that men had more opportunity to use and develop their intellect either in the struggle for survival or in the struggle to obtain a mate. As a result, Darwin concluded in *Descent of Man and Selection in Relation to Sex* (1874) that men were superior to women in the realm of "deep thought, reason, or imagination."

Even the first law of thermodynamics, the principle of the conservation of energy, was drawn upon to support women's inferiority to men. This principle states that although energy can be converted into work or into new types of energy, the amount of energy present does not increase or decrease. This physical theory of a fixed amount of energy was adopted in discussions about the energy needed to work both the body and the mind. Energy directed to one purpose in the human mind left less energy for the body or for other intellectual activities. Women were believed to need a significantly large amount of energy for their bodies at puberty. In his *Sex in Education; or, A Fair Chance for the Girls* (1873), Edward Clarke, former professor at the Harvard Medical School, noted: "It is . . . obvious that a girl upon whom Nature, for a limited period and for a definite purpose, imposes so great a physiological task, will not have as much power left for the tasks of the school, as the boy of whom Nature requires less at the corresponding epochs." Clarke's English counterpart, Henry Maudsley, wrote in his article "Sex in Mind and Education," published in *The Fort-*

nightly Review (1874), that it was "not a question of two bodies and minds that are in equal physical conditions, but of one body and mind capable of sustained and regular hard labour, and of another body and mind which for one quarter of each month during the best years of life is more or less sick and unfit for hard work." According to these scientists, energy was at a premium, and women's need for significantly more energy to be directed toward reproduction resulted in her weaker physical and mental state when compared with men.

Scientists expected to find evidence in nature that would prove the inferiority of women. Religious, social, legal, and political frameworks had already established the lower status of women in mind and body and had encoded these findings in law. Scientists assumed that nature would support contemporary notions of men's and women's roles in society. Not surprisingly, their science was a self-fulfilling prophecy. They saw in nature what they expected to find, and when nature contradicted their expectations, they were readily willing to believe in errors of experimentation and observation to discount them. The science of woman established the scientific fact, based in nature, that women were inferior members of the human race whose minds and bodies naturally relegated them to a different life from men. According to the scientists of the time and those who followed their creed, absolutely nothing could be done to change this fact.

Science and Gender: Biology as Social Destiny

Scientists' views and representations of women's bodies and minds legitimized their roles in society as wives, daughters, and mothers on biological grounds, and therefore, scientific men helped keep women out of intellectual pursuits, including science. Moreover, the study of women's minds and bodies clearly indicated through observation and experimentation the need to maintain the health of women's bodies for the good of society and civilization. Women who took part in excessive intellectual activity not only ruined their own bodies physically and mentally but were thought to put at risk their chances for reproduction and motherhood, which were deemed their natural roles. Although these scientific and social ideologies had long been working in tandem, they took on increasing importance in the eighteenth and nineteenth centuries as economic, political, and social unrest caused men and women in society to worry about the future. Scientists "proved" woman's limited capacity, and in so doing fitted her perfectly to the gendered norms that society, politics, and culture deemed necessary for the advancement and smooth running of civilization.

During the eighteenth century as the middling class began to grow in size and in social influence, women's social role was elevated in importance and pres-

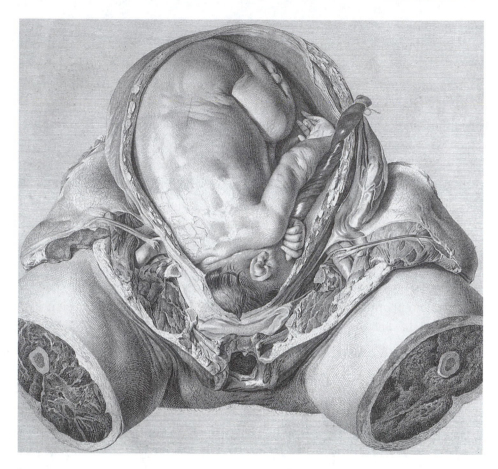

Copperplate engraving from The Anatomy of the Human Gravid Uterus, *William Hunter, 1774. An eminent anatomist and obstetrician, Hunter confined himself to a specific topic (late pregnancy) and "subject" (the dissection of a woman who died near the end of term). (Courtesy of the National Library of Medicine)*

tige. Whereas men's useful social role was to work, women's social role was to reproduce. Increasing cachet was given to this role, as it was seen as an all-important contribution to society. As producers of men, women were seen as important contributors to society. However, this role relegated women to the private world of home and family and prevented them from involving themselves in any sort of public role either as workers or as citizens.

In this period too, women were increasingly dubbed the "moral sex." But this morality was not defined in relation to women's intellectual abilities, as morality was in men, but in relation to her sex-specific characteristics. Thus, women's morality was defined in such terms as complaisance, gentleness, forbearance, and sensitivity. The French physician Pierre Roussel in his *Systematic Overview of Women as Physical and Moral Being* (1775) noted women's inability to think in the abstract due to the fact that a woman faces the "difficulty of shedding the tyranny of her sensations [which] constantly binds her to the immediate causes which call them [the sensations] forth, preventing her from rising to those heights which would afford her a view of the whole." This state of affairs was not seen in a negative light by such writers; rather, women's moral role demonstrated just how suited women were to their role as mothers, wives, and daughters. They had the innate ability to pay attention to the specific needs of individuals under their care, including husbands, parents, and children.

The French Revolution in particular instigated a contest over the control of public space. In carving out a position for themselves in government, middle-class Frenchmen attempted to eliminate the contestation of women for a voice in the public arena as citizens. The public body was represented as male, a body which could be corrupted by women. Revolutionaries pointed to the monarchy as characterized by this kind of corruption, justifying its removal. They attempted to relegate women to the private sphere, their morality in part defined in terms of their sexuality. Salons and politics were circumscribed as immoral, separating women from opportunities to learn, think, or participate in any way outside of the private realm. Women could only remain virtuous if they kept out of the public sphere. However, in discussing the private and public life of Madame Roland, Dorinda Outram has argued in *The Body and the French Revolution* (1989) that women were not always fitted to fulfill their roles as wives and mothers and that they could enter the salons and the public, political sphere despite attempts to keep them out. The lines between masculinity and femininity that both scientists and men and women in society tried to draw so carefully could be blurred and traversed.

Ideas about what it meant to be a man or a woman in society were embedded in the very scientific ideas they claimed were untainted by social ideas. Historian Londa Schiebinger has demonstrated how pervasive definitions of what it means to be a woman have been in science itself. She posited the question as to why Carolus Linnaeus, famous for providing a classification system based on the reproductive characteristics of life forms in his *Systema naturae* (1758), decided to adopt the name *mammals* for the groups of animals now associated with this label. Why did he draw attention to the fact that mammals feed their young milk from their mammae as an identifying feature of this group, when only the females are capable of milk production? Schiebinger asserts that Linnaeus was driven, probably unconsciously, into this choice of a name by cultural and political trends. Fears had been raised in Europe over wet-nursing, the tradition of aristocratic mothers to have their children breast-fed by other women. Linnaeus was involved with many others in attempting to encourage women to breast-feed their own children.

Breast-feeding fulfilled many functions. First, it fulfilled a biological function of feeding a child successfully and raising it to adulthood. Wet nurses, usually lower-class women, were notoriously sickly and/or often accused of taking on too many babies to nurse at one time. Babies often died in their care due to lack of sustenance, something that it was assumed would not occur if the baby remained with its middle-class mother. In this way, women would be contributing to society via reproduction. Second, breast-feeding fulfilled a social function, emphasizing the importance of women's role as mothers in society. Third, breast-feeding fulfilled a political role in helping to undermine women's ability to participate in politics in the public sphere. They were encouraged to remain in the home to nurture their children. Women achieved political power and social importance in this period indirectly through their roles as mothers, not as full citizens. Linnaeus believed he had found a universal term for this group of animals, the mammals, but his formulation had "infused nature with middle-class European notions of gender" (Schiebinger 1993, 74).

Eighteenth-century medical drawings and models provide another example of how science was imbued by contemporary social notions about gender. William Hunter (1718–1783), a leading obstetrician and professor of anatomy, published his *The Anatomy of the Human Gravid Uterus* in 1774, which contains large, dramatic plates of the anatomy of pregnant women. Though more anatomically accurate than many other earlier renditions, Hunter's work reflected his society's preoccupation with the importance of mothering. Just as contemporary artists painted mother-and-child scenes that showed the closeness and connection between the two, so did Hunter's anatomical drawings: rather than a small, floating fetus, Hunter drew full-term fetuses pressed tightly within the mother's uterus—"the two lives being portrayed as a single interconnected system" (Jordanova 1999, 199). The negative side to this depiction, though, was scientists' intimate exploration of women's bodies. The models used for anatomical instruction, mostly constructed in Italy, were often highly feminized and sexualized. With flowing hair and long eyelashes, their recumbent positions and facial expressions possibly denoted sexual or religious ecstasy. Such models implied an aspect of sexual titillation in anatomists' investigation of the dissected female body, and their ability to investigate and know women's bodies in the first place enabled them to define and thus attempt to control them. In these representations women's bodies could be simultaneously idealized and violated.

Even male scientists who believed that women could and should study science or who created approaches to science that included women in some way nevertheless, in the end, depicted women as inferior creatures to men or as human beings destined for a particular social role based on their biological sex. Erasmus Darwin (1731–1802), grandfather to Charles Darwin, wrote the very popular verse *The Loves of the Plants* in 1789. In this lengthy poem he outlined the Linnean classification system, hoping that by doing so, women might better understand the system. Linnaeus's system of classifying plants involved examining and counting the sexual organs of plants, the stamens and pistils. Erasmus Darwin was fascinated by the idea of evolution, and he believed that reproduction was at the heart of the variations that arose.

Nevertheless, despite the close attention he paid to botanical accuracy, his poem anthropomorphized the plants he described and in so doing depicted the social and sexual characteristics he and others of his time associated with the relationships between men and women. Darwin depicted plants with a larger number of male (stamens) than female parts (pistils) as either a seductive woman or a woman needing protection. Plants with more than five or six female parts were described as "seductive harlots." Plants with one male part and two female parts were described in familial terms, as maids or sisters caring for the male without sexual connotations. Plants that had an equal number of male and female parts were described in terms of a marriage partnership. Although Darwin recognized the sexuality of women and believed that the root of sexual relations resided in physiology he nevertheless could only see female roles in his plants that mirrored those in his society. There were no intellectual women in his floral settings, and certainly no divorcées. Janet Browne has argued in "Botany for Gentlemen" (1989) that though Darwin supported women's education in writing his *Loves of the Plants* in the first place, he nevertheless envisioned women's education as a support for her appropriate social roles in life and in terms of the benefits of women's education to men.

Johann Wolfgang von Goethe (1749–1832), an important German poet in the Romantic movement who was also interested in science, formulated a scientific methodology known as Naturphilosophie, which drew on Romanticism and rejected Linnean classification. He believed that the scientist had to be in sympathy with nature in order to study it. As sympathy was a characteristic particularly associated with women, Naturphilosopes included ladies in their community. However, Goethe established his science in the private, not the public, sphere toward which modern science was moving and excluded women as practitioners of science, seeing them instead as helpmates, audience, and muse. He did this not only through his published work but in practice, with his wife, Christiane Vulpius, who inspired his *Metamorphosis of Plants* (1788–1790) but who was never more than a lover, housekeeper, and gardener, largely ignorant of scientific botany. In a short poem he wrote for his wife in 1812, he actually turns Vulpius into a plant. Lisbet Koerner (1993, 470–495) has argued in this context that women and plants must be silent so that the male voice remains as the only authority on nature.

In the nineteenth century, emphasis was placed on the complementarity of the sexes. Liberalism and the idea of the equality of all men had raised the problem of the inequality of women. Because women, like men, were human beings, logically they also deserved political and social equality within the framework of liberal ideology. Scientists of the time responded by continuing to maintain as fact that gender distinctions based on the sex of the individual were natural, though they attempted to downplay the idea of the inferior female and superior male. Instead, they argued that the differences between male and female *complemented* one another. Each person contributed equally to society, but men and women filled different roles and thus were interdependent. According to the ideal of complementarity, women's roles were concentrated in motherhood and the care of the home. Extensive education, scientific practice, and intellectual disputation were outside the realm of femininity. Even in the context of this effort to raise women to equal status, women's biology more often than not still relegated them to the position of their sex—without legal rights as a person and dependent on men for their economic and social survival. Thus, complementarity did not result in equity in day-to-day living.

Science Turned Upside Down: Contesting the Science of Woman

Although scientists' theories about the differences between men and women were never completely overthrown, the contesting of gender norms and the contribution of new research presented continuous challenges. Occasionally, new investigations turned science upside down, overthrowing old theories and proving them false.

From the Renaissance on, works listing and extolling the achievements of women in the arts and sciences attempted to counteract the argument that the lack of female achievement provided supporting evidence for the biological and behavioral inferiority of women. Jacques Du Bosc in his *L'Honneste femme* (1632) argued that women's bodies and minds were actually better suited to the study of the arts and sciences than men. Some commentators manipulated religious arguments to provide evidence of women's superiority, arguing, for example, that the fact that woman was created last actually makes her superior to man.

Other theorists explicitly addressed sex differences. Marie le Jars de Gournay in her *Égalité des hommes et des femmes* (1622) argued that physical differences between the sexes simply formed the basis of human reproduction and nothing more. Marguerite Buffet in her *Nouvelles Observations sur la langue francoise* (1668) argued that souls have no sex and, drawing on ancient

Bucknell University women basketball players, ca. 1896–1900 (Library of Congress)

ideas, noted that the sex of male and female bodies are quite similar. The differences that do exist have nothing to do with the mind or the soul but are simply a function of reproduction. On differences in brain size between men and women, Buffet noted that men's larger heads do not prove their intellectual superiority, but only that they have something in common with "stupid animals and large beasts." She went on to claim that women's minds, beauty, and virtue are superior to that of men and argued that female babies take longer to come to term than male babies, indicating that female babies are more complex organisms than male babies.

Ex-Jesuit Poullain de la Barrie drew on Cartesian philosophy to argue that whereas the body was sexed, the mind was not. Women could just as easily as men become reasoned, scientific thinkers. He argued in *De l'égalité des deux sexes* (1677) that the lower numbers of women who practiced science compared to the numbers of men had more to do with the fact that they were relegated to "housewifery" and that their social and political subordination to men had made them "languish in idleness, softness, and ignorance, or otherwise grovel in low and base employments." Education was all that was needed for women to attain intellectual equality with men.

By the beginning of the twentieth century, the Victorian gender paradigm based on scientific evidence was eroding. Mendelian genetics proved recapitulation false. Discovery of human chromosomes showed that a larger sex chromosome was needed to make a female than a male, and the discovery of sex hormones showed that male and female hormones were equal in activity and chemical strength. An American study of 3,350 menstrual cycles in 1910 by Clelia Duel Mosher, associate professor of personal hygiene and resident physician for women at Stanford, demonstrated that women's periods were not naturally incapacitating to women but rather a culturally induced response due to lack of exercise, restrictive clothing, constipation from poor diet, and lack of muscle tone. The brain theories that had attempted to prove women's inferior intellect fell by the wayside as it became increasingly clear that the evidence on brain size continually contradicted itself.

In 1912, Jean Finot, a naturalized Frenchman, published *Problems of the Sexes*, in which he argued against almost all of the scientific arguments presented to prove the innate inferiority of women and superiority of men. Although not a scientist himself, he put forth well-reasoned arguments using recent scientific findings. For example, with regard to the differences between egg and sperm, he noted that new evidence proved the nuclei to be identical in both whereas the egg had a greater amount of cytoplasm. If more is better, then the egg could be considered superior. He questioned the assumption of sperm's superiority based on its variability and speed due to its small size. Finot asked whether it would be equally legitimate to postulate that the egg's stability suggests "seriousness and weight," whereas the sperm's smallness and variability could suggest "the fickleness and the weakness of man." Finot believed that any extrapolation from the biological to the cultural roles of men and women is ridiculous.

In 1913, taking yet another approach, the Reverend John A. Zahm (1851–1921) published his *Woman in Science*. Like other commentators he took the time to outline the long struggle women had faced since Greek and Roman times in obtaining access to "things of the mind" and contradicting scientific ideas that claimed to establish women's inability to do science. But he also counteracted the notion that there were not any women scientists as, chapter after chapter, he recounted specific women's achievements in all areas of the sciences. Even today, his research is considered to reveal an impressive number of women scientists from Greek times to the early twentieth century.

There were other ways, besides confronting the scientific evidence, in which women could contest the gendered meanings placed upon sex difference. Denying altogether the very foundation of the power and authority of science, they turned instead to alternative practices and the spiritual world. Mesmerists often used women as subjects for their experimentation, and women thus assumed a position of power through their ability to have "otherworld" experiences, often seeing the future or the heavens while in a mesmeric trance. As Alison Winters has argued in *Mesmerized: Powers of Mind in Victorian Britain* (1998), mesmerism brought into question classed, gendered, and racial assumptions about power and authority in society: "The mesmerist demonstrated the essence of influence; the subject displayed amazing new feats of perception and cognition" (5). Though male mesmerists sought to exercise power over their largely female subjects, some women successfully resisted their influence, some by becoming the analysts while supposedly in a trance. Women also became mesmerists themselves or chose to use mesmerism as a medical treatment, thereby asserting women's ability of self-control, their moral and intellectual superiority, and the power and authority they could wield.

Similarly, women also turned to spiritualism, the practice of utilizing psychic powers to communicate with the spirit world, a phenomena that reached its height in the 1860s and 1870s on both sides of the Atlantic. Women were again disproportionately represented in spiritualist circles because of their apparent capacity to communicate with the dead. Spiritualism took on a scientific air as scientists attempted to prove or disprove, depending on their perspective, the legitimacy of contact with the spirit world. Again, as with mesmerism, traits associated specifically with women—such as sensitivity, intuition, and feeling— were characteristics that were more likely to enable a person to become a spirit medium. Such a role bound women to the traits socially and biologically associated with women, but it also gained them social, economic, and spiritual power. Moreover, whereas women would usually adhere to gender norms in their day-to-day lives, the séance room became a site in which they might transgress gender boundaries.

The force of the centuries-long debate over sex and gender nevertheless persisted, even affecting women's own scientific ideas. Isabelle Gatti de Gamond (1839–1905), a leading late-nineteenth-century Belgian feminist and educator, included scientific education for women in her school curriculum at both the primary and secondary level. She had been responsible for founding the first girls' secondary school in Brussels. Arguing that she was preparing girls to better fulfill their duties as women in society, she also posited that a decent education would allow women to attain independence and equality. Science, she argued, would give women a better sense of reality, teaching them how to reason clearly and make judgments of their own accord. The study of science would also allow women to become a part of the science of progress by entering the industrial world as workers. Gatti de Gamond believed women were capable of doing science, but she could not escape naturalizing feminine traits such as women's morality, maternalism, and their pacificism in her arguments, noting that these traits could be brought to science and society. Women's complementary traits would temper the masculine traits therein and achieve social equality for women. Even though Gatti de Gamond imagined that women could break through gender boundaries, believed women were capable of practicing science, and utilized scientific knowledge as at least one of the keys to women's emancipation, her attempt to use science as an ally of feminism could not be fully realized while she still held on, in part, to a belief in the biological foundation of masculinity and femininity.

Later, in the first half of the twentieth century, American women attempting to professionalize themselves as physical educators of women athletes found it necessary to accept to some degree the idea that physical and behavioral sex differences were biologically innate. In order to make sure women athletes were provided with the necessary equipment and gym accommodations, they emphasized the need for women's separate space from male athletes by stressing their different needs. Some educators asserted that women's physical athletic abilities were weaker than those of men and that their attitudes toward sports tended toward cooperation rather than competition. Women educators used the science of sex difference to their students' immediate and practical advantage yet, at the same time, might downplay women's biological differences, arguing that only cultural disadvantages had held them back in the field of sports. Such inconsistency was perhaps inevitable as women encountered the ambiguous challenges to their aspirations and worked within a strongly masculine playing field. Nonetheless, Hazel H. Pratt of the University of Kansas observed in *Spalding's Official Basket Ball Guide for Women* that "during the past forty years, increased exercise and outdoor life to which women have been admitted have added to their weight, height, lung capacity, and physical vigor." Interestingly, for both Gatti de Gamond and her students and for the female physical educators and their students, though the science of sex difference could disadvantage women in society, the scientific ideas and theories about sex that influenced cultural notions of gender could also be manipulated to gain women some small access to scientific and athletic experience.

Conclusion

For today's reader it is easy to be dismissive of wandering uteruses, multiple cranium measurements, and hysterics induced by mental or physical overexertion during menstruation. In the present day these theories and ideas appear to be nothing more than the naïveté, misjudgment, or misogyny on the part of scientists who ardently believed that the tools of the new science factually legitimated men's domination over women, knowledge, nature, and society. Nevertheless, the justification of women's social inferiority based on a scientifically perceived biological inferiority persists to the present day. Questions as to whether women are inherently inferior mathematicians due to differences in brain function or, even more generally, that women are the intellectual inferiors of men based on supposedly fair and objective intelligence tests that do not account for the race, class, or sex of the test-taker continue to be debated.

In medical literature, the egg and the sperm are made to reflect male and female social roles in society. The egg is constructed as a passive recipient and the sperm as an active pursuer. Though scientists have recently discovered a more active role for the egg in conception, the language of scientists' papers still implies the heroics, strength, and stamina of the sperm and now characterizes the egg as entrapping and dangerous to sperm. In addition, the discovery of male and female hormones has served to "prove" the aggressiveness of men and the emotionality of women as natural tendencies. Do the type of hormones or the size and speed of our gametes dictate our characteristics as men and women in society? Scientists still pose these types of questions and often come to the same conclusions as their forebears: biology dictates the body, the mind, and perhaps even the soul of male and female human beings and cannot be altered but must be accepted. Too often these "facts" continue to point to women's lesser status in nature and thus too in culture.

Bringing conscious awareness to the social construction of human bodies and their repercussions is an important step in instituting changed attitudes toward the power wielded by scientists in society and toward the implications of "sexual science" for men's and women's lives, roles, and relationships in society. In 1926, Charlotte Haldane, the British writer, feminist, wife of scientist J. B. Haldane, and advocate of the importance of motherhood, wrote *Man's World*. Her story is a dystopian novel about a society in which women were reduced to their biological functions and controlled by male scientists. Haldane feared the development of new technology, in which her husband was instrumental, to determine the sex of a child before birth. In her novel this technology is co-opted by the state to increase the number of boys born, perpetuating patriarchy, the patrilineal class system, and industrial and colonial projects. In *Man's World* women were categorized as mothers, "neuters" (workers), or entertainers. In this way the state determines the roles suitable to women who reproduce and those who do not, decides which women could breed, and used their bodies to reproduce the gendered status quo. The two main characters, Christopher (an effeminate man) and Nicolette (a woman who resists the control of the state over her body and social role), each attempt to contest these attitudes toward women, reproduction, and the gender ideologies of their society. Both fail, however, as neither of them can overcome the internalization of the biological and social imperatives of their society.

More recently, Ann Fausto-Sterling, a biologist, feminist, historian of science, and social activist, has studied societal and scientific attitudes toward intersexed humans in her work *Sexing the Body: Gender Politics and the Construction of Sexuality* (2000). She found that scientists' concern with defining a newborn baby as either male or female is firmly routed in the fact that men's and women's roles in society were dictated from the first day of their life based on whether they were born male or female. Children born with a mixture of male and female sex organs are defined by science as either more male or more female and are forced into our dichotomous world via surgical operations. Fausto-Sterling notes that "physicians believe that their expertise enables them to 'hear' nature telling them the truth about what sex patients ought to be" (27–28). A child whose sexual organs are neither fully male nor female but a mixture of both is deemed socially and scientifically unacceptable—biologically and culturally "unnatural." Fausto-Sterling introduces the idea that there may well be a third sex in nature, or even "shades of difference" (3) indicating multiple sexes. Such an assertion, if taken seriously by the general public, would throw a spanner into the orderly workings of the male/female divide that not only neatly addresses biological function but also legitimizes the male preeminence in the social order and the authority of the scientist to underpin the social order via natural laws.

From the eighteenth century on, scientists had the opportunity as never before to legitimize the social/sexual order via natural laws. Consequently, women, the intersexed, and homosexuals who live outside of social expectations have been forced to contest the normative values placed on their sex, which determine gendered, class, and racial roles and relationships in society. Haldane's novel about sex, sexuality, reproduction, and thus women's role in society and Ann Fausto-Sterling's work on attitudes toward the intersexed both warn against masculine science that has naturalized the relationship between men's and women's minds and bodies and their roles in society. They also implicitly suggest a need for change, a rethinking of the status quo that will permit all sexes to be more than their bodies suggest and to create a human world whose science explores and accepts difference, variety, and the possibility of being more than the sum of one's biological parts.

Bibliographic Essay

Early works that explore the relationship between the science of women's minds and bodies and the gender roles of society emphasize that the theories that have been produced were simply bad science. These writers' discoveries of the slides in logic were fueled in part by the new women's activism of the 1970s. They argue that this science was consciously constructed by men intent on maintaining their social, legal, and political authority over women, who were beginning to agitate for educational and legal rights, full citizenship, and entry into the professions. Susan Sleeth Mosedale in "Science Corrupted: Victorian Biologists Consider 'The Woman Question'" (1978) argues that men who posited scientific theories that undermined any hope of equality for women in society "betray, by their uncritical acceptance of popular opinion about women, their emotional commitment to the traditional concept of the female's place in society"(3). Elizabeth Fee makes a similar argument, focusing specifically on craniology, in "Nineteenth-Century Craniology: The Study of the Female Skull," *Bulletin of the History of Medicine* 53 (1979):415–433. In *Midwives and Medical Men* (1977), Jean Donnison argues that the professionalization and medicalization of midwifery by medical men resulted in an increased powerlessness for both the mother and the traditionally female midwife in the birthing room. Men's ability to define women's bodies and their functions led to the naturalization of male power in the social arena and the increasing loss of power for women over their bodies.

Though to some extent this approach remains a part of later works, current emphasis is more often placed on the inability of the scientist, and thus science, to disconnect itself from social ideologies embedded in the scientists' psyche and cultural milieu. Authors of such works argue for the importance of recognizing that science, consciously or unconsciously, is biased by scientists' gender ideologies. As a result of this bias, the normative moral and social values placed on "male" and "female" become social constructions, not matters of incontrovertible scientific fact. The two most important and influential works in this area are Londa Schiebinger's *The Mind Has No Sex? Women in the Origins of Modern Science* (1989) and Cynthia Russett's *Sexual Science: The Victorian Construction of Womanhood* (1989). Schiebinger's research encompasses the seventeenth through the nineteenth centuries, and Russett's analysis focuses on the nineteenth century. Both outline scientific theories that seemed to establish women's physical and mental inferiority and thus confirmed their gendered social status. Schiebinger's focus is on women's exclusion from science on the basis of biological explanations of women's mental and physical incapacity to do science. Russett examines the rise and fall of Victorian theories of sex difference, arguing that the authority of science in this period gave them the power to, at least temporarily, undermine women's struggle for equality. Both Schiebinger and Russett underscore the fact that scientists as members of the larger community cannot separate themselves from the cultural norms of their time and that their claims to authority based on unbiased investigation cannot be supported.

Other useful works that explore the development of scientific ideas about sex and how they have influenced ideas about gender over time include Thomas Laqueur, *Making Sex: Body and Gender from the Greeks to Freud* (1990); Nancy Tuana, *The Less Noble Sex: Scientific, Religious, and Philosophical Conceptions of Woman's Nature* (1993); and Joan Cadden, *Meanings of Sex Difference in the Middle Ages: Medicine, Science, and Culture* (1993). Both Laqueur and Tuana cover broad chronological periods. Laqueur's focus is an examination of the ideological shift in the eighteenth century from the idea that women were lesser or imperfect men to the idea of two completely different, contrasting sexes. He pays particular attention to the changing scientific and social attitudes toward the role of the female orgasm in reproduction. He argues that "sex, in both the one-sex and the two-sex worlds, is situational; it is explicable only within the context of battles over gender and power" (11). Tuana takes a thematic approach, examining how scientific theories about women's bodies influenced and were affected by religious and philosophical conceptions of woman as a "misbegotten man" (ix), an imperfect version of the perfect creation: man. Tuana demonstrates how these theories lead to the assumption that women need to be under the control of men. Cadden's chronological emphasis is the twelfth through the fourteenth centuries. She details the medieval perspectives on the difference between the male and female sex, sexuality, eroticism, and the role in reproduction and then identifies how the influence of ancient authorities resulted in a variety of eclectic medieval opinions on the subject. She examines the ways in which "scientific and medical ideas [about sex differences] developed in specifically medieval settings . . . and also . . . the ways in which those ideas, institutions, and genres of writing made up a part of a larger culture" (10).

Two works specifically focused on the naturalization of the "feminine" in the eighteenth century with emphasis on pivotal texts and theorists who were involved in this process are Lorraine Daston, "The Naturalized Female Intellect," *Science in Context* 5, 2 (1992):209–235; and Lieselotte Steinbrugge, *The Moral Sex: Woman's Nature in the French Enlightenment* (1995). Daston argues that the meaning of nature and thus of naturalization changed over time during the eighteenth and nineteenth centuries and varied between countries such as France and Britain. This change in meaning can be seen in writings about the female intellect that pay close attention to women's "nature" and the "naturalization" of female traits. Whereas during the eighteenth century, women retained some, albeit lesser, intellectual traits, by the late nineteenth century "the link between 'female' and 'intellectual' was all but severed" (221). This clean break was due to the fact that in the nineteenth century, "reasonable necessity of natural laws had given way to the physical constraint of laws of nature" (224). Pointing to the eighteenth century as a period in which the equality of the sexes was addressed, Steinbrugge argues that in France, women's intellectual abilities were cut off from rationality and redefined from that of "deficient man" to a sex-specific emotionality that relegated women to the private, moral realm. Sensitivity, empathy, and compassion became innately female traits. Though women became the "moral sex," they were deemed, as creatures closer to nature, biologically incapable of action in the public realm. Theorists from varying disciplines drew on biological arguments to support their claims about women's fundamental intellectual difference from and inferiority to men.

In other works, specific attention has been paid to the rise of gynecology and obstetrics as a specialized discipline that has focused on the male medical community's construction of women based on their reproductive roles. Ornella Moscucci has argued in *The Science of Woman: Gynaecology and Gender in England, 1800–1929* (1990) that whereas women's reproductive functions were believed to be so overpowering as to effect their mental, emotional, and behavioral state and thus to have social and moral consequences, men's reproductive functions were believed to play no part in their behavior. Such a distinction caused women to be socially bound by their reproductive capacities and for the normal functions of a woman's body to be classed as abnormalities—diseases that needed to be treated by the newly formed discipline of male doctors in the field. This new science resulted in what Anne Digby has referred to as "women's biological straightjacket," forcing women to live their lives by the dictates of their reproductive organs or risk anemia, stunted growth, nervous breakdown, nymphomania, hysteria, or insanity. See "Women's Biological Straightjacket," in Susan Mendus and Jane Rendall (eds.), *Sexuality and Subordination: Interdisciplinary Studies of Gender in the Nineteenth Century* (1989):192–220. Ludmilla Jordanova has provided an interesting extension of these arguments in *Sexual Visions: Images of Gender in Science and Medicine between the Eighteenth and Twentieth Centuries* (1989), showing how medical and scientific images promoted gendered ideas about men and women. Medical art in the form of wax models for anatomical study, depiction of the female body in medical texts, and artists' renderings of the doctors' profession all reinforce the notion that women's bodies are natural objects to be studied by male medical practitioners, reinforcing the closer association of women with nature and body, and men with reason, intellect, and rationality.

Works that examine how gender ideologies are embedded in science provide examples of how social expectations surrounding gender norms slide seamlessly into mainstream science and in turn naturalize socially constructed gender roles. The most ground-breaking work in this area is Londa Schiebinger's *Nature's Body: Gender in the Making of Modern Science* (1993), in which she demonstrates how ideas about gender pervade the very science produced in the eighteenth century. Early studies of primates reveal male scientists superimposing stereotypical gendered behavioral characteristics of human women such as modesty onto female primates, then arguing that such feminine modesty must be natural because the characteristic was found in primates. In addition, two useful articles published in *Isis* demonstrate how even scientists who supported some women's participation in science and constructed science on a more feminine model nevertheless reinforced the gender norms of their time through their scientific productions. See Janet Browne, "Botany for Gentlemen: Erasmus Darwin and *The Loves of the Plants*," *Isis* 80 (1989):593–621; and Lisbet Koerner, "Goethe's Botany: Lessons of a Feminine Science," *Isis* 84 (1993):470–495.

Recently some historians have begun to examine the impact of theories of sex difference on women's theories of general education, on science education for women, and on women's physical education. These articles demonstrate the advantages and disadvantages that sex differentiation posed for women in society. See Martha Verbrugge, "Recreating the Body: Women's Physical Education and the Science of Sex Differences in America, 1900–1940," *Bulletin of the History of Medicine* 71, 2 (1997):273–304; and Kaat Wils, "Science, an Ally of Feminism? Isabelle Gatti de Gamond on Women and Science," *Revue Belge de Philologie et d'Histoire* 77, 2 (1999):416–439. Ideas about the human body and their impact on women's role in society can be found in Dorinda Outram, *The Body and the French Revolution: Sex, Class, and Political Culture* (1989); and in Ludmilla Jordanova, *Nature Displayed: Gender, Science, and Medicine, 1760–1820* (1999). In addition, studies on women's role in mesmerism and spiritualism attest to women's ability to utilize the traits supposedly characteristic of their sex to attain social power. Interestingly, both mesmerism and spiritualism were rigorously investigated by the male scientific community, which assumed them to be fraudulent practices. See Alison Winters, *Mesmerized: Powers of Mind in Victorian Britain* (1998); and Alex Owen, *The Darkened Room: Women, Power, and Spiritualism in Late Victorian England* (1990).

Although the rise of the science of women took place in the eighteenth century and nineteenth century, many of the authors mentioned here make at least passing reference to the fact that similar biases continue in science today. Other authors have gone forward to study the ways in which the naturalization of gender roles is still promulgated by science. Anne Fausto-Sterling's *Myths of Gender: Biological Theories about Women and Men* (1985) demonstrates how the sciences of the brain, genes, hormones, and indeed much of the field of sociobiology produce theories that appear to establish the significant gender differences between men and women and thus continue to establish natural rationales for the gender divisions in society. Emily Martin's study of scientific language demonstrates how stereotypical male and female roles are echoed in explanations of the function of egg and sperm in human reproduction. Other similar works include Sandra Harding and Jean F. O'Barr (eds.), *Sex and Scientific Inquiry*, Part 2: "Bias in the Sciences" (1987); and Bonnie B. Spanier, *Im/partial Science: Gender Ideology in Molecular Biology* (1995). Fausto-Sterling's most recent work, *Sexing the Body: Gender Politics and the Construction of Sexuality* (2000), looks at how intersexed humans contest the dualism of male/female as a natural divide and yet how scientists continue to force this dichotomy via surgical intervention or by their choice or approach to their research in such areas as the anatomy of the brain or the study of sex hormones.

Women Doing Science: Multiple Avenues

Attention to a garden is a truly feminine Amusement.
If you mix it with a Taste for Botany and a knowledge of Plants,
you will never be in want of an excellent Restorative.
—Reverend John Bennett, *Letters to a Young Lady* (1795)

The Industrial Revolution began in England in the late eighteenth century. At that time, when the middle classes were establishing the possibility of the self-made man who could climb the social ladder if he worked hard enough, eighteenth-century intellectuals noted how much brighter the future appeared for "man." The new, rational science shed light on the natural world, freed human beings from religion and superstition, and emphasized that right reasoning would lead to greater knowledge that would, in turn, benefit mankind even further. The world known to Europeans was expanding as they explored and colonized the globe. These changes ignited new ideas in Europeans' minds about the role of government and monarchy, and new theories of political economy. Intellectuals were immersed in the Enlightenment project that sought to transform the world by asserting reason over emotion, rationality over the irrational. They increasingly ordered, categorized, and thus believed they controlled knowledge about the world. Confidence in progress was the order of the day.

Within this context of the rapid development of knowledge came an increasing dissemination of information via books, informal discussion groups, and formal lectures. Such information was now within the reach of a wider group within society who could read, understand, and incorporate new world views into their mindset. The rising middle classes recognized the importance of education for themselves and their children, not only to ensure personal success in the business arena but also so that they could move into the upper echelons of society and engage in civilized conversation, make business alliances, marry wealthy partners, and become members of the governing elite. They upheld the ideal of education for all members of society. As a result, Europeans saw them-

selves as more civilized than their predecessors, more cultured, more knowledgeable, and more self-assured.

The period, however, did not seem to have been one of "enlightenment" for
women. As the world of men expanded with new ideas, the world of women
seemed to contract. Even within their traditional spheres of authority, men
began to deem women's ways of knowing to be old-fashioned. Increasingly,
among the middle classes, women's work for income was no longer necessary to
family survival, and, ideally, men and women were relegated by their gender to
either the public or private spheres, the world of work or the world of home. Men
worked outside of the home, and women increasingly worked within it. Desiring
to attain the lifestyles of their betters, middle-class men believed that they
needed to protect the women of their family from the outside world. Moreover,
upper–middle-class men strove to ensure that their women did not have to do
manual work within the home, either; they increasingly became household managers or ladies of leisure. The economic success of the middle classes enabled
many families to hire servants and thus accomplish this goal to a greater or
lesser extent, which meant that many women were left with time on their hands.
Intellectuals debated the position of women in society at length during the eighteenth and nineteenth centuries. Increasingly, however, the prevailing view
among the middle and upper classes was that women should only fulfill their
gendered roles of daughter, sister, wife, and mother and should not trouble themselves with work for financial gain, or with politics, government, or any other
serious intellectual pursuits.

Nevertheless, eighteenth-century educators upheld the ideal of education
for all. Women were no exception to this dictate. Women had tended to be given
a "finishing" education that instructed them in the rudiments of art, music,
dance, and polite conversation with perhaps a smattering of modern languages.
Increasingly, society worried that such an education produced women who were
well mannered and who could be entertaining for a short period of time, but who
were not sufficiently educated to enable them to become either satisfactory
intellectual companions to their increasingly well-educated husbands or teachers to their younger children. Proponents of an improved education for women
argued that the rising ideal of the companionate marriage and the increasing
focus placed on the importance of nurturing the life and mind of children in the
eighteenth century suggested that women required a more rigorous and balanced education than they generally were receiving. Moreover, they pointed out
that ignoring such education could result in languid, promiscuous, or ailing
women whose lives were all-consumed by fashion, society, and the latest,
morally questionable novels.

Most women were not accorded an education equal to that of men. The idea of separate spheres meant that men and women had to be educated to those spheres. Providing women with a man's education was viewed as a waste of time and could even be dangerous—to the individual woman and to society as a whole—as it might distract them from their proper sphere. Yet whereas the masculinization of science in the seventeenth century had increasingly pushed women away from science, the eighteenth-century leisured lady found that a certain amount and a certain kind of scientific knowledge was socially acceptable for her to pursue. If society was going to progress, it needed to produce rational, well-ordered individuals who were able to think logically and reason correctly. Educators considered science a good discipline for the mind. Though no one believed that women could shake their essential nature as emotive beings, educators argued that this tendency of women could be tempered with a rational education. Educators thus encouraged women to engage in some minimal scientific education, especially in botany, chemistry, and astronomy. These activities, among others, were added to the list of acceptable female accomplishments. Women read popular science works written especially for the female beginner, works that taught lessons in a pleasant and accessible manner.

While women gained access to science in this way, a new line was being drawn in the sand as men began to professionalize science, slowly separating the discipline from informal amateur practice. As the nineteenth century progressed, science was being transformed into a formal, hierarchical organization that required its members to obtain university degrees, society memberships, and remunerative positions in the university, government, or private sector; to conduct original research work; and to attempt to achieve funding grants and awards that recognized scientific achievement. Certainly some male scientists could find themselves on the outside of the science profession, but there were far more opportunities for men to gain access to science professions than there were for women. Moreover, many male scientists consciously strove to keep women out of the profession. Some men continued to believe that women were not intellectually capable of doing science and thus that their attempts to practice science would bring the profession into disrepute. Some men believed that the simple presence of women would give the burgeoning science profession the appearance of a frivolous pursuit rather than a serious endeavor. Thus women were largely barred from scientific societies and educational institutions, making their advancement within the profession unlikely.

Still, just as there had been women in the seventeenth century who wished to enter into the company of the male scientists' inner circle or who wanted to question the new science, likewise in the eighteenth and nineteenth centuries there were women who wanted to join scientific activities or contest their ideological basis. Many women incorporated science into their lives as a pleasant pastime, but others utilized this path as an avenue to reach much farther. Certainly the majority of women seriously engaged in science during this period were relegated to the role of assistant or popularizer. But it was through these roles and through a domesticated, feminized education in science that women could fill scientific niches left unexplored or untended by male scientists—or even, on occasion, move into the increasingly well-established networks of masculine science. The fact that the professionalization of science was gradual and only began to firmly establish itself in the late nineteenth-century meant that the boundaries of scientific practice were still flexible and permeable. Women could enter science in this period more freely than the male gentleman scientists of the seventeenth had intended.

Science as a Feminine Occupation

In 1686 the natural philosopher Bernard Le Bovier de Fontenelle (1657–1757) published *Conversations on the Plurality of Worlds*. A Cartesian, Fontenelle was at the center of French intellectual life: He was elected to the Académie Française in 1691 and became permanent secretary of the Académie des Sciences after 1697. In 1701 he joined the Academy of the Inscriptions and Belles Letters. His *Conversations* was an attempt to popularize the Copernican worldview among polite society and its implication that humans are not at the center of the universe, and that life might indeed exist on other planets. To accomplish this goal, Fontenelle wrote his *Conversations* in the form of a dialogue between a male philosopher and a marquise, a titled noblewoman. Their six-part discussion takes place in her garden during six evenings of pleasurable entertainment. The philosopher teaches the marquise some science, but it is not the science of the earlier Cartesian women who sought a deep understanding of the New Philosophy and an active engagement with its male exponents. In Fontenelle's dialogue, the woman student becomes the passive recipient of simplified knowledge, a spectator and not an investigator of the natural world; the male philosopher, in contrast, represents the masculine view of nature. This is not a dialogue between intellectual equals, but a lesson to show the student, in this case female, the correct mode of modern thought. Fontenelle characterized the marquise as an innately subjective, feeling, and intuitive being whom he must introduce to the objective world. The philosopher enthusiastically encourages the marquise to understand and accept the New Philosophy. She is skeptical about the philosopher's "plurality of worlds" and resistant to his desire to think objectively, but eventually she is converted to his point of view. Fontenelle's goal

Outdoor class in botany, Washington, D.C., ca. 1899 (Library of Congress)

was to educate women to gain social prestige and acceptance for the new science within elite society, not to encourage women to actively participate in it or question it.

Other natural philosophers were as equally eager to dispel the notion that science was a solitary and thus antisocial activity, and they promoted the acceptability of polite discussion about science at soirees and in family drawing rooms. Raising the status of science improved the status of the scientist within society. Taking up Fontenelle's lead, many natural philosophers published polite science books to introduce elite society, women, and even children to the new science. For example, Voltaire published *Elements of Sir Isaac Newton's Philosophy* (1738), Francesco Algarotti published *Philosophy of Sir Isaac Newton Explained* (1765, English edition), Benjamin Martin published *Young Gentleman and Lady's Philosophy* (1755), and John Bonnycastle published *Introduction to Astronomy in a Series of Letters* (1786). These popular works avoided scientific language and mathematic formula considered incomprehensible to the layperson. They were often written in a dialogue or letter format and embellished with literary and poetic quotations. While avoiding both atheism and religious enthusiasm, these popular works of science might also emphasize God's creation of the wonders of the universe. Such works abounded in the eighteenth century.

As the century progressed, women increasingly engaged in this kind of polite science, which was largely considered to be an appropriate sphere of learning for women. Scientific instruments such as Orrerys (mechanical models of the solar system), telescopes, barometers, microscopes, and globes became important symbols of social prestige. Many eighteenth-century family portraits included such instruments along with other trappings, such as a grand country house, a thoroughbred horse, fine clothes, and jewels that established their status in society. Just as libraries of unread books might serve as indicators that the man of the house was well-educated and interested in learned pursuits, scientific instruments could serve the same function and never be used. However, the upper and middle classes increasingly believed that their ability to use these instruments and to converse occasionally on scientific subjects should be a part of mannered social conversation, serving as an indication of their social status. Moreover, men and women could converse about science with one another without any sense of impropriety. The middle classes also believed that science helped to promote good relations between the sexes. Women, with a little scientific knowledge, might ask the right questions and engage a gentleman in conversation, enabling him to exhibit his own superior knowledge. Women educated in this way might keep their husbands entertained at home so they would not be tempted into immoral activities outside the domestic sphere. Science could also keep women out of trouble. James Ferguson's *Young Gentlemen and Lady's Astronomy* (1748) finds a brother teaching his sister astronomy. When the sister asks whether science is really a suitable activity for a lady, her brother reassures her. Not only would it promote domestic harmony and family values but, he argued, if more women practiced science "the consequence would be, that the ladies would have a rational way of spending their time at home, and would have no taste for too common and expensive ways of murdering it, by going abroad to card-tables, balls, and plays: and then, how much better wives, mothers, and mistresses they would be, is obvious to the common sense of mankind." Science was an acceptable mixed-gender, familial, and social activity that increased in popularity throughout the eighteenth century.

Women could associate science with other appropriately feminine activities such as drawing, modeling, writing, collecting, gardening, herbalism, home decoration, floral arrangement, and taking walks in the fresh air. The science of botany in particular became a woman's domain, as it did not involve dissecting, killing, or any unseemly physical exertion. In their homes, women were often educated in drawing and water-color art in the eighteenth and nineteenth centuries by their mother, governess, or, a male artist. Women's art included such genres as landscapes as well as floral and fruit arrangements and might include modeling flowers and fruit in wax. Mary Delaney (1700–1778) became famous for her botanically accurate paper mosaics of flowers. Other women stitched

Jane Webb Loudon (1807–1858)

Jane Webb was home-educated and had some literary and artistic talent in her youth, but she had no interest in or knowledge of science. Her parents died while she was still a child. Because her father, Birmingham businessman Thomas Webb (?–1824), had been having financial difficulties before his death, he left his daughter with little means of financial support. While living with friends, she wrote a science-fiction romance, *The Mummy: a Tale of the Twenty-Second Century* (1827). The novel received positive reviews for its optimistic depiction of the future and its technological innovations.

John Claudius Loudon (1783–1843), a well-established and well-traveled horticulturalist and editor of the very popular *Gardeners Magazine*, read and reviewed *The Mummy*. He was so fascinated by the book that he asked to meet the author, who he assumed was a man. He was surprised to learn that the author was a woman, but clearly their first meeting went well, because later that year, in 1830, they married. Jane Loudon immediately began learning botany both through instruction from her husband and by attending the public botany lectures of John Lindley. As John Loudon had suffered the amputation of his right arm, Jane assisted him in his horticultural work by writing, taking dictation, and copying manuscripts for him. She also accompanied him on his numerous tours of English gardens. Unfortunately, by 1838 the Loudons found themselves heavily in debt. John Loudon's important work *Arboretum Britannicus* had been lavishly illustrated and was priced so high that it did not sell as well as had been expected.

Jane then started writing her own publications to raise money. They ran the full gamut of women's popular science works, although Jane paid only minor attention to the natural theology tradition. Some of her works, such as her *Instructions in Gardening for Ladies* (1840), were largely horticultural and emphasized the "how to" of various gardening projects. She wrote her *Young Naturalist's Journey; or, The Travels of Agnes Merton and her Mama* (1840), republished in 1863, for women and children and was only very loosely scientific. Her aim was to bring the natural world of the British Isles into their homes and engage their interest in natural history. Her series of eight volumes entitled the *Ladies Flower Garden*, published beginning in 1838, specifically addressed her love of flowers and how to grow them, but interwove into this discussion an extensive and in-depth botanical knowledge. Her *Botany for Ladies* (1842), later published under the title *Modern Botany* in 1851, avoided the Linnean system of classification, teaching women Augustin-Pyramus de Candolle's natural system of classification instead. In this work she took up a textbook format, eschewing the feminine conversational format that was losing fashion by midcentury. The book exhibited Jane Loudon's belief in the importance of women's science education and in the need for books that made science accessible to women. Their debt still remained upon John Loudon's death in 1843, despite some small amounts of money received from the government and from the Literary Fund. Jane continued to write until her death in 1858, in a continual effort to support herself and her daughter Agnes.

flowers in embroidery or crochet. Such skills not only filled women's spare time but enabled them to decorate their homes.

It was a short step from observing the natural world for artistic purposes, to collecting and pressing flowers and plants for personal herbarium collections and exchange. Botanical activities were considered to invigorate the mind as women were taught the Linnean system of ordering plants to categorize and name them. (Some men and women found the Linnean system's ordering of plants by their sexual parts to be a risqué subject for women, and some chose other systems by which to order their collections.) Moreover, the gathering of specimens for collections necessitated spending long periods outside and walking fair distances. Though it was not considered appropriate for women to travel alone very far from their homes, the association of plant studies with exercise was clear, and women needed appropriate apparel for pursuing such activities to their full extent. Jane Loudon (1807–1858) refers to a lady's gauntlet for gardening in her *Instruction in Gardening for Ladies* (1840), and Margaret Gatty (1809–1873) in her *British Seaweeds* (1872) outlines for women suitable dress for collecting seaweeds, which included a strong pair of boy's oiled shooting boots and woolen petticoats that remained above the ankle. Practicality in this instance, argued Gatty, ought to outweigh fashion if not propriety. As well as collecting and preserving specimens, women's increased interest in plants added to their herbal knowledge and encouraged a desire to plant gardens to watch the growth and life cycles of plants as well as to further embellish their homes, activities women herbalists and midwives had practiced for centuries. These middle-class women kept conservatories, window gardens, and aquariums of plants and also planted gardens outdoors.

Women flocked to botany in such numbers that by the mid-nineteenth century, Victorians feared that botany "crazes" were denuding the landscape, and some scientific men began to see botany as "femininized." In 1887 an article by J. F. A. Adams entitled "Is Botany a Suitable Study for Young Men?" in the journal *Science* addressed this very point. Scientists such as John Lindley (1799–1865) worked to recover botany as a masculine pursuit by consciously defining it separately from botany for ladies. In 1865 he wrote a popular botanical work entitled *Ladies' Botany; or, A Familiar Introduction to the Study of the Natural System of Botany*, consciously indicating that women's botany involved the general knowledge of the student. Although he contributed to the popular botanical literature and encouraged women's scientific learning, he nevertheless argued that botanical research was men's work and consequently attempted to exclude women from it.

Middle-class and upper-class women could also obtain scientific knowledge through public lectures. Established scientific societies gave public lectures that included women as well as men in their audiences. Women in the audience sometimes did not understand these lectures, making it more of a social event than a learning experience. Nevertheless, it was Jane Marcet's (1769–1858) inability to follow Humphrey Davy's (1778–1829) public lectures at the Royal Institution on the concepts of the new chemistry that prompted her to write her popular *Conversations on Chemistry* (1839). Able to learn the chemistry from "familiar conversations" of scientific friends, Marcet then found Davy's lectures far more enjoyable. The dialogue format of her *Conversations* facilitated women's chemistry learning, for as Marcet writes in her book, female "education is seldom calculated to prepare their [women's] minds for abstract ideas, or scientific language." Marcet wanted her women readers to be engaged by science, not just for the sake of science itself but also because she believed the nature of science would improve their minds more generally. Moreover, Michael Faraday (1791–1867), famous for his work on electricity and magnetism, was led to a life of science in part through a reading of Marcet's *Conversations on Chemistry*. Historian Susan Lindee has highlighted the success of Marcet's book, noting the extensive career of this volume in many forms throughout the first half of the nineteenth century as a chemistry textbook in U.S. girls' schools. Lindee (1993) argues that while the teaching of chemistry was justified by women's schools as an appropriate subject for women because of its domestic associations such as cooking, in fact women's schools "promoted feminine interest in scientific theory at a level that exceeded that required for domestic efficiency or religious gratification" (155). Public science lectures could be more than a social function for women and could promote women's scientific learning much further and more deeply than the lecturer intended.

As a result of women's increasing interest in science, the number of popular science books directed at them proliferated. These works were how-to manuals for the uninitiated, itemizing what to wear, where to go, what equipment was needed, the basic elements of the science under examination, and how to identify and understand what they observed. In this way, women throughout the eighteenth and nineteenth centuries were *students* of science, as was Fontenelle's marquise in the late seventeenth century, but they also were *teachers* of science—to their children. Popular science writers often designed their works with this fact in mind and with a desire to encourage and provide the skills for a mother to educate her children in science. Until the middle of the nineteenth century, mothers could read popular works to their children that were written in a conversational form, often representing a woman teacher or mother answering the basic scientific questions of the children in her charge and encouraging them to carefully observe and record the natural world. Nevertheless, historian Greg Myers has argued that as readers of popular science works, women and children became "the consumers, not the producers, of scientific knowledge" (1989, 198). Whereas philosophers in the seventeenth century created a masculine science, in the eighteenth century they promoted the creation of a feminine science that encouraged women's interest and enthusiasm for studying the natural world, giving science social respectability. However, male scientists attempted to maintain their authority over the nature of the science learned and to restrict the role women could play within science via science popularization.

Women Writing Science

Though men also wrote popular science books, this kind of writing was particularly attractive to women for a number of reasons. First, women read popular science works to their children as part of their education. Women writers reasoned that in writing such works they were extending, but nevertheless still remaining within the confines of, the role of mother. Historian Ann Shteir has shown that from the turn of the century until the middle of the nineteenth century, women science writers made extensive use of "the narrative voice of the maternal educator" (1996, 62). The first woman to write a botany book for children was Priscilla Wakefield (1750–1832), with her *An Introduction to Botany, in a Series of Familiar Letters* (1796). In this work a young girl, Felicia, is taught botany and the Linnean system by her mother and governess to cheer her during her sister's absence from home. Subsequently, Felicia writes about her lessons to her sister. Wakefield held out botany as a particularly useful study for girls as it encouraged them to engage in physical exercise and self-improvement beyond the vacuous "accomplishments" of fashionable society. Working within the constraints of women's private world and within the behavioral and intellectual expectations for women, Wakefield nonetheless wanted women to understand that they could find enjoyment in scientific study.

Similarly, a year later, Maria Elizabeth Jacson (1755–1829) published *Botanical Dialogues: Between Hortensia and Her Four Children* (1797). Hortensia, the mother, was portrayed as an expert authority on botany who educated her children in the Linnean system and decried works that attempted to oversimplify science to match the supposedly limited intellectual capacity of ladies. Wakefield's work was so influential on subsequent women writers that in Sarah Atkins Wilson's (1801–1863?) book *Botanical Rambles, Designed as an Early and Familiar Introduction to the Elegant and Pleasing Study of Botany* (1822), one of the children in her story turns to Wakefield's *Introduction to Botany* as an authority. Wilson also had her character bring to bear the same rationales for women's pursuit of science, praising her mother, who "wishes us to pursue this

Margaret Gatty (1809–1873) (Hunt Institute for Botanical Documentation, Carnegie Mellon University)

fascinating science, not merely as an amusement, but as a pleasing change for some of the trifling objects that too often occupy the time of many young ladies." Within the Enlightenment tradition of teaching and learning, and moral, physical, and intellectual self-improvement, women writers could not only learn science but could take up their pen to teach others. Though still restricted by the gendered conventions for women to the popular genre and the conversational format set in the domestic context, women writers like Wakefield and those who followed her exhibited their own extensive scientific knowledge and in turn cultivated a desire to learn science in women who read their works.

In the United States, women science writers' desire to educate other women and children was clear, but their goals were often slightly different from those of their European counterparts. According to historian Nina Baym, British female popularizers encouraged women to be better educated within their gendered sphere, whereas American female popularizers viewed science education as a way for women to rise above their present state and achieve intellectual parity with men. Ironically, in presenting women's lack of scientific learning they highlighted and often eventually accepted the secondary status of women in their society.

Almira Hart Lincoln Phelps (1793–1884) was an American science writer and educator of women who promoted scientific knowledge among women with her *Familiar Lectures on Botany* (1829). This work saw twenty-eight editions over forty-four years and sold approximately 350,000 copies. Subsequently she wrote other textbooks on chemistry, geology, and natural philosophy. Phelps worked with her sister, Emma Willard (1787–1870), at Willards' Troy Female Seminary in New York and later became principal of the Patapsco Female Institute in Maryland. Phelps believed that the responsibilities of a republican citizenship demanded an educated populace that included women. She argued that science promoted analytical thinking; improved the powers of observation, memory, and reasoning; and overrode superstitious and primitive ways of thinking about the world. Yet, despite her experience as an educator and her success as a science popularizer, Phelps believed that men had greater natural abilities in the realm of science than women. She was dependent on male support for her work and she, along with other American women science popularizers, celebrated the achievements of masculine science. In some sense she was working against herself, writing popular works of science for women yet believing that women could not attain the same accomplishments as men in their study of the natural world. Nevertheless, she was a strong proponent of scientific education for women, and her published works did facilitate women's introduction to science.

Another aspect of popular science writing that made it an acceptable province for both American and British writers was its religious component. Women had long been considered more spiritual beings than men and thus were able to publish as religious writers without social censure. Natural history was the perfect subject for promoting faith. In the natural theology tradition in which many of these women wrote in the late eighteenth and nineteenth centuries, they argued that the study of the natural world brought the student of nature to worship at the feet of God. Examining the grandeur of the planetary system or the design of a tiny insect's wings implied in most peoples' minds the idea of a designer. William Paley (1743–1805) in his *Natural Theology; or, Evidences of the Existence and Attributes of the Deity, Collected from the Appearances of Nature* (1802) wrote perhaps the best and most well-known explanation of the argument about design. Design in the natural world, he posited, implies a creator, just as a watch implies a watchmaker. Writers like Paley showed how the awe and wonder that most men and women experienced when studying the natural world was easily transformed into veneration for the designer of the universe. Thus natural history provided a service for theology, bringing an ever-increasing materialistic and rational world back to the faith of their ancestors.

Some women writers only nodded in the direction of natural theology as a cover for their real interest in science and simply included a perfunctory note in the introduction or conclusion of their work that pointed out the connection between the natural world and the creator. Other women took up the religious implications with vigor. As part of their educational and maternal role they believed it was important to nurture faith through the study of nature. Margaret Gatty (1809–1873), although the author of the popular science work *British Seaweeds*, was most well-known in her day for her series of children's books entitled *Parables from Nature*, published from 1855 to 1871. Parables convey religious or moral lessons in simple stories. Gatty, working within the natural theology tradition, used her scientific knowledge about the natural world to teach children lessons that would instill in them a faith in God. In her story "Red Snow," a young boy is told by his mother and his pastor that the mountains surrounding his Swiss home are cold and lifeless like a hardened human heart without God. There was no point traveling to the mountains for there was nothing to see there. One day, a naturalist passed by and met the boy, who was gazing wistfully at the mountain full of pity for it because it was an outcast of God where no flowers grow. The naturalist quickly informed the boy that in fact there is a red flower that grows only in snow and lives there in abundance and presents the moral: "God makes the very wilderness to burst forth and blossom like a rose: that there are no outcast ends of the earth, uncared for by Him; no desolate corners where His goodness is not shown forth." The boy and his mother learn through their discovery of the red plants on the mountain that no place, and no individual creature or person, is ever deserted by God. The natural history lesson contains a religious message and thus provides a religious rationale for studying nature.

In the United States, Almira Phelps in *Familiar Lectures on Botany* (1829) explicitly instructed her readers to look for evidence of God in nature as part of their program of self-improvement. She wrote, "those who feel in their hearts a love for God, and who see in the natural world the workings of His power, can look abroad, and adopting the language of a Christian poet, exclaim, 'My Father made them all.'" Though male popularizers of science certainly also wrote within the natural theology tradition, it was particularly important for women writers to do so, as it linked the female sphere of influence with science. The natural theology tradition helped to legitimize women's science popular writing as an educational sphere open to women because of its moral and religious context.

As male scientists worked to professionalize science they made every attempt to disconnect their science from religion and to separate popular writers who were grounded in the natural theology tradition from professional scientific research. They believed that religion and religious men could both undermine and supersede scientific authority if the interconnection between science and religion remained. In order to establish themselves as a distinctive voice of authority, these men had to separate themselves and their knowledge from writers who persisted in maintaining the connection. Despite their best attempts, the popularity of natural theology continued well into the second part of the nineteenth century as many male and female popularizers of science continued to reject the separation of science and religion.

Another way in which women became involved in science was through their empathetic attempts to protect the natural world. Taking up the argument of masculine science that women were closer to nature than men, these women argued that in that case, they were clearly the best spokespersons to protect the natural environment. Many women writers took up the cause to protect the natural world from the encroachment of civilization, industrialization, and masculine science. Within the first year after the founding of the Society for the Protection of Birds in England in 1889, all of its officers were women. Women were also heavily involved in protecting the natural landscape from urban and industrial encroachment. They believed that in order to maintain their humanity, human beings had to be in harmony with the natural world. Women such as Octavia Hill (1838–1912) in London petitioned for more open spaces in urban areas, arguing on the grounds that such spaces would improve the moral, physical, and spiritual health of the poor.

More sinister in the minds of these women writers was the vivisection of helpless, feeling animals by heartless medical men. Vivisection, a form of dissection that examines the observed creature while it is still alive and conscious, allows students and researchers to observe firsthand the functioning of various bodily systems. Frances Power Cobbe (1822–1904), a particularly strong opponent of this practice, argued that not only was this cruelty against animals, it was a degradation of all the characteristics civilized human beings held sacred. She believed that sympathy for one's fellow creatures was an indication of a higher state of civilization than was the lack of sympathy exhibited by the medical profession. How, she wondered, would such medical men go on to treat their patients with kindness and respect if they were educated in this manner? Moreover, she connected men's mistreatment of animals and their mistreatment of women, arguing that animals and women both were victims of a patriarchal society that had no respect for the natural world. As a strong advocate for women's rights, she did not let an opportunity slip by that might draw society's attention to the inequality women suffered. She was instrumental in forming an antivivisection society, the Victoria Street Society in London, that worked in concert with the Royal Society for the Prevention of Cruelty to Animals to try to bring about an end to vivisection.

Women not only contested the treatment of and attitudes toward nature by a masculine and industrializing world but also examined the new theories of science, contested them, and sometimes were impelled to rewrite science. Margaret Gatty railed against Charles Darwin's theory of evolution by natural selection in her letters to William Henry Harvey (1811–1866), a well-known botanist and authority on algae. She believed not only that Darwin's theory of evolution was scientifically incorrect, but that in presenting such a theory to the public he undermined the religious foundations of society. She called upon Harvey to write against Darwin, and in a couple of her *Parables* she pointed out how ridiculous some theories could appear to be, berating the smug scientists who believed they knew everything. In 1883 Arabella Buckley (1840–1929), secretary to the famous geologist Sir Charles Lyell, went a step further than Gatty and wrote *Winners in Life's Race*, a popular work recounting the development of the vertebrate species. Buckley both drew upon and extended Charles Darwin's theory of evolution in her work. However, unlike Darwin, who described the indifferent force of "natural selection" as responsible for evolution, Buckley added the evolution of sympathy. She argued that the caring of one animal for another had developed: "amidst toil and suffering, struggle and death, the supreme law of life is the law of *self-devotion and love.*" Buckley's contemporaries and present-day scientists could deem her views representative of the typically feminine traditions of cooperation and even sentimentality, but historian Barbara Gates has argued in *Kindred Nature: Victorian and Edwardian Women Embrace the Living World* (1998, 169) that "this important woman popularizer's . . . vivid and sympathetic work" should be seen as "something more visionary, and also more original—a significant rewriting of Darwin for the future." Gates referred to Buckley's and other women's unique approaches to understanding the natural world as "womanist visions of nature" (1998, chapter 5). Though these women science writers worked within the popular tradition of science writing that was deemed socially acceptable for women, they nevertheless broke free of the traditions masculine science tried to impose on their view of nature and instead, like their forerunners in the seventeenth century, constructed and published their own opinions and theories about the natural world. Women could be constrained in their science writing by gender norms, but such written work also taught women about science. With this knowledge in hand, women could conceptualize a different science for their readers, one constructed from a woman's view of the world rather than a man's.

Women Assisting Science

Women did not only write popular science texts, they also practiced what they preached. Men often provided introductions and instructions in science practice to their wives and daughters. On the Ormerods' Gloucestershire estate, Sedbury Park, the Ormerod children had eight hundred acres of agricultural land, woods, pasturage, and water sites to explore. Eleanor Anne Ormerod (1828–1901) was the youngest of the ten children but did her best to keep pace with her older sisters and brothers' long and sometimes treacherous walks. Her brothers engaged her in natural history lessons, including instruction on how to use a microscope. In the United States in the colonial period, Jane Colden (1724–1766) learned the Linnean system of plant classification, Latin, and botany from her scientist-politician father, Cadwallader Colden. Jane conducted her study in the gardens of their three-thousand-acre family estate at Coldengham in New York. Cadwallader believed that "botany is an amusement which may be made agreeable to the ladies, who are often at a loss to fill up their time. Their natural curiosity, and the pleasure they take in the beauty and variety of dress, seems to fit them for it" (quoted in Bonata 1991, 5).

Some scientific men understood that educating women in science could take them beyond its entertainment value and become for some women more than a hobby. Such men were well aware that a reasonably well-educated female family member might provide valuable unpaid assistance to their own scientific endeavors. Though some historians have argued that the role of assistant is a passive and subordinate one, women saw such a role as a socially acceptable avenue to scientific studies. They could fulfill roles as gatherers of specimens, data collectors, note takers, copyists, observers, laboratory assistants, and illustrators. An excellent example of a woman acting in these capacities can be seen in the life and work of Elizabeth Gould (1804–1841), who was a talented illustrator of the ornithological works of her husband, John Gould (1804–1881). In her short life she produced an enormous number of illustrations, including those for John's work on Australian birds, painting most of the 681 original watercolors and engraving eighty-four of the plates. It was on their trip to Australia that Eliz-

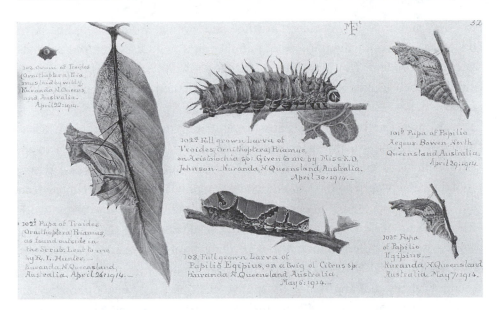

Lepidopetra larva from Queensland, Australia, drawn by Margaret Fountaine (1862–1940). (Natural History Museum, London)

abeth began to develop a personal interest in observing birds as well as in painting them. Unfortunately their partnership was cut short when she died at thirty-seven of puerperal fever after the birth of a daughter. Another well-known example of a woman fulfilling the role of assistant is that of Caroline Herschel (1750–1848), sister of famous astronomer William Herschel (1738–1822). She became a valuable assistant to her brother in his astronomical observations, as it was necessary for one person to observe and another to fetch instruments or take notes. She became so adept at observation that she discovered a new comet in 1786, and seven more after that.

In the United States, Elizabeth Cary Agassiz (1822–1907), wife of Swiss naturalist and Harvard professor Louis Agassiz (1807–1873), was her husband's interpreter and translator when they first moved to Massachusetts. The couple started a girls' school together in their home to supplement Louis's income, and here Elizabeth began to learn natural history. She attended his class and recorded his lectures, took accurate notes on field trips with him, and received an education from him equal to any one of his male students. In 1859 Elizabeth published her own teachers' guide, *A First Lesson in Natural History*. In 1865 she published *Journey in Brazil*, her account of daily experiences on their field trip to Brazil. In the same year she also published *Seaside Studies in Natural History*. From 1869 on she published scientific articles in the *Atlantic Monthly*, which demonstrated she was an active participant on Louis's field expeditions. Louis ignited Elizabeth's interest in science that resulted in both her assistance to Louis and in her own science practice and writing. Women could be introduced to science for the purpose of fulfilling the role of assistant to husbands, fathers, or brothers. However, this experience allowed them to participate extensively in science and even sometimes to succeed in their own personal, scientific achievements.

Women also became assistants to nonfamily members. In an era where men considered gathering specimens to be of paramount importance to the advancement of their field, many male scientists developed systems of correspondence networks so they could obtain a large collection of specimens from many different locations. Keeping in touch with individuals all over the world who were interested in science meant that scientists could put many eyes to work in multiple locations for that elusive, undiscovered specimen. For example, Dawson Turner (1775–1858), a banker interested in cryptogams (flowerless and seedless plants that reproduce by spores, such as fungi, algae, mosses, and ferns), corresponded with Ellen Hutchins in Ireland. Hutchins, an amateur botanist, sent Turner many useful specimens and drawings. William Henry Harvey not only corresponded with Margaret Gatty (noted earlier) on scientific matters but received specimens from another amateur botanist, Amelia Warren Griffiths. He found Griffiths so helpful that he dedicated his *Manual of British Algae* to her in 1849. Dawson Turner likewise admitted how important Ellen Hutchins's assistance was to him. He bewailed her early death in his *Fuci* (1819), noting that he had "been deprived of a most able assistant" and that "Botany has lost a votary, as indefatigable as she was acute, and as successful as she was indefatigable, this work bears abundant testimony; and the *Lichenographica Britannica*, should it ever be published, will do so no less unequivocally."

Mary Anning (1799–1847) was a gatherer of a different sort. An impoverished woman living with her family in Lyme Regis, an English coastal town, she became a self-educated paleontologist (studying fossils and ancient life forms) and a fossil dealer. Uncovering many significant fossils, such as a complete plesiosaurus and a fossil flying reptile named *Pterodactylus macronyx*, Anning sold her findings to prominent gentlemen scientists including Adam Sedgwick (1785–1873), William Conybeare (1787–1857), and Henry De la Beche (1796–1855). These men made their scientific names, in part, by buying, naming, and displaying Anning's finds. They passed her name on in private correspondence as a useful resource, making her a local celebrity in her hometown. Despite the fact that Hutchins, Griffiths, and Anning never published anything in their own names, the written testimony of the men they assisted speaks volumes about their contributions to science.

Some women managed to go beyond the garden and the family estate and venture abroad. As the nineteenth century progressed, men and women interested in science founded innumerable local natural history societies. Although these organizations limited women's participation, women were permitted to accompany parties in rambles across the countryside in search of specimens, in part because organizers often considered these events to be semi-social occasions. These groups consisted mostly of amateurs who enjoyed dabbling in natural history and the socializing that accompanied it. The Providence Franklin Society and other American societies were particularly interested in encouraging women to participate in their field trips, courses, and lectures. Harnessing the excitement of exploring a new world and the competitive spirit of establishing scientific concerns equal to those in Europe meant that scientific men and scientific societies had to rely heavily on amateurs, including women, to establish American science on a firm footing.

Other women had the opportunity to travel further afield than their own gardens. Whether emigrating from Europe to the New World, traveling as a family member of colonial government appointees, or as a companion to a naturalist, women gained experience in strange new natural landscapes. Some women took up natural history as a pastime while abroad. Others, already interested in science before leaving home, took the opportunity to study their favorite science abroad. Louisa Twamley Meredith (1812–1895), an established English author, emigrated as a colonist with her husband, Charles Meredith, to Tasmania in 1839. She had already published *The Romance of Nature; or, The Flower Seasons Illustrated* (1836) and *Our Wildflowers Familiarly Described and Illustrated* (1839). Meredith decided to educate British audiences, including scientists, about her new land, believing her firsthand experience gave her valuable information about the lives of the plants and animals there. Her works included *Notes and Sketches of New South Wales during a Residence in that Colony from 1839–1844* (1844), *My Home in Tasmania* (1852), and *Tasmanian Friends and Foes: Feathered, Furred, and Finned* (1880). Meredith was not a lone woman naturalist, as is evident in the correspondence of Ferdinand von Mueller (1825–1896), government botanist of Victoria. Of his 2,800 correspondents, 240 were women, many of whom were scientific correspondents or who at least shared a mutual interest in science with von Mueller.

Catherine Parr Traill immigrated with her husband to Upper Canada in the early part of the nineteenth century. She turned to the study of flowers to elevate herself above the misery of pioneer life, writing that the wild plants were like "dear friends, soothing and cheering by their unconscious influence hours of loneliness and hours of sorrow and suffering" (quoted in Bennett 1991, 84). In 1854 Traill wrote a guidebook, *The Female Emigrants' Guide, and Hints on Canadian Housekeeping,* so that new immigrants would more quickly understand the benefits and usefulness of certain plants than she had and know how

Mary Henrietta Kingsley (1862–1900)

Mary Kingsley was the product of a dissolute middle-class father and an invalid working-class mother. While she and her mother remained confined to their little house and garden in Highgate, England, her father, George Kingsley (1827–1892), traveled the world collecting, shooting, and note taking. Mary did at least have the opportunity to read books from her father's library, and when he was at home, she learned the basic tenets of natural history from him. When ill health forced George into retirement in

Cambridge, his self-educated daughter assisted him by running the household and working with him to put together his notes on a comparative study of worldwide sacrificial rites. He died before the work was completed, in 1892, and his wife, Mary's mother, died the same year. For a short period, Mary looked after her brother, Charles, but when he left for a trip to China she was, at age thirty, finally free to choose how to fill her own time.

First she traveled to the Canary Islands. Thereafter, in 1893, she decided to travel to Africa to collect fish specimens for her sponsor, Albert Gunther of the British Museum of Natural History, and to conduct the research that would allow her to finish her father's book. In practice she

(Hulton Archive/Getty Images)

was more interested in studying African culture and the natural world on her own terms. Insisting upon dressing in a curiously old-fashioned European garb, even in the midst of the West African heat, she traveled alone with one bag and some collecting boxes, living among the wild animals and tribes of Africa. Despite finding herself in such a completely incongruous setting for any proper middle-class Victorian woman, her familiarity with her mother's working-class family facilitated her comfortable communication with all types of people and enabled her to barter and trade with some

to prepare and utilize them. Her work also included descriptions of animal behavior and ecological relationships. She went on to publish botanical studies including *Canadian Wild Flowers* in 1868, the first Canadian botany book that was both illustrated and written by women, and *Studies of Plant Life in Canada* in 1885. She wrote both works for a general audience, so they were unencumbered by scientific terminology, and drew not only upon her own knowledge of Canadian flora but that of native women and old settlers. Traill struggled to make a living from her writing, constantly undermined by publishers who paid her a pittance and then made a fortune from numerous editions of her work. They

fair degree of skill. Though experiencing the full brunt of the primitive living conditions in Africa, neither her physical nor her mental health suffered. She relished her
travels collecting fish. Gunther praised her powers of observation when he found that
she had returned with sixty-five species of fish, three of which were previously undiscovered. She also kept a diary of her adventures and recorded the ethnographic
knowledge she collected along the way, not only about African society but also about
the imperial relationship between the British and Africans. She made a second trip to
West Africa in 1895.

When she returned to England later that year, she stayed with her brother in London so that she could write her travels, published as *Travels in West Africa* (1896)
and *West African Studies* (1899). She toured the lecture circuit, giving talks on her
adventures in Africa. Her lectures were a huge popular success, sometimes drawing
audiences of up to two thousand people. Kingsley was entertaining, informative, and
opinionated. Although often depreciating the worth of her work compared to that of
male scientists, Kingsley clearly saw herself at the same time as a heroic adventurer
and a knowledgeable field worker and gatherer of data. She also formulated her own
view of African society. She believed that African people ought to be studied in the
context of their own society and culture, not in comparison with European culture.
She spoke with respect about all aspects of their culture from polygamy, to cannibal-
ism, to trading practices. She pointedly disagreed with many aspects of British imperial encroachment such as hut taxes, the ethics of the liquor trade, and Christian mis-
sionaries, even though she considered herself an imperialist.

Although criticized by scientists for her lack of objectivity and theoretical clarity,
she persevered in her multifaceted presentation format, symbolic of the complexity
of the human culture she was discussing. In this way she contested the objective
approach to the study of the natural and the human world. The popularity of her public lectures allowed her voice to be heard despite the concerns and doubts of men in
the scientific community as to whether or not she ought to define herself as a scientist and her work as science. She took up her travels only once more, in 1899, this time
to South Africa to nurse British soldiers during the Boer War. The war was an imperial cause she ardently supported, but she died from enteric fever at the age of thirty-
eight in 1900 before the war had ended.

took advantage of her often dire financial situations and her naivety about the
publishing world. In turn, she had no choice but to accept whatever meager sum
they offered her to keep her family housed and fed (Gray 1999).

Sarah Wallis Bowdich Lee (1791–1856) experienced her own share of financial difficulties. She married naturalist Thomas Edward Bowdich Lee and
accompanied him on field trips to Africa. After Thomas died during the second
trip, in 1824, she had learned and observed enough natural history to complete
his unfinished work, *Excursions in Madeira and Porto Santo*, to which she
appended a natural history of zoological and botanical descriptions. Lee went on

to publish her own works in natural history, including *The Freshwater Fishes of Great Britain* (1828–1838); *Anecdotes of the Habits and Instincts of Birds, Reptiles, and Fishes* (1853); and *Trees, Plants, and Flowers: Their Beauties, Uses, and Influences* (1854). These works enabled her to pay off her husband's debts, support her children, and gain for herself a solid scientific reputation. For women of that era, science was not only a way to pass the time but an intellectually engaging activity that occupied their minds and hearts, assuaged loneliness, and also paid their bills when necessary. Despite prescriptive strictures against work for middle-class women, many had no choice but to make their own living. Middle-class women's employment options were limited, but from the late eighteenth century on, writing became a socially acceptable form of work for middle-class—as well as upper-class—women. Putting pen to paper to support themselves and their families could still be done within the confines of private, domestic spaces, and certain types of writing, including popular science writing, were by then considered by society to be within women's purview.

Other women naturalists traveled alone, much like their male counterparts, purely with the intent to explore the natural world. Marianne North (1830–1890), Margaret Fountaine (1862–1940), and Isabella Bird Bishop (1832–1904) lived lives that were not at all in keeping with gender norms for Victorian women. They braved the harsh climates and conditions of foreign lands and lived and associated with people outside of their class and race, personally financing their explorations of the globe. Neither North nor Bishop published strictly scientific works, but their travelogues recounted their experiences in nature as they traveled around the world. North painted an enormous number of canvases in oils representing the natural world in an ecologically harmonious and scientifically identifiable manner. She built a gallery at Kew Gardens with her own funds where she exhibited her voluminous botanical artwork. Her art is still on display at her gallery today. Fountaine amassed a sizeable collection of twenty-two thousand butterflies, which she willed to the Norwich Museum, and a large collection of scientifically accurate watercolors of butterflies, which she left to the British Museum of Natural History upon her death. Whereas many women were introduced to the natural world beyond Europe through family circumstances, others sought the life of an adventurer unconstrained and unencumbered by gendered restrictions, free to pursue the life of a naturalist at large.

Women were very much a part of science practice in this period. In an ironic twist of fate, though, women were, in the words of historian Debra Lindsay, "immersed in a community that formally excluded them" (1998, 632). Nevertheless, though women were often silent partners in their husbands' scientific work, the boundaries between home and work were still flexible. Women worked in the production of masculine science, assisting men in the pursuit of natural knowledge. But they could also use the training they received working on others' projects to their own ends, producing their own science and often becoming known and respected in scientific circles.

Women Scientists in a Man's World

Fictional female characters engaging in science in popular works could establish the plausibility of women scientists, argues historian Paula Findlen (2003). "Women could become philosophers in conversation with men. A bookish idea helped to solidify a social reality" (61). But entering the professionalizing world of science was not easy for women. As the professional scientific community grew, male scientists refused women admission to scientific societies. In 1880 at the dinner of the first meeting of the recently founded American Chemical Society, the evening entertainment included the performance of anti-female songs and poems that were subsequently published as *The Misogynist Dinner of the American Chemical Society*. Rachel Bodley (1831–1888), the first woman elected to the society, resigned in protest at these activities. A less obvious exclusionary tactic was for the men to organize "smokers." Before the 1920s it was considered socially unacceptable for women to be in the same room while men smoked. Therefore an organization that labeled a particular meeting a "smoker" was, consciously or unconsciously, keeping women out. The American Geographers, for example, deliberately held "smokers" to discourage women from attending their meetings. Official scientific societies have a long history of excluding women. However, societies such as the Botanical Society of London, the Zoological Society, the Royal Entomological Society, and the Royal Institution freely admitted women from their inceptions in the first half of the nineteenth century. In Britain and the United States, women had been admitted to local clubs and societies that were often social as well as scientific clubs and that often required the women to keep up their numbers. Attempts of the men in the British Association for the Advancement of Science to keep women out of their membership and annual meetings failed from its inception in 1831. Although told they were not welcome, the ladies gate-crashed on a number of occasions until the male members relented, realizing at the very least that women were a financial asset to the association.

However, by the second half of the nineteenth century the debate over the admission of women to other scientific societies intensified as the increasingly professional body of male scientists began to feel uneasy about the presence of women amateurs in their midst. When renowned geologist Charles Lyell suggested in a letter to Thomas Henry Huxley in 1860 that women be admitted to the

Maria Mitchell and Mary Watson Whitney at the observatory at Vassar College (Special Collections, Vassar College Libraries)

Mary Fairfax Somerville (1780–1872) (Getty Images)

Geological Society, Huxley drew a clear line between the status of men and women in science. He noted that the Geological Society was not "to [his] mind, a place of education for students but a place of discussion for adepts: and the more it is applied to the former purpose the less competent it must become to fulfil the later—its primary and most important object." Women were but students, amateurs in the scientific endeavor. Huxley, keen on professionalizing science, wanted to keep the "Naturalists of the Boudoir" out of the serious scien-

Maria Mitchell (1818–1889)

Maria Mitchell was born in Nantucket, Massachusetts, the third child of nine. Her father, William Mitchell, introduced her to science. He was a teacher and bank clerk with an interest in astronomy who worked for the U.S. Coast Survey on the island of Nantucket

(Library of Congress)

and was responsible for checking the accuracy of whaling ship chronometers by means of stellar observation. Maria demonstrated a high level of skill in making observations and in mathematics. From 1836 to 1856, she became the librarian for the Nantucket Athenaeum, following in her mother's footsteps. There, she began to teach herself advanced mathematics.

During this same period, she continued making astronomical observations with her father, who placed his telescope on the roof of his bank. In 1847, at twenty-nine years of age, she discovered an unknown comet, later named after her, that won her the gold medal offered by the King of Denmark. This discovery and subsequent award bought her some fame in scientific circles in Europe and America. As a result, she was elected to the American Academy of Arts and Sciences in 1848 and the American Association for the Advancement of Science in 1850. Her calculations of the comet's orbit were published in the *Monthly Notices of the Royal Astronomical Society* in 1848. She would later be presented with honorary degrees from Hanover College in Indiana (1853),

tific meeting and played a pivotal role in keeping women out of the Ethnological Society and the Anthropological Society in the 1860s (Richards 1989, 257).

Yet the boundaries of science remained permeable during this period, so that despite the lack of formal education for women in science and the fact that they could rarely be full members of the science community, some women nevertheless attained a respected place in the male scientific community. These women's works were often considered equal to that of any man's scientific productions. In the nineteenth century Mary Somerville (1780–1872) became famous in scientific circles and beyond for her grasp of mathematics and for her ability to understand the full range of scientific knowledge of her day. Though she engaged in some experimental research of her own, she became most

Rutgers Female College (1870), and Columbia University (1887) for her work in astronomy.

In 1849 Mitchell herself joined the U.S. Coast Survey to assist in taking part in a survey of the coast of Maine, for which she published a report in 1852. At the same time she was hired by *American Ephemeris and Nautical Almanac* to compute the positions of the planet Venus for the astronomical ephemeris for the meridian of Washington. She held this part-time assignment for nineteen years. However, in 1865 Mitchell was offered the position of professor of astronomy and director of the observatory at Vassar College, the second science chair in an American college to be filled by a woman. Taking the position, she moved to Poughkeepsie, New York, with her father. At this point in her life she cautiously began promoting college education for women and published a pamphlet on the subject in 1881. She was also a woman's rights advocate concerned with improving women's lives and became founder and president of the Association for the Advancement of Women in 1875, where, among other causes, she promoted science education for women. As well as teaching students astronomy, she continued her own observations and became interested in sunspots, becoming one of the first astronomers to use photography in this area. She published numerous papers in scientific journals on such subjects as double stars, the total eclipse of the sun, Jupiter and its satellites, and the planet Saturn and its rings and satellites. Her interpretations about her observations can be found among her notes, indicating that she was interested in more than data collection. Like Mary Somerville, however, Mitchell believed herself to be an "industrious" worker but no genius. She also recognized that with a few exceptions, women were not welcome in the masculine science community. While American science was professionalizing in the second half of the nineteenth century, Mitchell continued to promote the collaborative approach to doing science rather than the image of the heroic scientist working alone. The collaborative model of science found in such work as coastal surveys and astronomical observations opened a space for women's participation in science. Mitchell worked and taught up until a year before her death in 1889.

famous for her translations, synthesis, and explanations of others' scientific work. In 1827 she was asked to undertake an English translation of Laplace's *Mechanism of the Heavens*. This was a long, mathematically complex work that utilized Newton's theory of gravitation to show how the solar system was a self-regulating mechanism. Somerville spent four years writing a straightforward translation of Laplace's work, and added to it a "Preliminary Dissertation" that provided an explanation of the basic mathematics required for understanding the ideas in the work, a history of the subject, and an explanation of his work using her own mathematical derivations and proofs. Somerville's work became a standard text for the rest of the century. In her second work, published in 1834, *On the Connexion of the Physical Sciences*, Somerville demonstrated her knowledge not only in mathematics but also in astronomy, mechanics, magnetism, electricity, heat, and sound. Even more successful than the *Mechanisms of the Heavens*, her second work demonstrated the interdependence of the various branches of the physical sciences. In 1848 she published her most successful work of all, *Physical Geography*. Drawing upon the new geology of Charles Lyell, she described the "successive convulsions which have ultimately led to its present geographical arrangement, and to the actual distribution of land and water." In subsequent editions of this work, though, she avoided any discussion of Darwin's theory of evolution. She may have disagreed with Darwin, or she may have shied away in her advanced age from the kind of denunciations she had received against her in the House of Commons and from the pulpit of York Cathedral. They had branded her a godless woman after the publication of *Physical Geography* for her discussion of the great age of the earth, which contradicted the much shorter biblical dating.

Male scientists were amazed by Somerville (Laplace told her that she was the only woman who had ever understood his work), and she received memberships in many scientific societies around the world, including the Royal Astronomical Society in 1835. However, they were all honorary memberships, and her status relative to men in the scientific community remained unequal. That same year, though, she was awarded an annual pension from the British government of £200, which two years later was increased to £300, and in 1870 she won the Royal Geographical Society's Victoria Gold Medal. In her *Personal Recollections* (published posthumously in 1873), Somerville recounted that according to Robert Peel the government's purpose in giving her the pension was to "encourage others to follow the bright example which you have set, and to prove that great scientific attainments are recognized among public claims." This was an amount comparable to that given to men of science and reflected her place among them. Her mathematical facility was recognized by her male scientific contemporaries as making her their intellectual equal. Historians have debated whether her scientific productions were equal to those of men. She never made any original scientific discoveries, although she certainly participated in the activity of scientific research and published a few papers in scientific journals. Thus, it could be argued that Somerville was predominantly an intelligent popularizer of science rather than a scientist per se. Yet, despite contemporary and present-day arguments to elevate her as an equal to men, Somerville herself believed, in accordance with the gendered norms of her day, that women could never be geniuses. She wrote in her *Personal Recollections*: "Although I had recorded in a clear point of view some of the most refined and difficult analytical processes and astronomical discoveries, I was conscious that I had never made a discovery myself, that I had no originality. I have perseverance and intelligence, but no genius; that spark from heaven is not granted to the sex. We are of the earth. Whether higher powers may be allotted to us in another existence, God knows; original genius, in science at least, is hopeless in this." Interestingly, though Charles Lyell noted to his wife that had Somerville been married to a mathematician, "she would have merged it [her work] in her husband's, and passed it off as his," in actuality an unusual gender-role reversal took place in her marriage. Moreover, Mary's second husband, William Somerville, acted as her assistant in science editing, copying, compiling information, and overseeing her scientific correspondence.

Eleanor Anne Ormerod (1828–1901) and Agnes Mary Clerke (1842–1907) found themselves in a similar position to Somerville. Like her, they both believed that they could not quite attain equal scientific status with men because of their sex. Unlike Somerville, though, both Clerke and Ormerod believed they ought to be included as equals with the scientific men of their respective disciplines. Ormerod established herself as a popularizer, communicating to the agricultural public the recent scientific findings that might assist them in protecting their plants, crops, and livestock from injurious insects. Her work was well-respected by scientific entomologists around the world, with whom she corresponded on a frequent basis. She engaged in her own experimental work with insects, and although she did not make any original discoveries, she certainly attempted to do so, frequently reporting her insect observations to scientific journals as well as published textbooks and handbooks. She was a member of both the Meteorological Society and Entomological Society and had even been permitted to read her own paper at the Entomological Society in 1879. She was an honorary member of numerous entomological societies worldwide. She was also the Honorary Consulting Entomologist for the Royal Agricultural Society, was agricultural adviser to the Board of Agriculture, and for three years was co-examiner in agricultural entomology for the University of Edinburgh, beginning in 1896. In 1900 Edinburgh University awarded her an honorary LLD degree; she was the first woman to be so awarded. However, despite all of her public scientific achievements and positions held, Ormerod remained an outsider. Though Edinburgh University had called upon her to be an examiner at Edinburgh in 1896, earlier, in 1889, when the university set up a lectureship in agricultural entomology, its administrators consulted with Ormerod about who ought to be chosen for the position. They automatically assumed she could not be considered for the position herself because she was a woman. Even after receiving her honorary degree from Edinburgh, she did not feel as if she truly belonged. Ormerod wrote in a letter to her entomological colleague Professor Wallace at Edinburgh in 1900, "though I do not really hold any post among you, yet I like to think of myself now not wholly separate." By the end of her life, Ormerod seemed all too well aware that women could not hope to attain equality with men in their scientific endeavors simply because they were women.

Agnes Clerke was an astronomical popularizer and as such was praised for her work. Her major works were *A Popular History of Astronomy during the Nineteenth Century* (1885), *The System of the Stars* (1890), *The Herschels and Modern Astronomy* (1895), *Astronomy* (1898), *Problems in Astrophysics* (1903, coauthored), and *Modern Cosmogonies* (1905). In 1903 her work was recognized by the Royal Astronomical Society with an honorary membership. Like Ormerod and Somerville, Clerke was not merely popularizing science for a lay audience of women and children. She went one step further than Ormerod by also writing for scientific men. Following professionalizing trends, she wrote her works on historical and current astronomical findings in an impersonal voice and reflected upon the larger meaning of the scientific work that other scientists had introduced, indicating which lines of research she thought ought to be pursued next. In keeping with women's popular science writing, Clerke introduced religion into her works, believing that the new astronomy would support Catholicism. She pointed to the design and harmony she saw as inherent in the solar system. When she attempted to move outside of the realm of popularizer by criticizing and analyzing the science of the men whose work she popularized, they quickly put her in her place. Reviews of her work questioned her scientific expertise while admiring her writing style, drawing attention to her natural abilities as a woman and the worth of these attributes to a female science popularizer. Richard Gregory (1864–1952), assistant editor of *Nature*, denied Clerke's credibility as a scientific critic because she was a woman. In a 1906 review of her work in *Nature*, he wrote, "In preparing a statement of the position of fact and theory in any branch of science, great care must be exercised, and not a single assertion should be made without substantial reason for it. A cynic has said that it is a characteristic of women to make rash assertions, and in the absence of contradiction to accept them as true. Miss Clerke is apparently not free from this weakness of her sex." In attempting to construct the heavens to suit her own religious beliefs and critiquing the works of the expert male astronomers, Clerke was vulnerable to attacks upon her abilities on the basis of her sex. While reformulating the popular science text to give her space to move beyond the conventions of popularizations, male scientists nevertheless attempted to force her back into the category of popularizer, based purely on the fact that she was a woman.

Throughout the eighteenth and nineteenth centuries, women demonstrated their ability to do masculine science, and their work was recognized and respected by the male scientific community. But such women still had to know their place in the social and scientific hierarchy of the sexes, and keep to that position or risk public censure and ridicule. Somerville, Ormerod, and Clerke all recognized this fact. Somerville accepted it and worked within it, Ormerod complained about it privately and chafed against being held down, and Clerke believed in her own scientific abilities and her right to contest men's theories and refused to acknowledge that her place on the scientific ladder was automatically lower simply because she was a woman.

Conclusion

For every Mary Somerville, Eleanor Ormerod, or Agnes Clerke who attempted to set themselves on a par with the men in their discipline and to various degrees succeeded, there were many more women who could not break free from the gender norms of their day. In some cases women internalized these norms to such a degree that they believed that the only place for them in science was, at best, that of popular science writer. Other women who tried to gain access to the upper echelons of science failed completely and so sought refuge in traditional spheres of women's science work.

Beatrix Potter (1866–1943) is a figure as well known as a children's author today as she was one hundred years ago. Peter, a naughty young rabbit, along with such favorites as the hedgehog Mrs. Tiggy-Winkle and the frog Jeremy Fisher, encapsulated in their water-colored environments, are still a part of many children's imaginations. But in the 1890s, Potter, working alone, had accomplished important research in mycology, the study of fungi. She had discovered molds accompanying higher fungi by examining germinating spores under high magnification. She took her work to William Thiselton-Dyer, the director of the Royal Botanic Gardens at Kew and a family friend, only to have her work dismissed so rudely that she wrote in her journal that he must certainly be a misogynist. Although a year later her paper "On the Germination of the Spores of *Agaricineae*" was read at a meeting of the Linnean Society, it was never published. Not having read her paper to the society herself or even been at its presentation, she was only told that her paper needed more work. After that, she became wary of sharing her experimental investigations with men in the scientific community and gradually pulled away from the experimental discovery that was the epitome of the masculine construction of science. Instead, she moved toward writing light-hearted tales for children. Her stories often reflected upon childhood mischief and its repercussions, and they allowed a woman mortified and enraged by her rejection by the scientific community to draw and paint the natural world she had investigated with delight and to share it with children who appreciated her efforts. Potter dressed her animals in smart clothes, converted them into little people, and followed their adventures while teaching children a little about nature and even more about human nature. In *Kindred Nature* (1998), Barbara Gates has noted that Potter's animals often, in one way or another, escape the confines of their little outfits by the end of her tales, allowing their true natures to be revealed. Perhaps Potter was expressing her desire to escape her own "clothes" that represented the confines of her gendered social role, which, she believed, resulted in her being ostracized from the male stronghold of scientific research.

Even women who managed to find a scientific niche that ostensibly included them in the research community of science often felt, as had Eleanor Ormerod, that they could never be "one of the boys." Nevertheless, the work of women like Potter, Somerville, Clerke, and Ormerod continued to demonstrate, throughout the nineteenth century, that women were intellectually capable of scientific research and scientific popularization. These women's work, and the work of others like them, lay the foundation for women's gradual acceptance into scientific institutions as professionals in their own right. In April 1997, on the one-hundredth anniversary of the reading of Potter's paper at the Linnean Society, the society offered Beatrix Potter an official apology.

Bibliographic Essay

Faced with their ostracization from masculine science and all-male scientific institutions from the seventeenth century on, women nevertheless negotiated space for themselves in scientific practice. Historians have long engaged in "collecting" women scientists. One of the earliest of such works is Reverend John Zahm's *Woman in Science*, originally published in 1913 but republished in 1991, an impressively extensive recounting of women's participation in and contribution to science from Greek times until the early twentieth century. Since the 1970s, the second wave of the Feminist Movement prompted historians to explore women's past accomplishments. A significant number of women have been identified who practiced science in the eighteenth and nineteenth centuries. Margaret Alic provided an important starting point for such work in her *Hypatia's Heritage: A History of Women in Science from Antiquity through the Nineteenth Century* (1986), as did Marilyn Ogilvie with her *Women in Science: Antiquity through the Nineteenth Century: A Biographical Dictionary with Annotated Bibliography* (1986). Maria Myers Bonta has written biographies of twenty-five significant American women scientists in *Women in the Field: America's Pioneering Women Naturalists* (1991), spanning the eighteenth century to the present. Elizabeth Wagner Reed, a teacher and researcher in biology, has drawn particular attention to American antebellum women in science in *Ameri-

can Women in Science before the Civil War (1992), gathering a list of twenty-two biographies of such women, some well-known and some obscure figures, but all of whom published in science before 1861. One other useful biographical work is the dictionary by James T. Edward, *Notable American Women, 1607–1950* (1971). Two recent single biographies of women scientists, Kathryn Neeley's *Mary Somerville: Science, Illumination, and the Female Mind* (2001) and M. T. Bruck's *Agnes Mary Clerke and the Rise of Astrophysics* (2002), attest to the availability of sources for extensive studies of individual women and reveal the extent of the contribution that women have made to science both as popularizers and as scientists in their own right.

At the same time, studies that reflected the difficulties women faced in achieving success in the male-orientated world of science and their struggle to balance scientific work with family life contributed to the growing historical literature on women and science. Three important collections of essays that explore these issues are Pnina G. Abir-Am and Dorinda Outram (eds.), *Uneasy Careers and Intimate Lives: Women in Science 1789–1979* (1987); Marianne Gosztonyi Ainley (ed.), *Despite the Odds: Essays on Canadian Women in Science* (1990); and Helena M. Pycior, Nancy G. Slack, and Pnina G. Abir-Am (eds.), *Creative Couples in the Sciences* (1995). Other more recent works have explored how women practiced science within the socially acceptable spheres of home and marriage, including Paul White, "Science at Home: The Space between Henrietta Heathron and Thomas Huxley," in *History of Science* 34 (1996):33–56; and Debra Lindsay, "Intimate Inmates: Wives, Households, and Science in Nineteenth-Century America," *Isis* 89, 4 (December 1998):631–652.

Another entirely different approach to the history of women and science seeks to recognize the significance of women scientists' work that until this point, historians had deemed as mainly peripheral to the masculine world of science. Ann Shteir has written some of the most important and ground-breaking works to date. She has brought women's forms of science practice in the eighteenth and nineteenth centuries to the center of historical research. Her earlier articles demonstrated how botany in particular provided women with access to science while allowing them to remain firmly ensconced within the confines of the social norms for women. Polite society considered botany to be an appropriately feminine occupation for genteel ladies. These women practiced science within the domestic sphere and wrote science for women and children using conversational formats. See for example, Shteir's "Botanical Dialogues: Maria Jacson and Women's Popular Science Writing in England," *Eighteenth Century Studies* 23 (1990):301–317; "Linnaeus's Daughters: Women and British Botany," Barbara J. Harris and JoAnn K. McNamara (eds.), *Women and the Structure of Society: Selected Research from the Fifth Berkshire Conference on the History of Women* (1984):67–73; and "Botany in the Breakfast Room: Women and Early Nineteenth-Century British Plant Study," Pnina G. Abir-Am and Dorinda Outram (eds.), *Uneasy Careers and Intimate Lives: Women in Science, 1789–1979* (1987), pp. 31–44.

Shteir's earlier work culminated in her *Cultivating Women, Cultivating Science: Flora's Daughters and Botany in England, 1760 to 1860* (1996), which outlines the rise and fall of botany as a female domain from the eighteenth century through to the late nineteenth century. In this work Shteir explores the many routes women could follow to establish feminine spaces within botanical science and exist without social censure. She also documents the demise of these niches for women with the rise of the male professional scientist at the end of the nineteenth century in "Gender and 'Modern' Botany in Victorian England," *Osiris* 12 (1997):29–38. Elizabeth Keeney has noted a similar role and position for American women prior to scientific professionalization in her *Botanizers: Amateur Scientists in Nineteenth-Century America* (1992), chapter 5.

As equally important as Shteir's work is Patricia Phillips's more chronologically expansive work, *The Scientific Lady: A Social History of Women's Scientific Interests, 1520–1918* (1990), which recognizes the existence of, and explores women's various interests in, outlets for participating in science. Suzanne Le-May Sheffield's work, *Revealing New Worlds: Three Victorian Women Naturalists* (2001), explores the largely positive intellectual and personal impact of science on women's lives and the fluidity that often existed in this period between popularizer and professional that facilitated women's participation in science. Shteir, Phillips, and Le-May Sheffield explore science as an appropriate sphere of activity for women, but also highlight women's agency through their ability to do significant scientific research and/or writing while working within and sometimes around the social constraints of their time.

In *Kindred Nature: Victorian and Edwardian Women Embrace the Living World* (1998), Barbara Gates explores how women utilized their naturalized association with nature not only as justification for their study of the natural world but to establish their affinity for doing so. She argues that women created their own feminized niches for scientific practice. Finally, joining forces as editors, Gates and Shteir brought together a collection of essays that demonstrate how British, North American, and Australian women constructed "distinctive discourses as they narrated the story of science" in popular texts in *Natural Eloquence: Women Reinscribe Science* (1997).

Nina Baym, writing *American Women of Letters and the Nineteenth-Century Sciences: Styles of Affiliation* (2001), has noted that in the American context science could fill the same place in women's lives as in the British context, and that women were popularizers of science in the United States who constructed femininized approaches to science. However, she argues that these science writers' intent in involving women in science could also be different from those women writers in the British context. According to Baym, American women writing science wanted to learn science so they could benefit fully from the technological progresses that science afforded in their lives, and they wanted to promote science among women to "pry women out of nature and move them into culture" (20). Increasingly the life of the mind was becoming more important than brute strength, and American women science writers believed that, as a result, women could at last stand on an equal footing with men if only they availed themselves of scientific knowledge.

Women's Education in Science

The Woman of the Future! She'll be deeply read, that's certain,
With all the education gained at Newnham or at Girton;
She'll puzzle men in Algebra with horrible quadratics
Dynamics and the mysteries of higher mathematics.
—"The Woman of the Future" in *Punch* (1884)

In 1833 Albertine Necker de Saussure (1766–1841) published *L'Education Progressive* in which she argued for the importance and necessity of a scientific education for girls. Necker de Saussure had an extensive education as a result of her upbringing in Geneva, Switzerland, and her scientifically well-connected family. Her great-uncle Charles Bonnet and her father, Horace-Benedict de Saussure, were both famous naturalists. Her father was professor of philosophy at the Academy of Geneva and supported the modernization of college and secondary school education. Necker de Saussure was immersed in science as part of her education and as part of her family's daily life. She went on geological and botanical expeditions with her father and had the opportunity to discuss science with people such as the famous French chemists Monsieur and Madame Lavoisier. She married Jacques Necker, whom she supported as he worked toward a scientific career, initially writing his botany lectures for him when he entered his position at the academy. She went on to educate her own children in a wide range of subjects, including science.

In her *L'Education Progressive* she stressed that girls were just as intellectually capable of scientific learning as were boys and argued that in their primary education they ought to be schooled together. She went on to say that girls' education in science would make them strong mother-educators and would bring them closer to God. She did concede, however, that boys would go off to public school while girls remained at home to receive the rest of their schooling. But she did not consider that the location of their education undermined her general belief in the equal capacity of girls and boys to do science. She did not suggest that women practice science as a feminine accomplishment nor that they learn scientific facts by rote. Rather, she argued that women should be taught the scientific method and be grounded in the spirit of curiosity and research that was at the heart of science. She believed that such an education would lead girls away from vanity, simplicity, and excessive imagination. Moreover, a scientific education would provide women with the skills to engage in lifelong learning and to occupy their minds in old age.

Yet though she had forward thinking ideas about science education for girls, Necker de Saussure believed that women could only "imperfectly fill the place of men" in scientific research, largely as a result of their lack of institutional support. Despite the fact that she associated with some of the central figures of French science and that the strong push for the professionalization of science had not yet taken place, women were still on the margins. They could not fully enter the scientific community, but she understood that education was the answer to overcoming this limitation. Her ideas provide evidence that at least in some minds there was a basic understanding of the importance of equal education in general and the importance of a comparable scientific education between boys and girls if girls were to obtain an even chance of success in scientific research.

Although the amateur and professional worlds remained blurred in the first half of the nineteenth century, women's private, home-based education provided them with access to a scientific education. In part, this kind of education was equal to that of men's, at least with respect to science, because the public schools and universities did not take up science as a formal subject of education to any great extent until the second half of the century. The lack of a divide also meant that women could situate themselves in various ways as science writers and even as scientific researchers. But the nature of science and scientific practice changed in the second half of the nineteenth century, requiring more equipment, laboratory experimentation, and research. The focus of scientific study also changed, requiring more in-depth training as students were expected to turn their minds to physiology and ecology over taxonomy. Women could not keep pace with men, who were now gaining access to the new approaches in science through all-male educational institutions and who were increasingly specializing in their fields.

Women such as Necker de Saussure and Mary Wollstoncraft (1759–1797) had been calling for substantial and meaningful women's education since the second half of the eighteenth century but met with no success. Only in the second half of the nineteenth century did women begin in earnest to seek access to educational institutions. The same pedagogical issues arose, but now a discussion took place around women going to college and to graduate and medical school. What kind of education were women capable of obtaining? Should men and women be educated together or separately? Should men and women be taught the same subjects in the same manner, or should education be sex-differentiated based on biological considerations and the ensuing social roles they would be expected to fulfill as adults? What was the purpose of a woman's education? There were mixed responses to these questions, both for education generally and in the context of scientific education specifically. Some men and women felt that women should be educated differently and separately from men, and others believed absolutely that women needed to gain an equal education.

Between the mid-nineteenth century and the early twentieth century, women in Europe, Britain and its colonies, and the United States had made significant strides toward achieving high standards for women's scientific education. But barriers still remained as women struggled to achieve equal access to educational institutions and instruction and respect for their ability to learn and contribute to intellectual enterprises, including science. Moreover, though some women did receive scientific education equal to or greater than that of some men, few women could find as equally fulfilling career paths. Prejudice against women's education in science continued, and even the opening of doors to scientific education did not mean that they would automatically remain open without contest. At the same time that women gradually began to enjoy the experience of institutionalized education, obtain access to the best teachers, share their learning experiences with like-minded women, and to achieve university degrees that firmly established their scientific credentials, activists at the forefront of gaining access for women to institutional education remained vigilant in the face of continual opposition to their cause.

Girls' Elementary and High School Education in Science

Many women had, for centuries, studied science for the love of it, for the intellectual challenge, and for the personal development and growth it provided to them. The difference from the 1870s on was that women were gradually included in the institutional curriculum of elementary and high schools, raising the question among educators about whether science should be included as a subject for women and, if so, what approach to the subject ought to be taken. Science education was introduced into girls' education at these levels; nevertheless, the majority of girls did not receive the proper background training necessary to study science at the university level. So, though women wanted to improve their minds with science and some wished to become scientists, they were stymied by inadequate preparation.

Public school science education was often seen within the context of

Group of young women studying electromagnets, Washington, D.C., ca. 1899. (Library of Congress)

women's gender. Middle-class pedagogues often saw in a girl's scientific education a way to ensure against the extremes of femininity that were considered as detrimental to her own health and the future well-being of her family, as was the loss of her femininity through education. Whereas in the prerevolutionary years many American girls had little education, perhaps just enough to read the Bible, in the postrevolutionary period they benefited from the republican spirit. Benjamin Rush in a lecture delivered before the Young Ladies Academy of Philadelphia in 1794 stressed the importance of women's education so that they might instruct their "sons in the principles of liberty and government." Such republican motherhood opened public schooling for women in the United States much sooner than in Europe. In both the North and South, in large and small towns alike, many private female academies and seminaries were founded after the War of Independence that taught a wide range of subjects including history, geography, mathematics, and natural sciences. Between 1790 and 1830 about four hundred such schools were founded. In Britain and in European countries, public schooling for girls began to increase only after about 1850.

The study of science had taken place in the home throughout the early days of the United States, although it was supplemented with scientific publications for family consumption. It was a natural extension for schoolteachers to include science in their curriculum for both boys and girls. Later, American women's schools at all levels invested heavily in scientific equipment, laboratories, and even astronomical observatories, and they proudly outlined their scientific offerings in the advertisements for their schools. In Hartford, Connecticut, a Mrs. Value opened a boarding school for young ladies in the 1810s that in its advertisement listed the subjects to be taught including "orthography, reading prose and verse, writing, arithmetic, parsing English grammar, the elements of astronomy on the celestial globe, geography on the terrestrial globe with a correct knowledge of the atlas and maps, history, Blair's lectures and composition." Mr. Value would also teach the students French, music, dancing, and "polite manners." Science lessons were a part of this school's curriculum as it was in many others. In addition, many seminaries and colleges included botanizing trips as part of their science studies program.

Though these advertisements do not attest to the quality of science education girls received, examining committees and foreign visitors noted how proficient American women were in mathematics and science. Frances Trollope in her *Domestic Manners of the Americans* (1839) reported that the Englishwoman Harriet Martineau (1802–1876), independent intellectual, feminist, and writer on women's education, among many other subjects, noted her surprise when visiting a public exhibition of a girls' school in the United States, that "the higher branches of science were among the studies of the pretty creatures assembled there." The fact that many women's science courses were taught by prominent male scientists further suggested the quality and the seriousness of such study.

In Britain, public education for boys and girls did not become prominent until after 1870. Investigations by the British Association for the Advancement of Science in 1867, 1888, 1889, and 1890 noted the decided lack of scientific education for girls in public schools. Still, this period witnessed the rise of high schools for girls, such as the North London Collegiate School for Girls and the Cheltenham Ladies' College. Untrammeled as they were by educational traditions, many were keen to introduce science into the curriculum. At the North London Collegiate School, science lessons included the properties of matter, the laws of motion, the mechanical powers, simple chemistry, and electricity, with outlines of geology, botany, natural history, and astronomy. Initially such courses were taught as a means of broadening girls' knowledge, but increasingly science was seen as a subject for mental training. Through science lessons, argued Sophie Bryant, author of *Educational Ends or the Ideal of Personal Development* (1887), girls would be taught the "ability to observe and compare, to measure and calculate, to infer and demonstrate, to inquire into the causes, to invent means, to choose ends." Just like Albertine Necker de Saussure so much earlier, Bryant argued for the equal education of girls and boys in science. Such an education would prepare girls for university science for those who wished to pursue further education. For those women who did not go on to higher education institutions, such science lessons educated girls in the rhetoric of the day and gave them an interest in science.

British schools struggled, however, against a lack of equipment and limited school hours. In 1894 the *Journal of Education* reported "the almost universal lack in girls' schools of suitable equipment for the experimental teaching of science owing to which girls are severely handicapped in the competitions for the scholarships tenable at Local University and Other Colleges." As a result, few girls reaching high school had a solid elementary school foundation in either mathematics or the sciences, which disadvantaged them still further in their higher-level science studies. Despite the opportunity for educational innovation, girls' high schools aimed to provide their students with an equal education to that of boys so that they might do well in external examinations. Unfortunately, this meant that the opportunities to escape the narrow classical education long established for boys were rarely taken up by girls' high schools.

By the 1870s, nearly everyone agreed that women ought to be educated for the good of the nation in most countries. The question that was pressing was in what ways women should be educated. Some educators argued that women should not be encouraged to become overly intellectual as they were not training to practice in a profession. Even those who believed in the equal training of girls in science did not think that they would go on to become research scientists. As a result, many decided that it was best for women to be trained to their domestic role, explaining the rise of home economics in the 1890s. American Anna Cooley in her *Teaching Home Economics* (1919) stressed that domestic sciences "afford just the right opportunity for the training of the girl as a member of society in her own home and in her community associations. They contributed, as all studies rightfully pursued should, toward the development of the social efficiency of the girl." Women were as equally intelligent as men but had to use their intelligence for different purposes, thus the practical applications of science instruction for girls would be different than that for boys. The rise of domestic science not only fulfilled the need to educate women to their sphere but also offered science lessons that were specific to women's roles as housekeeper and mother. In Britain from 1878 onward, domestic science was compulsory for girls in elementary schools. In the United States, domestic science rapidly expanded in the 1890s and was institutionalized as a field for women after 1910. By 1914 a survey of 288 U.S. high schools found that though many of these schools offered domestic science, most focused largely on food, nutrition, and sewing.

A class in mathematical geography studying Earth's rotation around the Sun, Hampton Institute, Hampton, Virginia, 1899. (Library of Congress)

The value of such courses has been debated by historians and contemporaries alike. Originally this subject had been introduced as a middle-class attempt to educate working-class girls (and by extension their mothers too) in the correct and scientific management of their households. But by the beginning of the twentieth century it was believed that domestic science would make the subject of science attractive to all classes of female students by making it relevant to their everyday lives. Though the 1904 Education Act in Britain stated that students were to receive seven and a half hours of science instruction a week, increasingly domestic science took the place of general science for girls over age fifteen. In the United States, domestic science became vocational, focusing on practical skills and not on training in scientific method.

Whatever their worth for the budding housekeeper, clearly such a gendered division as "domestic science" could keep women out of the "masculine" sciences. Moreover, the scientizing of women's traditional skills undermined their individual authority and knowledge base within the home. Peter Asbjornsen published *Sensible Cooking* in 1864, in which he noted that "popular diet and methods of cooking must be changed to accord with the principles of natural science,

Home economics lab (Courtesy Division of Rare and Manuscript Collections, Carl A. Kroch Library, Cornell University)

and that women must learn everything anew." Many male scientists supported the notion of a domestic science to minimize female competition in their own science fields. Yet Arthur Smithells, chemistry professor at the University of Leeds, in his "School Training for the Home Duties of Women," published in the *Annual Report of the British Association for the Advancement of Science* (1906), argued for the importance of a "solid, scientific foundation" for domestic science for women. Another chemistry professor, Thomas Cartwright, wrote in his preface to *Domestic Science* in 1900 that "his experience with girls has taught him that they soon become adepts at manipulation, and that they are not one whit behind boys in their power of grasping the truths that the experiments are intended to make manifest." Some male teachers clearly believed in the idea of science for girls whatever the application might eventually be.

In Britain, the substitution of domestic science for general science lessons was opposed by women teachers and feminists because it was often bad science and because it was believed that the two subjects taught completely different lessons. Ida Freund, science tutor at Newnham College, Cambridge, published an indictment of domestic science in a 1911 volume of the feminist journal *The Eng-*

lishwoman's Journal. She argued that it was "erroneous to think that through the study of the scientific processes underlying housecraft and especially cookery, you can teach science, that is, give a valuable mental training that should enable the pupils in after life to judge whether an alleged connection between effect and cause has been established or not." Moreover, women educators argued that such a subject as domestic education undermined the hard-won opportunities for equal education. As a result, the substitution of domestic science for general science was not established in the British education curriculum, although certainly domestic science has persisted as a form of technical education. Nevertheless, Dena Attar has argued that in Britain, girls have certainly perceived that domestic science was the science for them, while boys have determined that they were not expected to take such courses and that the general sciences were their domain. Attar notes that from the advent of domestic science to the present day, boys who have taken domestic science courses do so with a career goal in mind, whereas girls rarely do so. Rather, girls take domestic science to prepare them for their future lives as wives and mothers.

Many men and women who founded educational institutions in both the United States and Britain worked hard to convince the community that women's intellects were equal to those of men. They knew that their students were going to need to be able to support themselves financially and that education could provide a route to independence for widows and spinsters. Others believed, however, that it was more practical for women to be educated to their gendered role, or that this type of education was all women were capable of achieving. Some women begrudged the fact that they did not have access to the same kind of education as men, and in response they developed equal education for themselves and for their students. In the United States, Emma Willard opened the Troy Female Seminary in 1821 and Catharine Beecher established the Hartford Female Seminary in 1823. Both of these women educators included science as part of the curriculum in their schools. Education could, of course, broaden women's horizons. Most obviously, women had increasing opportunities to be paid teachers outside of the home. Although teaching has generally been considered a socially acceptable role for women, historian Mary Kelley has argued that the growth of women's educational institutions allowed women to "create female-dominated spaces" in which they and their female students could flourish intellectually. Women's schools often hired their own graduates as teachers, thus providing important role models for girls of the next generation. As well as teaching, medicine was also a possibility for women as the caring aspect of the medical profession made it an appropriate sphere for women. By the 1860s three hundred American women were qualified physicians. In Britain, medical schools were among the first institutions of higher education to open their doors to women. Nevertheless, the fact remained that though girls were often introduced to some science education at either the elementary or high school level that could open certain gendered career paths, such education did not prepare them for higher education and thus significantly restricted their ability to become practicing scientists.

The Struggle to Enter the Ivory Tower: Gaining a University Education for Women

While girls were entering public education at the elementary and high school level and taking some science classes and while women's medical education was, sometimes grudgingly, accommodated within the social framework in both the United States and Europe, the question of higher education at the universities was still being resisted in the second half of the nineteenth century. What was the point in educating women beyond their sphere either at university or to subjects outside of the domestic purview? In the end, women would be expected to work at home and their education would be largely wasted. University authorities considered a higher education in science as useless to women because it had no direct utility in women's sphere. Moreover, they believed that the study of science was a masculine activity and that it went against women's natural character to practice science. For women to attend a university, sitting in classrooms with young men and working beside them in laboratories, even fraternizing with them during social activities on campus, went against all the social Victorian tenets of femininity and women's morality. Besides, most male educators continued to believe that women did not have the intellectual capacity or physical rigor necessary to study the sciences in higher education, partly because their mathematics training at the secondary level was not sufficient, but just mostly because they were women.

Proponents of the equal education of women and the entry of women into the universities continued to posit the mother-educator role as a good reason to educate women, and they also made the argument that women were suited to certain kinds of knowledge. They recognized that women's ability, for instance, to note detail and to have the patience and take the time to do so recommended them to scientific endeavors. Others argued simply that women as members of the human race, equal in their attainments to their brothers, were entitled to the same cultural advantages of education and intellectual fulfillment. Feminists saw women's education as a first step toward their more general emancipation both politically and socially. Moreover, it was generally recognized in society by the late nineteenth century that there would be some surplus women in the popula-

Students protesting the admittance of women to Cambridge University, 1897. (Hulton Archive/Getty Images)

tion who would never marry, so it was considered acceptable for such women to pursue a career for financial reasons. Grant Allen, evolutionary naturalist and novelist, wrote in an 1889 article for *The Fortnightly Review* entitled "Plain Words on the Woman Question" with some sense of relief that "Out of every hundred women, roughly speaking, ninety-six have husbands provided for them by nature and only four need go into a nunnery or take to teaching the higher mathematics." Of course not all women would benefit equally from such circumstances. The majority of women who attended university had middle-class backgrounds, had been educated in high schools, and had a great deal of personal determination and often, but not always, family support. Many women who attended university in the second half of the nineteenth century not only had the financial means necessary to pay for a university education but often came from families where intellectual support and stimulation for women was voluntarily forthcoming.

London University was the first British university to admit women, beginning in 1878. But London University did not have the prestige of Oxford or Cam-

Sofia Kovalevskaia (1850–1891)

Although Sofia Kovalevskaia benefited greatly from the growing access to European universities for women in the second half of the nineteenth century, she nevertheless suffered from barriers to women's higher education. Tutored at home in Russia, she was interested in mathematics in her early years and found her father supportive of her interests. When she was fifteen years old he found her a tutor who could train her in university mathematics. It was through this tutor that she was introduced to nihilist

(Courtesy of Joan Spicci)

ideas and became a nihilist for the rest of her life. Obtaining what was called in Russia a "fictitious marriage" to escape the control of her father or a husband, she married an amateur paleontologist named Vladimir Kovalevskii. In Russia at that time, women did not have separate legal identities and could not live apart from their father or husband without written permission. It is uncertain, considering the support her father gave her as a young girl, why she felt the need to marry. She may have believed that her father would withdraw his support for her education when she announced she wished to move away from home to study.

Though Kovalevskii understood the platonic nature of their marriage and supported Sofia in her studies, he nevertheless believed that eventually he and Sofia would become lovers. Sofia's intention, however, was to marry in name only so that she would be free to study. From 1869 to 1871 she studied mathematics in Heidelberg, Germany. Then from 1871 to 1874 she lived in Berlin to further her studies. Because as a woman she was not allowed to attend lectures at the university, Karl Weierstrass, a great mathematical analyst and her Berlin adviser, tutored her privately. In 1874 she received her doctorate degree from the University of Göttingen for three treatises. Although all three were considered work of a high order,

bridge University, and advocates of higher education for women set their sights on women's entry into these bastions of male education. Oxford and Cambridge both had women's colleges early on. Girton College was founded in Cambridge in 1869, and Newnham College followed in 1871. In 1878 and 1879 respectively, Lady Margaret Hall College and Somerville College were founded at Oxford. As was the case at the elementary and high school levels, some women's colleges upheld equal opportunities for women, and others advertised a female education especially suited to women. Women were allowed to attend the university's lectures, including science lectures, although often only if the lecturer gave special the first, which added to work already completed on the solution of partial differential equations, was considered the most important.

Between 1874 and 1880, unable to obtain a university teaching position, she decided to live as husband and wife with Kovalevskii and gave birth to a daughter, also named Sofia. During these years Kovalevskaia gave herself over to the gendered roles of society and did not study mathematics at all, but she was unhappy. In 1880 she separated from her husband and returned to Moscow, where she resumed correspondence with Weierstrass and moved in with her best friend. Sofia was preparing to take the Russian Masters degree with the support of several members of the mathematics department at the Moscow University to enable her to teach at the university level in Russia. Unfortunately, even with the department's support, the university refused to allow her to sit the exams. Instead, a friend of Weierstrass, upon his request, managed to find Kovalevskaia a professorship at the University of Stockholm in Sweden. Anna Carlotta Leffler noted in *Her Recollections of Childhood* (1895), her biography of Kovalevskaia, that despite her shy personality, Sofia had nevertheless seen her university position as not only a way to pursue her love of mathematics but as a way to "open the universities to women, which has hitherto only been possible by special favour—a favour which can be denied at any moment." Here she completed her other significant work, published in 1889, on the revolution of a solid body about a fixed point, known as the "Kovalevskaia top," which linked complex function theory and analytic mechanics. For this work she won the Prix Bordin, the highest award of the French Academy of Sciences. Broadening her work in this field, she was later awarded another monetary prize by the Swedish Academy of Science.

Kovalevskaia was the first woman in modern times to receive her doctorate in mathematics. She became the first woman outside of eighteenth-century Italy to be awarded a chair at a research university, the first woman to be on the editorial board of a major scientific journal (*Act Mathematica*), and the first woman to be elected a corresponding member of the Russian Imperial Academy of Science. Her educational credentials along with her brilliance as a mathematician allowed her to be viewed as a professional mathematician, and she was accepted as such by her male colleagues. Tragically, she died of pneumonia in 1891 at the age of forty-one.

permission for women to attend. When women could attend, often they were only admitted within the confines of separate spaces in the classroom so that there was no risk of distracting their male colleagues. By 1873, twenty-two out of thirty-four Cambridge professors allowed women to attend their classes. The resistance of male faculty to admitting women to their lectures was, they claimed, due to the "indelicate" subject matter their classes covered, which they deemed unsuitable for discussion in mixed company. On one occasion when a brain was passed around the lecture hall for observation, the male students all turned to the back of the room to observe the women's reaction to this anatom-

Female undergraduates at work in the laboratory of Girton College, Cambridge University, ca. 1900. The college, founded in 1869, was the first for female undergraduates. (Getty Images)

ical sight. Other lecturers argued that male students would be afraid to ask questions for fear of appearing ignorant before the ladies. Women university students often suffered from both the subtle and explicit backlash of male students and faculty against their presence in lectures and on campus.

Women's permitted attendance at lectures was not officially recognized by Cambridge University, so they could not write exams or take degrees. In 1865 special exams were set for women who wanted to become teachers and governesses, and women who passed them were given certificates but still could not take degrees. Women were granted the right to sit for the degree exams for the first time in 1874, including those in the natural sciences. However, if they passed the exam their names were not allowed to appear on the list advertising their standing in the exams in comparison with their peers. Many women received the top places of honor in their exams, but they were not allowed to receive degrees no matter how well they did. In fact, Oxford University did not allow women to attain degrees until 1922, and Cambridge held out until 1948. One of the main fears of the university governing body with respect to giving women degrees at

Oxford and Cambridge was that doing so meant that women would become full members of the university, which would include a say in the governance of the university. Male faculty and students alike wished to retain the university as a male stronghold.

Despite all the apparent drawbacks, numerous women sought a university education, and a significant number went to university to study science. Prior to the opening of the Cavendish Laboratory for Physics to women students at Cambridge in 1882, women had to utilize the laboratories in the women's colleges. The forerunner to Girton, Hitchen College, was running its own science class beginning in 1871. Students at women's colleges worked in less than ideal circumstances, often in makeshift laboratories with insufficient equipment and instruction. There were many advantages to getting a place in the Cavendish lab to work and, according to historian Paula Gould, women focused less on the barriers they faced in obtaining a university education and more on the activities in which they came to participate. In fact, their scientific partnerships in the Cavendish lab were often likened to the family unit. Male/female partnerships sometimes became marriage partnerships, the one subsumed within the other. Some professors were willing to admit women to the laboratory on an equal basis with men, but others believed women researchers needed a special program. As a result, women's research interests and achievements were determined to a fair degree by the work of those professors who most readily welcomed their equal participation. Women did not always find themselves in antagonistic relationships with male faculty and students, but, on both sides, accommodation was required to integrate women into university life and the science lab therein. Similarly, historian Marsha Richmond has argued that the Balfour Biological Laboratory for women at Cambridge, built for the students at Girton and Newnham and in use between 1884 and 1914, provided a similarly high quality research laboratory equal to that of the men's while allowing women to form their own scientific subculture. The politics of gender were alive and well in late nineteenth- and twentieth-century Britain, but clearly this did not mean that the story of women gaining access to higher education was solely a battle of the sexes. Accommodations and compromises were made on both sides.

In contrast to the slow, gradual development in women's university education in Britain, the 1860s and 1870s was an era of significant gains for Russian women in science because it was both the golden age of Russian science and the beginning of the Russian revolutionary movement. Tsar Alexander II was initially believed to be more liberal than his father, and so there was hope of political and social reform and thus change and progress for women. The government sought to reform science education to promote technical, military, and medical progress and sent many young people to German and Swiss universities for a scientific education. As a result, Russian women were among the first in the world to receive doctorates in science. Women also followed this path and were allowed to attend Russian medical schools and take up technical studies as unofficial auditors. The government was taking a risk. Nihilists, a group of men and women who sought to replace tsarist rule with a more egalitarian society, saw science as a route to accomplishing this end. They believed that science overrode superstition and religion and that it provided a direct route to truth and progress. Science and social activism went hand in hand in Russia. Many of them brought scientific ideas to the revolutionary cause.

When Russian women were studying at European universities they suffered, like their British counterparts, from a lack of formal secondary education. They also found themselves studying largely in the more "feminine" areas of natural sciences and in medicine. They experienced social censure from the Swiss population and even from the non-Russian students in the universities, sometimes seen as little more than prostitutes by the public because they fraternized with male colleagues. Many of them resisted eye contact with Russian male students for fear of being accosted, harassed, and insulted. Even those men who were not openly hostile simply did not understand women's desire to obtain a university degree, especially one in mathematics or science. Still, Russian women were among the first women to attend universities in Europe, and they opened doors for others to follow. The first woman doctorate in medicine was Russian, Nadezhda Suslova (1843–1918). Her dissertation from the University of Zurich was on reflexes in the heart and lymph glands of frogs. She returned to Russia, where she obtained permission to practice medicine. As a result of Suslova's achievements and other women who followed in her footsteps, Russia had more licensed women physicians than any other country in Europe in the 1860s and 1870s. In 1873 there were over a hundred Russian women studying at the University of Zurich and at the polytechnic institute. These Russian women comprised more than 40 percent of the Russian degree aspirants in Zurich and about 85 percent of all women at the university.

As more women began to study science in the universities, student uprisings became more frequent in the universities. Thus, science began to be seen by the Russian government as a threat to the establishment. With the assassination of Tsar Alexander II in 1881, the government resorted to closing Russian universities and calling Russian women home from the University of Zurich. The government announced that any woman who stayed on at the university would not be given permission to work in her profession when she returned. Most women found ways to stay despite the governmental edict. Some women students compressed their studies into the time remaining, others stayed on believing the government would not enforce the proclamation, and still others left Zurich and went elsewhere— to the United States and Italy—knowing they would be denied access to the universities in Russia. One exception to this increased exclusion was a course for midwives that had opened in St. Petersburg in 1872. However, graduates were not allowed to call themselves doctors, and these courses were not of the same caliber provided in the medical schools. Professional status was naturally lower for those who worked in women and children's health.

Late in the nineteenth century, Russian women were attending universities in Europe and studying for graduate degrees, and British women were slowly making headway gaining access to university lectures and exams. In Spain women had since 1868 been expected to obtain a letter from the academic authorities granting them individual permission to attend university. But after 1880 they were allowed to attend without special permission. Some men, through the Free Education Institution, fought for women's access to education on the grounds that they would be better mother-educators and less susceptible to the conservatism of the Catholic Church. Unlike in other European countries and the United States, there were no women's colleges in Spain, and only fifteen women graduated from university between 1880 and 1910. Spanish women were not allowed full access to university until 1910, but once they were given access, their numbers in science increased faster than in other disciplines, suggesting that institutional barriers and not a lack of interest or intellectual ability had been keeping women out of science in Spain.

Higher education for women came slowly in Germany. Science education in the elementary and secondary level had long been omitted from the German girls' school curriculum. During the 1880s and 1890s some American and European women attended German universities to study science, albeit with difficulty and often without obtaining degrees. Ironically, German women were refused entry. Foreign women went back to their countries of origin, but German authorities assumed that university-educated German women would disturb the gendered social order and would start competing with German men for teaching positions. Still, the admission of foreign women certainly played a part in the gradual admission of German women to the universities beginning in the 1900s.

In the United States access to a college education was much easier for women, and it occurred much earlier in the United States than it did in Europe. By 1890, 56,000 women were enrolled in colleges in the United States. Doctors' concerns about ill health, sterility, and mental degeneracy in educated women had persisted throughout the nineteenth century but were largely ineffectual. Dr. Edward Clarke, a prominent Boston physician, argued that women's education had damaging repercussions for their minds and bodies. His arguments were well-received by those opposing women's education. Equally strong were men and women who argued against his assertions. Educators fighting for women's right to an equal education with men, or any education at all, explained why a woman's education was necessary. Anna Brackett in *The Education of American Girls* (1874) forwarded the assertion that women were equal in the eyes of God and that "to God, the brain of a woman is as precious as the ovary and uterus." God, she believed, had intended women to use both. The republican rationale and the mother-educator rationale went hand in hand and certainly provided fuel for the fire against Dr. Clarke.

Most powerful and successful of all the arguments against Clarke was the one posited by the 1882 all-women's Association of Collegiate Alumnae's (ACA) study of 705 women graduates. Calling Clarke out for his lack of statistics, they fired back with statistics of the personal health histories of their membership and published a 78-page pamphlet entitled *Health Statistics of College Graduates.* Seventy-eight percent reported themselves in good or better health after graduation. John Dewey responded positively to the ACA's 1886 report in *Popular Science Monthly,* writing that "for the first time the discussion is taken from the *a priori* realm of theory on the one hand, and the haphazard estimate of physician and college instructor on the other." Still, seventeen percent of the women graduates said that they were in bad health, and the ACA felt obliged to provide an explanation for this significant number. The women graduates' ill health could be explained, the association wrote, by their foremothers' ignorance of the laws of sanitary science, which once understood would enable them to raise daughters who would not suffer the same ills. But of course Clarke had argued that women's education would result in women's infertility, and even John Dewey acknowledged that the birth rate among educated women was lower than the national average. The first generation of college-educated women failed to reproduce themselves biologically. Women faculty were not permitted to marry, and if they did so they would forfeit their position at the university. When Harriet Brooks (1876–1933), a physics instructor at Barnard College, suggested to the dean that she should be allowed to remain on staff after marriage, she was told not only that a woman was expected to see homemaking as her profession once married but that if she believed home duties came second, she could not possibly be a woman of good character. Brooks resigned despite pleas from her department that she be kept on. Persistent calls were made for women's education to be directed toward women's gendered roles, which, it was claimed, were only the social manifestations of biology.

Women's colleges improved their educational offerings throughout the nineteenth century, including science courses, as state universities opened their doors to women. Among the institutions that educated women in science and invited them back as faculty members were Cornell University and the University of Michigan. At the same time, more women's colleges opened, including

Smith College in 1871, Wellesley College in 1875, and Bryn Mawr College in 1885, expanding women's opportunities still further as these well-endowed institutions supported the laboratory and equipment needs of science students. Such opportunities resulted in better-trained women educators who by the 1890s and 1900s could pursue the advanced degrees required to teach at women's colleges.

At the same time that women's colleges increasingly expected their faculty to possess Ph.D.'s, American women found it difficult and sometimes impossible to gain admission to doctoral programs. Many universities admitted women at this level only as exceptions, and most women were denied access on the grounds that there was "no precedent" for women earning doctorates. Of course if no women were ever admitted, there would never be a precedent for admitting women. Though a few institutions began to accept women as doctoral candidates in the 1880s, the high-status institutions, such as Harvard and the new Johns Hopkins, refused to do so. As a result many American women sought their doctoral degrees from European universities, particularly German institutions and the University of Zurich, which were held in high esteem in the States. These women went on to mentor subsequent generations of American women students to carry on after them. Still, even in the second and third generations, teaching duties were emphasized over research, providing little time or money for women science faculty to produce papers. These women already had the best jobs they could attain; other private and public institutions hired men over women. Moreover, as women's institutions attempted to improve their reputations, they also began to hire men over women in an effort to accomplish this aim. There was no real professional purpose, other than personal motivation, for women to pursue further research.

Between 1870 and 1914, women in Europe, Britain, and the United States fought for and entered the coeducational universities in order to study the "masculine" subjects of mathematics, science, and medicine offered there. For each woman, though, the experience of a university education and the attainments she reached could be quite different. Although Leeds University opened all its degree options to women students in 1904, it was 1922 before the university had its first two female math graduates. Math seemed to remain an unwomanly science. Highly successful math student Philippa Fawcett (1868–1948) gained higher marks than any man in the 1890 mathematical exam at Cambridge when she was a student at Newnham College. But she was discouraged from pursuing math by her mother, Millicent Fawcett (1847–1929), a renowned women's rights campaigner in Britain, on the grounds that law would be a better avenue for her to pursue—despite the fact that women had not at that time been granted access to law. Philippa Fawcett turned down a postgraduate scholarship in mathematics as a result. Nevertheless, she went on to teach mathematics at Girton and

Ellen Henrietta Swallow Richards (1842–1911)

Ellen Swallow was born into a modest American household, but educational opportunities opened whole new worlds to her. In turn she would open educational and scientific doors to other women. Still, her science education and career were dogged by barriers to women's education and science practice that many women of this period faced. Her parents, although not wealthy, moved to Westford, Massachusetts, where Swallow attended the Westford Academy in 1859 and graduated from the school in

*(Special Collections,
Vassar College Libraries)*

1863. The Westford Academy provided a rigorous education, but she still found time to help her father in his grocery store and to assist her mother with household chores. Her interest in science began early as she collected plants and fossils and, through her experience in her father's store, became aware of and interested in product purity issues and air and water contamination.

Working at various jobs and suffering under the strain, she nevertheless saved enough money to attend Vassar College, in Poughkeepsie, New York, beginning in 1868, a women's college that had just opened. Continuing her science interests at Vassar, she decided not to study astronomy with Maria Mitchell but instead to pursue chemistry under Charles Farrar, whose work interested

her for its application of chemistry to everyday life. She received her BA in 1870. That same year she went to study chemistry at the Massachusetts Institute of Technology, becoming the first woman admitted to the school and the first woman to be accepted at a scientific school. She was admitted without charge and assumed that this was due to her strained economic situation. She was later to find out, much to her dismay, that her nonpaying status enabled the institute to claim that she was not an official student. Although initially behind in her mathematical studies, she worked hard to catch up, and in 1873 she received a BS from MIT and a master's degree from Vassar for a thesis on the chemical analysis of iron ore. Though she stayed on at MIT for two years as a postgraduate student, she never obtained a doctorate degree. The chemistry department heads refused to grant the first doctorate in chemistry to a woman and so discouraged her doctoral studies.

obtained the position of council member there, which she held between 1905 and 1915. In 1918 she became a fellow at the University College of London, remaining as such until her death in 1948.

Despite the increasing number of women seeking university degrees, only a small percentage finished their studies, and of those who did, few continued in

After marrying Robert Hallowell Richards, professor of mining and metallurgy at MIT, in 1875, Ellen, with the support of her husband, spent the next ten years of her life opening up science education to women. She established a Women's Laboratory at MIT, where she taught chemistry until 1883, when women were permitted to join the men in MIT classrooms. She also organized the science section of the Society to Encourage Studies at Home. Her work with this society led her to think about the importance of teaching women science that they could then apply to many aspects of their lives, including diet and physical exercise, stressing the importance of education to women. Richards believed that such education would improve family life in all aspects.

In 1884 Richards was appointed instructor in sanitary chemistry in a new MIT laboratory dedicated to this study. As such, Richards has become known as the founder of home economics or domestic science. In her laboratory, she taught students how to analyze air, water, soil, and sewage samples. Her lab took on government and industrial work to test for harmful substances and thereby contributed extensively to public heath. Although she never taught domestic science, she was interested in applying science to the problems of women's everyday lives. For instance, she was interested in teaching women how to provide nutritious meals for their families at a reasonable cost. Her experiments in this field led to the publication of *The Chemistry of Cooking and Cleaning* (1880) and *Food Materials and Their Adulterations* (1885). She went on to open a demonstration kitchen in Boston to demonstrate how to prepare cheap and nutritious meals, an idea that quickly spread to other cities, schools, and hospitals. The Boston School of Housekeeping, established in 1897 and revamped by Richards, demonstrated the importance of science to women's future careers and began to see domestic science as a career in itself. In 1908 she was involved in the formation of the American Home Economics Association and its *Journal of Home Economics*, both of which emphasized the professional aspects of domestic science. Richards believed that in order to be competent homemakers, women had to be educated in chemistry and mechanical and physical laws. She also believed that such an education would move women's sphere of influence beyond the home and into the community.

Though Richards's work in home economics was important, her path to this profession was directed by the chemistry department at MIT. It seems that staff members were happy to have her involved in applied science projects that were considered appropriately feminine and that did not encroach on their scientific domains. Despite her educational accomplishments and, eventually, her teaching position within the university, she was discouraged from pure research in chemistry. Higher education provided many new opportunities for women, but it did not necessarily open all doors.

their specialties. For some women, family pressures or the internalized conviction that science was "unwomanly" was the cause of their departures from the university. Other women felt guilty about engaging in selfish pursuits. Many Russian women were following the path of science studies because of their radical political convictions. Yet when they returned home to Russia, other paths seemed more important politically than scientific research, so they engaged instead in direct political activism. For other women the lack of appropriate role models and the overt prejudice of male professors and students drove them away. In the United States, despite these types of barriers, the availability of women's science education resulted in many women practicing science in the 1870s and 1880s. For example, Sarah Frances Whiting (1846–1927) became the first professor of physics at Wellesley College after graduating from Ingham University for Women and having taught at the Brooklyn Heights Seminary. Lydia White Shattuck (1822–1899) became a naturalist after studying at Mount Holyoke Seminary, and Rebecca Pennell Dean (1821–1870), a graduate of the Lexington Normal School in Massachusetts, taught most of the science courses at Antioch College during its early years. In Britain, women's access to classrooms and laboratories, to exams and degrees was a slow process of give and take. British universities gradually opened their doors, but there were still few women students in science. In 1878 only thirty of five hundred women in British universities took science subjects. The Royal College of Science in South Kensington had sixty-six women students between 1881 and 1916. Between 1879 and 1911, Manchester had only seventy women science graduates. But where the welcome was warmer, more women showed an interest in and aptitude for science. During that same thirty-two-year time span, just over six hundred women received the University of London's bachelor of science degree.

A Case Study in Women's Higher Education in Science: Obtaining a Medical Degree

One of the best-documented struggles for women's education in a scientific discipline was their attempt to gain access to medical education. This case study encapsulates many of the challenges women faced when attempting to enter medical schools. It also highlights what could be achieved by women with appropriate education and training. Many of the women who were the pioneers in women's medical education went on to promote such education among the generation to follow, expanding and improving women's access to medical education and the nature of the education they received.

Medical schools in Switzerland and France were the first to open their doors to women. In 1864, University of Zurich, by then known for its exceptional medical education to its students, gave Russian student Maria Kniazhnina permission to attend lectures in anatomy and microscopy. The following year Nadezhda Suslova (1843–1918), another Russian student, joined her. Both were initially auditors, but then Suslova's request to take the medical examination to

British doctor Elizabeth Garrett Anderson (1836–1917) (Hulton Archive/Getty Images)

obtain a degree in 1867 was accepted. The liberal reforms of Zurich, together with the largely German medical faculty who had come to Zurich to escape the restrictions of the German universities, set the stage for the medical faculty to agree to Suslova becoming a formal student and writing the exam. Suslova passed the exam and in the same year defended her thesis on the physiology of the lymphatic system, which impressed all of those in attendance. In 1867 she wrote, "I am the first but not the last. After me will come thousands" (quoted in

Bonner 1992, 37). Zurich continued to accept women students into its medical program without question. Other women from Britain, Germany, Russia, and the United States followed the two Russian pioneers.

The medical school in Paris was open almost as early as the one in Zurich, admitting women from the early 1870s. Unlike at Zurich, there was strong opposition from the medical faculty and the general public on the grounds that a medical education would affect women's moral purity, but women were nevertheless admitted. Once again, Russian and British women were the first in attendance. Mary Putnam Jacobi (1842–1906), born in England to American parents, received her doctorate degree in Paris in 1871 and won a bronze medal for her thesis on fatty degeneration in various clinical disorders. She returned to the United States to teach at the Woman's Medical College of the New York Infirmary and in 1873 founded a medical dispensary for women and children at Mount Sinai Hospital. French women did not receive the necessary secondary education for admission to medical schools, as secondary education for girls in France did not begin until the 1880s. As a result there were fewer French women in the Paris medical schools than there were foreign women. By 1874, though, 150 women were engaged in medical studies in Paris and Zurich.

Elizabeth Blackwell (1821–1910) played an important role in opening the medical profession to women. Arriving from Britain in 1832, she received a solid education at a New York school until, at the age of seventeen, the death of her father prompted her and her sisters to open a boarding school. Blackwell did not enjoy teaching and had no inclination toward medicine in her early years. However, she keenly felt the social inequalities women suffered. As she thought more about these injustices, the idea of obtaining a doctor's degree appealed to her on moral grounds. She applied to medical schools in Philadelphia and New York and at Harvard and Yale but was turned down at all of them. She was only accidentally accepted at Geneva College in New York when the college put her admission to a student vote and the students, believing the application to be a joke, accepted her.

She received her degree in 1849 but sought additional training in Europe between 1849 and 1851, as she was dissatisfied with the training she had received at Geneva College. Women's medical colleges had been established in response to the exclusionary policies of male schools, but there was a growing feeling that the education women received at these all-female institutions was inferior to that of male institutions. During the period of the formation of American women's medical schools in the 1860s, there was a great deal of discussion about the need to raise the medical standards in male institutions. These schools were giving out medical degrees after only six or nine months of study, without any instruction in dissection or clinical training. These discussions led to the assumption that women's medical school education must be inferior and was used as a reason by medical men to keep women out of the medical profession.

Blackwell returned to New York in 1851 but found it impossible to practice medicine until the publication of her *Laws of Life* the following year. This work brought attention to her cause, made clear that she was a qualified medical practitioner, and expressed her conservative views on motherhood, which disassociated her from the usual view that "woman doctor" was a euphemism for abortionist, and instead stressed the naturalness of women's involvement in the care of the suffering. As a result, in 1853 she was able to open a dispensary in a tenement district, which later became the New York Infirmary for Women and Children. In 1868 she founded the Woman's Medical College. Both her medical teaching and practice emphasized the importance in society of educating women as doctors and of educating women about their health and about hygiene to correct the imbalance of gender inequities in Victorian society. Women educators, including Blackwell, made every attempt to improve the training they offered to their female students, and by the 1880s the standards had improved considerably. However, during this period, clinical opportunities were opening for women, and in 1892 the Johns Hopkins Medical School began admitting women. Through the Women's Fund Committee, $100,000 had been raised in support of the medical school, the funds contingent upon women's admission to the school on the same terms as the men and that the school maintain a rigorous admissions policy. The school accepted the money and the conditions. As women gained access to coeducational institutions, female-only medical schools in the United States began to close. Yet discrimination against women medical students meant that enrollment of women in these coeducational schools remained low.

On March 2, 1859, Elizabeth Blackwell gave a lecture at Marylebone Hall in London on her education, her work, and the suitability of medicine as a profession for women. Little did she know at the time that Elizabeth Garrett (1836–1917) was in the audience and that her speech would motivate Garrett to pursue a medical career and become the first licensed woman physician in England. Garrett had been frustrated by the finishing education she had received at a ladies school in Blackheath, near London, and was annoyed by her lack of mathematical and scientific training. She had become interested in the women's movement in the 1850s and had read an article about Elizabeth Blackwell and her work in *The Englishwoman's Journal.* Upon hearing Blackwell speak and talking with her, Garrett decided to become a physician. As a surgical nurse from 1860 to 1861 at the Middlesex Hospital in London, Garrett attempted to further educate herself by attending lectures and examinations at the hospital. Sympathetic physicians allowed her to follow them through the wards, but other students petitioned against her presence because her superior test scores in comparison to those of her male colleagues had irritated them.

Garrett attempted to find a school that would admit her but was turned down by the universities of London, Edinburgh, and St. Andrews, despite intervention by Emily Davies (1830–1921), a proponent of women's education and founder of Girton College. Garrett did manage, though, to gain her Licentiate of the Society of Apothecaries in 1865. Finally, hearing in 1868 that women in France would be allowed to pursue degrees, she applied and was admitted to the University of Paris, receiving her MD from that school in 1869.

Garrett returned to Britain to become the visiting medical officer for a children's hospital, married James G. Skelton Anderson in 1871, and had three children, but continued her medical work at the same time. As an advocate for women's rights, Garrett Anderson responded to Dr. Henry Maudsley's published assertions that supported the infamous Dr. Clarke's argument about the effect of higher education on women. In her 1874 reply in the *The Fortnightly Review* she wrote that in comparison to the small number of cases that Clarke noted, there were thousands of young women "in which break-down of nervous and physical health seems . . . traceable to want of adequate mental interest and occupation in the years immediately succeeding school life." Garrett Anderson herself was living testimony to the fact that women suffered neither physically nor mentally from higher education, even in the sciences, and that their fertility remained unaffected by such work. She saw no conflict between marriage, motherhood, and her medical career.

Garrett Anderson went on to promote medical education for women, helping to establish the London School of Medicine for Women, to which she was elected dean in 1883. From 1886 to 1892 she oversaw the building and staffing of the New Hospital for Women, the teaching hospital connected with the school. She saw her daughter, Louisa, follow in her footsteps and attend the London School of Medicine for Women, where she received an MD in 1905. Elizabeth Garrett Anderson not only set a precedent for women doctors in Britain through her own struggles but also enabled British women to obtain their education without leaving the country. Like Elizabeth Blackwell before her, Garrett Anderson understood the importance of gaining education for herself as well as the importance of educating women of the next generation.

Initially women had more luck gaining entry to medical school in Scotland. In 1869, the year Garrett Anderson received her medical degree from the University of Paris, Sophia Jex-Blake (1840–1912) and four other women gained access to medical education at the University of Edinburgh. When they had proven that they could learn the course work, the university set up special segregated classes for women. Soon after, some male members of the faculty and students began an ardent campaign to have the women removed from the university. Their protests escalated to physical confrontation, blocking the women's access to the lecture halls, verbally abusing them, and, on one occasion, throwing peas at Jex-Blake. She wrote in her work *Medical Women: A Thesis and a History* (1886) that Dr. Laycock, a medical professor at Edinburgh, told her that "he could not imagine *any decent woman* wishing to study medicine—as for *any lady*, that was out of the question." At one point women were refused permission to study at the Royal Infirmary, which was necessary to obtain a medical degree.

After four years of expensive education, the University of Edinburgh refused to grant them the degrees they deserved. University administrators argued that the women's instructor was not qualified, and then refused to allow him to prove his qualifications by examination. In response, Sophia Jex-Blake petitioned Thomas Henry Huxley (1825–1895), well-known evolutionist and supporter of science education as well as of women's emancipation, to examine their instructor and certify him. Huxley refused. The women sued the university for breach of contract and lost. Sophia Jex-Blake and the other women were forced to accept certificates that did not allow them to practice medicine. However, Jex-Blake's exam papers were read by Huxley, who failed her. She sought her degree eventually at the University of Dublin and obtained it in 1877. Meanwhile the University of Edinburgh women took their case to Parliament, where in 1876 the Russell Gurney Enabling Act allowed medical examining bodies to test women. From then on, medical schools began to admit women more readily. Jex-Blake, alongside Garrett Anderson, was involved in founding the London School of Medicine for Women, practiced medicine in Edinburgh from 1878 on, and founded the Women's Hospital there in 1885, organizing the Edinburgh School of Medicine for Women in 1886. Once again, education was the foundation of her life, and she became involved in the promotion of higher education for the women who followed her.

The importance of women's access to universities and colleges to obtain a respected position in the medical and scientific communities in the late nineteenth and early twentieth centuries is clear, recognized not only by historians today but also by contemporaries of the period. Middle-class men and women in Britain, Europe, and the United States realized the importance of women's education to the family, to society, and to women themselves as individuals, and they started schools and agitated for women's admission to colleges and universities. The late nineteenth century saw the beginning of institutionalized science education, and it was necessary for women to obtain access to lectures, exams, and degrees in order to keep pace with the advances men were making at that time. Individual women experienced higher education in science differently from one another and utilized their education to different ends. But all, each in their own way, played a part in opening the doors of institutions for themselves and for the women who would follow.

Conclusion

Although men and women argued that women were capable of and deserving of a higher education, some remained concerned that education "unsexed" women. *Punch*, a satirical British magazine, described the "Woman of the Future" in 1884, outlining this very concern:

> O pedants of these later days, who go on undiscerning,
> To overload a woman's brain and cram our girls with learning,
> You'll make a woman half a man, the souls of parents vexing,
> To find that all the gentle sex this process is unsexing.
> Leave one or two nice girls before the sex your system smothers,
> Or what on earth will poor men do for sweethearts, wives and mothers?

Punch also had much sport with the figure of the "Girton girl," a fictional pretty young woman, tall and slender, who it was difficult to believe would be much interested in obtaining a university degree. The "Girton girl" stereotype denigrated university women's high accomplishments and undermined them by referring to them as girls. Intellectual women who had sought a higher education, particularly in the sciences, were depicted as physically unattractive and characteristically unfeminine, often symbolized by an unhealthy, pale look and the fact that they wore glasses. The "New Woman" was often associated with the negative ramifications of the "Girton girl." The educated woman created the New Woman, who was intent on earning her own living, avoiding marriage and motherhood, and supporting women's emancipation. The New Woman was a popular and well-recognized figure in fiction at the end of the nineteenth century and was often portrayed as a "Girton girl," such as Herminia Barton in Grant Allen's *The Woman Who Did* (1895).

Thus, as historian of science Paula Gould has noted, women science students themselves were often obsessed with dressing correctly and comporting themselves in a feminine manner in public. Both women's educational supporters and detractors speculated about these women by commenting upon their "visible bodies," and as a result women science students attempted to make themselves as inconspicuous as possible by their dress so as to draw the least attention to or criticism upon themselves. Yet underneath cinched waists, lace collars, long skirts, and suitably fashionable leg-of-mutton sleeves were hidden athletic and energetic bodies and minds that could compete on an equal playing field with the male students. The desire to be perceived as acceptably "feminine" while engaging in the unfeminine activity of studying science at a university or college was a delicate balancing act. Some women gave up the pretense altogether and accepted reputations that were at best eccentric and at worst "unsexed." Ida Freund, who did research work in physical chemistry at Cambridge and taught practical science at Newnham College, was described by one student as "a jolly stout German, whose clothes are falling in rags off her back." A woman who failed to pay attention to her appearance could no longer be perceived as womanly.

This apparent conflict between beauty and brains, and the concern that education might damage a woman's beauty and thus her potential to marry—and possibly even affect her fertility—was only part of the fear of those who contested women's entry into educational institutions. *Punch* magazine drew attention to the deeper fears of the male detractors themselves, whose concerns, on the surface, appeared to be for women's health and the future of civilization. What did they stand to lose by women's education? These men feared the encroachment of women into their all-male enclaves in the classroom and the laboratory. Adam Sedgwick (1854-1913), head of the School of Morphology at Cambridge at the end of the nineteenth century, wrote in a letter to the editor of the *Standard* in 1897 that if the university granted women degrees, "the glorious career of this University as a producer of great men will receive a most serious check." Many of them did not want to share the advantages of their sex and did not think they should have to do so, believing firmly in the biological and intellectual superiority of men over women. Moreover, many men did not want to share the benefits of their professions. They began to see educated women as competitors not only for positions at university but for teaching posts, medical positions, and scientific research opportunities. Such competition was seen by these men as an affront to their masculinity and to their status as breadwinner and patriarch. These fears, however, were largely unrealized, as women remained a minority of numbers in the university, especially in mathematics and the sciences, well into the twentieth century and therefore did not encroach heavily on the male strongholds of universities and the professions.

Bibliographic Essay

Research into women's scientific institutional education is not extensive. General works dealing with the rise of women's access to higher education in the late nineteenth and early twentieth centuries only peripherally address women's scientific and medical education. Still, such works provide a solid foundation in the history of women's institutional education and serve as a basis for further research in the specific area of scientific education. Elizabeth Seymour Eschbach's work, *The Higher Education of Women in England and America, 1865–1920* (1993), compares and contrasts women's education in England and the United States. She argues that whereas higher education for American women advanced in leaps and bounds and English higher education took a slow and cautious route to providing such education, in the end women in England built a firmer educational foundation that more fully resembled that offered to men. Carol Dyhouse in *No Distinction of Sex? Women in British Universities 1870–1939* (1995) focused on the British universities founded before World War II, aside from Oxford and Cambridge, that claimed to make "no distinction of sex" in their admission policies. Dyhouse examines patterns of distinction made against women at these universities and explores how women nevertheless constructed spaces for themselves within these institutions. James C. Albisetti in *Schooling German Girls and Women: Secondary and Higher Education in the Nineteenth Century* (1988) studies the development and nature of women's education. Albisetti examines how even when educational institutions' aims were to inculcate women in the ideals of womanhood, education itself could create different expectations for women's lives among students and their teachers. He argues against the feminist interpretations vilifying government officials who protected the status quo and instead looks at the efforts certain government officials made to assist in the progress of women's education. Finally, Christine Johanson in *Women's Struggle for Higher Education in Russia, 1855–1900* (1987) provides a balanced study of Russian women's access to higher education, studying not only those women who sought an education to overthrow the government but also those women, far more numerous, who worked with the government to attain a university education. Johanson dedicates a chapter to female physicians in Russia during the Russo-Turkish war (1877–1878).

Useful works that specifically address women's medical education are Regina Markell Morantz-Sanchez's *Sympathy and Science: Women Physicians in American Medicine* (1985) and Thomas Neville Bonner's *To the Ends of the Earth: Women's Search for Education in Medicine* (1992). Both works focus on women's efforts to secure a medical education and thus obtain a respected standing in the medical profession. Bonner's work compares and contrasts the accessibility and nature of education that American and European women received in the nineteenth century and the reasons for the educational differences that occurred. Morantz-Sanchez's work concludes that women medical students and doctors justified their entry into medical schools and the profession by arguing that women were natural-born healers, that they alone "could combine sympathy and science—the hard and soft sides of medical practice" (5).

Ironically, women's efforts to obtain equal status with men in medical education and as doctors resulted in the decline of female values in medicine.

Specific studies of women's science education have so far produced only articles or chapters on the subject. However, the scholarship in this field has been wide-ranging and in-depth. By far the most important book in the history of American women's science education in the universities is Margaret Rossiter's *Women Scientists in America: Struggles and Strategies to 1940* (1982). In the first two chapters of this work Rossiter outlines the rise of women's colleges and the strength and extent of women's education in the United States and goes on to discuss women's struggle and eventual access to gain doctorate degrees. These chapters set up Rossiter's larger discussion of women's scientific employment in the first half of the twentieth century. Two other notable works on American women's science education are Deborah Jean Warner, "Science Education for Women in Antebellum America," *Isis* (1978): 58–67; and Sally Gregory Kohlstedt, "Parlors, Primers, and Public Schooling: Education for Science in Nineteenth-Century America," *Isis* (1990): 425–445, both of which examine the pervasiveness of science education for women in the United States.

Regarding Great Britain, studies of women's access to science education at the university level include Catherine Manthorpe, "Science Education in the Public Schools for Girls in the Late Nineteenth Century," in Geoffrey Walford (ed.), *The Private Schooling of Girls: Past and Present* (1993): 195–213; Roy MacLeod and Russell Moseley, "Fathers and Daughters: Reflections on Women, Science, and Victorian Cambridge," in *History of Education* (1979): 321–333; and Chris Wills, "'All Agog to Teach the Higher Mathematics': University Education and the New Woman," in *Women: A Cultural Review* (1999):56–66. These three works establish the struggles women and girls faced in obtaining a scientific education. Paula Gould, "Women and the Culture of University Physics in Late Nineteenth-Century Cambridge," *British Journal for the History of Science* (1997):127–149; and Marsha L. Richmond, "'A Lab of One's Own': The Balfour Biological Laboratory for Women at Cambridge University, 1884–1914," *Isis* (1997):422–455, have provided important new work in the field of women's education in science. Gould emphasizes women's involvement and inclusion in university science rather than their battle against an unwelcoming male presence. Richmond demonstrates how the Women's Laboratory provided a female space for women to study, to work, and to establish a subculture that allowed them to function well within the university setting.

Studies of European women's education in science include Ann Hibner Koblitz, "Science, Women, and the Russian Intelligentsia: The Generation of the 1860s," *Isis* (1988):208–226; and Elena Ausejo, "Women's Participation in Spanish Scientific Institutions, 1868–1936" in *Physis* (1994):537–551. Ausejo's article outlines the development of Spanish women's access to scientific institutions in the late nineteenth and early twentieth centuries, and Koblitz's article outlines the rise and decline of Russian women's access to university. Both provide interesting contextualized studies of women's science education, on the one hand constrained by the conservative political and religious backdrop of Spain and on the other hand spurred by the radical politics of Russian nihilism. More work on the relationship of science to radical movements in the nineteenth century would be useful.

Biographical works on women educators, women who benefited from a scientific education, or women who fought for access to scientific education contribute the individual stories of women's experiences. See, for example, Ann Hibner Koblitz, "Career and Home Life in the 1880s: The Choices of Mathematician Sofia Kovalevskaia," in Pnina G. Abir-Am and Dorinda Outram (eds.), *Uneasy Careers and Intimate Lives: Women in Science 1789–1979* (1987):172–190; and her *A Convergence of Lives: Sofia Kovalevskaia: Scientist, Writer, Revolutionary* (1993), along with Lois Barber Arnold, *Four Lives in Science: Women's Education in the Nineteenth Century* (1984).

One area of study in women's science education that has been fairly well-researched is the advent of domestic science as a science discipline for women in the period from about 1850 to 1950. In the British context, Catherine Manthorpe in "Science or Domestic Science? The Struggle to Define an Appropriate Science Education for Girls in Early Twentieth-Century England" in *History of Education* 15 (1986):195–213, argues that domestic science never really took off in England as a form of science for women because women at the forefront of the educational movement felt that such education fed into the separate spheres education that they were fighting against. However, Dena Attar argues in *Wasting Girls' Time: The History and Politics of Home Economics* (1990) that though domestic science failed as a science for girls, a gender-segregated subject was created that not only wasted girls' educational time but also succeeded "in establishing the idea that ordinary science was too masculine for them" (104). Most recently two chapters in Sarah Stage and Virginia B. Vincenti (eds.), *Rethinking Home Economics: Women and the History of a Profession* (1997) examine women's domestic science education in the United States: Rima Apple's "Liberal Arts or Vocational Training? Home Economics Education for Girls" and Margaret Rossiter's "The Men Move In: Home Economics in Higher Education, 1950–1970." Apple demonstrates how in the first few decades of the twentieth century, domestic science moved from a liberal arts-and-science tradition as preparation for future home life in girls' education to that of a skills-orientated vocational training. Both approaches emphasized women's role as professional homemaker and the need for training to this purpose in the schools; however, vocational training focused on practical skills whereas the earlier science-based approach emphasized problem solving and critical thinking. Male governmental officials and school heads, and even some female home economists, preferred to see domestic science education in light of the traditional women's sphere. Ironically, Rossiter's article examines the male backlash in universities against the embarrassment of home economics as a discipline between 1950 and 1970. Male administrators pointed to the field's female faculty majority and its vocationalism, equating these characteristics with a lack of intellectual and scientific rigor. They managed to reinvigorate the field by moving in men and money and renaming the field "nutritional science," "human development," or "human ecology."

Professionalizing Women Scientists

I have no doubt at all that had [Dr. Stephenson] been a man she would have been elected to the Fellowship [of the Royal Society] some time ago. I was particularly impressed with the accuracy of her work, and the very large amount of research which it has directly inspired.

—J. B. S. Haldane, member of the Royal Society, in a letter to Henry Dale (November 15, 1943)

In 1881, Edward Pickering (1846–1919), American astronomer and physicist at the Harvard College Observatory, had become frustrated with his male assistants. He announced to them that he believed even his maid could do a better job of the copy and computing work than they were doing. Such work had multiplied considerably due to the changing nature of astronomy at this time. Astronomical photography produced an increasing number of photographs, creating significant amounts of data that had to be examined, identified, and classified. Moreover, "computers" were needed, women who could write legibly and had a knowledge of arithmetic. Pickering decided to hire Williamina P. Fleming (1857–1911), a teacher with only a public school education, who outshone Pickering's male assistants to such a degree that he instructed her to hire a staff of women assistants. He paid them the meager salary of twenty-five to thirty-five cents an hour to sort and classify photographs of stellar spectra. Fleming was remembered in her 1911 obituary by Pickering as being a woman of "energy, perseverance, and loyalty," and she herself reflected in an address at the 1893 World Columbian Exposition in Chicago that other women hired as science assistants by observatory directors "would undoubtedly devote themselves to the work with the same untiring zeal." Female assistants had, for the most part, nowhere else to go and thus once trained became highly experienced and dedicated workers. Meanwhile, male astronomers were freed from tedious work and could seek better positions in the growing field, where new observatories offered larger telescopes and career opportunities for research and promotion. This process, whereby women were permitted to gain some access to the science profession, has been termed the *feminization* or *ghettoization* of women's work in science, relegating them to the low-paid, low-status occupations within the scientific hierarchy. This was the drudge work of science that the majority of men were not interested in doing. Women who wished to enter science could do so at this subordinate level, but any hope of gaining higher ground within the profession was extremely unlikely. The feminization of the lower ranks of the profession was considered perfectly acceptable. This was only one of the numerous strategies used by male scientists to keep women in socially acceptable positions within the scientific community.

Another strategy was to exclude women altogether. Some individual scientists refused to educate women, and others refused to hire women scientists. Individual male scientists and scientific organizations were also responsible, both intentionally and unintentionally, for creating a climate that made women feel uncomfortable, ostracized, and unwanted. Suffering in such circumstances, women often left science altogether, and then male scientists could point to the fact that women dropped out of the profession as evidence either of the uselessness of educating women who would only marry, have children, and leave the profession, or of the inability of women to succeed in a masculine profession. The lesson to be learned here by other women was that they did not belong in science, they were not welcome, and those who did attempt to encroach on the profession would suffer in an often unfriendly and unsupportive environment.

Despite the barriers that existed to women's participation due to the professionalization of science during this period, women scientists in the first half of the twentieth century achieved graduate educations in science, contributed to scientific team-work, made important scientific discoveries, entered university positions and scientific societies, and won awards for their achievements. Whereas many male scientists employed various strategies to keep them out, women scientists employed their own strategies to assist them in moving into the science profession. In some cases, women accepted lower-status and lower-paid work as a stepping-stone in their fight for access to a career in science. Other women scientists established female subcultures that allowed them to participate in the male world of science while engendering the support of their female colleagues. Some women filled positions in new areas of science in which men were not interested, and other women pursued careers in science by networking with male scientists who supported women's right to participate in the profession. Such men could provide avenues to jobs, research opportunities, and career advancements where otherwise the door was firmly shut against women. There were also women who fought directly for their rights as full and equal partners in the science profession, who demanded equal pay for equal work, recognition for their contributions to research, and entry into the all-male clubs that facilitated scientists' networking. Though for some women this was a lonely and disappointing life path, for others the success they gained reflected the possibility that women could achieve equality with men within the science profession.

"Women's Work" in the Science Profession

The term *women's work* has long referred to the unpaid work of housekeeping and caring for children and the elderly undertaken by women. This work has been considered absolutely necessary to daily life but at the same time thought to be work beneath the dignity and intellect of men. The phrase "a woman's work is never done" suggests the endless and often thankless tasks that women fulfill on a daily basis. Such work is often repetitive, tedious, and mind-numbing, but it is also viewed as requiring certain skills and personality traits. "Women's work" needs patience, attention to detail, organization, dexterity, concentration, and even loyalty, commitment, and empathy. Such work is often referred to as "a labor of love." Love becomes a motivating factor to which a woman will submit her body and mind to others' needs and desires. Scientists as well as social commentators have long argued that women's biology naturally suits them to the occupations designated "women's work." Women do not need to be educated to "women's work," for the skills and personality attributes they need have been considered innate.

As a result of this pervasive ideology, paid occupations have long been gendered to reflect the perceived complementary skills and attributes associated with men and women. As science increasingly became a professional occupation in both the United States and Europe, so the inclusion of women became more and more problematic. Male scientists did not want their profession defined as "women's work." Such a label would enable employers to rank science as a low-brow occupation that could therefore be poorly paid. As a result, they believed they had no choice but to restrict women's role. Moreover, the changing nature of science in this period, which encouraged scientists to work in teams in laboratories funded by institutions, made professional science intricately hierarchical. The jobs in science at the bottom of the hierarchy were low-paid, low-status positions that could involve tedious, repetitive labor or social service work that appeared to require precisely those skills and traits associated with "women's work." However, many male scientists often realized that utilizing women on the bottom rungs of the hierarchy and never allowing them to climb the professional ladder could be advantageous for all concerned. Women who wished to work in the field of science could do so without affecting the reputation of the profession, and male scientists would be quickly freed from the more onerous work of science to engage in the more intellectually challenging research positions with commensurate responsibility, peer recognition, status, and pay.

Certain fields of science were feminized, including such areas as domestic science, nursing, child psychology, librarianship, and social work, because they offered low-paid positions and were considered suited to women's innate ability to care for others. Men generally avoided such fields of employment. Women were also relegated to positions that addressed the needs of female students, such as dean of women students, and as college physicians who oversaw women's health and hygiene concerns. Women filled these positions with the hope that they would provide a starting point for entry into other posts, but such women were rarely considered for high-level administrative positions elsewhere within the university. Positions unofficially classified as "women's work" did not further women's opportunities; rather, such work served to keep them relegated to these areas.

A feminized science field such as home economics, even when headed by a woman at the departmental level, was ultimately overseen by a body of administrative men at the university who could, without any notice, eliminate the department. In the case of the home economics department at the University of California, Berkeley, Dr. Agnes Fay Morgan (1884–1968) chaired a well-established department beginning in 1918. She fostered a scientific rather than vocational departmental identity, and the department achieved its own building in 1954. But only one year later, upon Morgan's retirement, the department quickly fell out of favor with university administrators. In the conservative atmosphere of 1950s America, the department was accused of offering too many science courses and not preparing women students for family life. It also was faulted for discouraging women from entering the program by offering too many science courses and thus not producing enough home economics teachers. Nutritional science, considered increasingly prestigious, subsequently was separated from the department of home economics and integrated into the male-dominated fields of food science and food technology at Berkeley, and the rest of the home economics department was moved to the Davis campus. The new female chair resigned because she had not been consulted or even informed that changes were being discussed, let alone that they were to be implemented. The elimination of home economics rid Berkeley of a department where women could advance professionally at the same rate and to the same heights as men. Had the department been more traditional, it would likely have been accused of not being scientific enough and would have been eliminated on the grounds that such a feminized department reduced the prestige of the university as a whole. Feminized fields that remained low-status and served a social purpose were tolerated, but feminized fields that attempted to accord themselves the same status as a masculine field of science had to be eliminated to guard the prestige of professional science. This meant keeping feminized sciences and the women who worked within them out of the professional inner circle.

The low pay and low status of "women's work" in science was rationalized in a number of different ways. Men feared that if women scientists filled well-paid positions in the upper echelons of science, some men would ultimately be unable to meet their responsibilities as the main or sole breadwinner in their family. They argued that women rarely had this responsibility and so did not need well-paying jobs to make ends meet. Efforts to obtain equal pay for equal work in the United States both by feminists and suffragists made little impact despite women's contributions, scientific or otherwise, to the war effort during World War I. Studies in the 1910s and 1920s clearly demonstrated the inequalities suffered by women doing the same job as men but being paid substantially less, but no one contributed suggestions for solving the dilemma. Researchers believed that by revealing the discrepancy, fair-minded people would work to correct the injustice. But in the 1930s, the Great Depression made it difficult for women to justify holding a job in the face of so much destitution, let alone continuing their call for equal pay. In such tough times, women had to measure themselves against the achievements of the "exceptions" to even be considered for a position. Their overachievement became known as the Marie Curie Strategy. Women understood that they had to have more degrees than men and work far harder for longer than did men, in order to be hired by a man who measured all women scientists against exceptions such as Marie Curie. Few women measured up, and the rest were confined to "women's work" in science with their attendant poor pay and low status. Often, in the end, men simply preferred to hire men as both professionally and socially more comfortable for all involved.

Although women engaged in "women's work" in science were grateful to obtain any sort of paid work within the profession, worked hard, and were often good at their jobs, they nevertheless felt the injustice of their situation. Williamina Fleming, of whom Edward Pickering had noted "occupied one of the most important positions in the Observatory," nevertheless felt keenly the affront of her poor pay, overwork, and the lack of opportunity to do her own research. In a diary she kept for two months in 1900, having by then worked with Pickering for twenty years, she expressed her frustrations: "Sometimes I feel tempted to give up and let him try some one else, or let some of the men do my work, in order to have him find out what he is getting for $1500 a year from me, compared with $2500 from some of the other assistants. Does he ever think that I have a home to keep and a family to take care of as well as the men? But I suppose a woman has no claim to such comforts. And this is considered an enlightened age!" (quoted in Rossiter 1982, 57). Ellen Hayes, a professor of mathematics at Wellesley College, noted with some disgust in a 1904 letter to James McKeen Cattell, professor at Columbia and editor of the journal *Science*, that women were welcome in science so long as they restricted themselves "to become experts in washing bottles and adding logarithms and dusting specimens." Many women felt keenly the unfairness of their position.

Moreover, once "women's work" had been established in a particular field or feminized fields had been constructed, it was difficult for women to move beyond them. Fleming managed to work within the constraints of overwork and low pay to expand her role. Though as a woman untrained in science she initially worked on simple tasks such as copying and ordinary mathematics, as time went by she expanded her work, which eventually enabled her to make her own scientific contributions. She discovered ten novae, more than three hundred variable stars, and fifty-nine gaseous nebulae. She had a paper entitled "Stars of the Fifth Type in the Magellanic Clouds" read for her by Pickering at a 1898 conference at Harvard. She became the first woman ever to receive a corporation appointment at Harvard as the curator of astronomical photographs from 1899 to 1911 and was made an honorary member of the Royal Astronomical Society in England in 1906, its fifth female member. In contrast, Pickering's niece, Antonia Caetana de Paiva Pereira Maury (1866–1952), who attempted to develop her own classification system and then expected to receive full public credit for it, ended up leaving the observatory in 1892, unable to work with Pickering any longer. Pickering gave her front page credit, but he did not adopt her system of classification because he believed it was too complex and thus could not be taken up by the women assistants who worked for him. Professional gender etiquette, sometimes more so than a woman's actual abilities, could make or break a woman's scientific career. Women whose personalities were considered abrasive or who attempted to assert any sort of authority over men were deemed "difficult" and often overlooked for even the most lowly positions.

Still, "women's work" gave women a space within the scientific community. At the Royal Observatory in Greenwich, England, between 1890 and 1895, four women were hired at the low rank of computer and for low pay; however, unlike the women working at the Harvard College Observatory, these women did more than tedious computations. They were trained in the use of astronomical instruments and made actual astronomical observations. After serving her term, one of the computers, Alice Everett (1865–?) went on to obtain a three-year position at the Astrophysical Observatory at Potsdam, Germany, in 1895. Returning to Britain after a year of work at Vassar College in the United States, she continued an active career. Another of the computers, Annie Russell Maunder (1868–1947), went on to collaborate with her husband in his work on periodicity of sunspots, albeit in a voluntary capacity, and on her own account analyzed more than 600 sunspot groups.

Some women managed to overcome the barriers against their becoming full-fledged professional scientists. A good example of a woman successful in achieving full professional status in the scientific community is Alice Wilson (1881–1964), who worked for the Canadian Geological Survey. Failing to complete her BA at the University of Toronto due to poor health, she had nevertheless obtained a position in the mineralogy division of the University of Toronto Museum in 1901, with the aim of eventually becoming a geologist. By 1909 she was working for the Geological Survey of Canada in Ottawa as a museum assistant, and she continued working there until 1946. Museum assistant was another category of "women's work," although Wilson was the first woman to hold such a position at that museum.

Wilson's male colleagues were not supportive of her ambition to become a geologist, and they blocked her from achieving her aim. While men around her were granted leaves to pursue additional education opportunities, Wilson applied and was refused such a leave for ten consecutive years, even when, in 1926, the Canadian Federation of University Women granted her a graduate fellowship. Finally her director relented, but only after she agreed to take only six months' leave and to do a full year's work in the preceding six months. Wilson earned her doctorate from the University in Chicago in 1929 under these trying circumstances, but her director refused to promote her or give her a salary increase. However, having achieved the Order of the British Empire in 1935, been the first woman elected a fellow of the Royal Society of Canada in 1938, and been the first Canadian woman elected a fellow of the Geological Society of America, she was promoted to associate geologist and finally to full geologist in 1944, a year before she retired. In 1991 the Royal Society of Canada, at the behest of its Committee for the Advancement of Women, established the Alice Wilson Award in her memory to honor the first woman elected to the society. The award is given yearly to "a woman of outstanding academic qualifications who is entering a career in scholarship or research at the postdoctoral level."

Though Alice Wilson was vindicated, albeit belatedly, few women had her capacity to work so hard for so long with so little recognition and among colleagues who were unsupportive and sometimes hostile. Some women were overqualified for the positions they held and longed for opportunities that would never come or were extremely late in coming. As Margaret Rossiter has argued, although a few women climbed to the top positions, "most were getting lost in labyrinthine passages that worked in a variety of ways to channel them into certain fields, keep them in low ranks, pay them low salaries, and direct them to adjunct positions as research associates and deans of women" (1982, 216). Alice

Wilson was an exception that breaks the general rule of women's low scientific achievement, which ironically contributed to the understanding among male scientists that only a few exceptional women could compete. Nevertheless, women such as Williamina Fleming and Alice Wilson demonstrated that even when relegated to positions construed as "women's work" and expected to never do more, these women could work within established boundaries and even expand those boundaries to become professional scientists despite the odds against them.

Beyond "Women's Work" in Science

Though significant numbers of women had been relegated to "women's work" in science, some women did attempt to move beyond this sex-segregated sphere. To enter the professional world of science, male and female scientists alike had to achieve certain goals. Increasingly in the twentieth century they had to have earned a university degree, preferably a graduate degree. They then had to find a paid position that would allow them to conduct their own research so they could subsequently publish their findings, contribute to the scientific enterprise, and receive career promotions. Professional career success depended not only upon the scientists' credentials and scientific achievements but also upon the contacts the scientists had in their field. Professional networking was therefore vital to initial placement in a paid position, to obtaining grant money and awards, and to any subsequent career advancement. Such networks included positive relationships with teachers, admittance to and acceptance within formal and informal scientific societies, and correspondence, friendships, and working partnerships with like-minded senior and junior colleagues. As in any profession, some individuals will be more successful than others based upon myriad interconnected factors. For women, though, the difficulty of effectively networking often stymied them in their attempts to move into the science profession or to move up within it.

Women's early socialization often put them at a disadvantage right from the start. Many did not receive familial support for a career in science, and others suffered from a lack of confidence engendered by a society that did not believe women were capable of any sort of professional career, let alone one in mathematics, science, or medicine. A qualitative study of faculty at the University of Minnesota in the early 1980s explored professional socialization. One respondent in the study reported, "I wanted to be a scientist, but I didn't decide that I wanted to be a faculty member because I thought that would be impossible because I was female." Moreover, many women worried about how they would balance career and family, a concern that men rarely perceived as an issue that would affect their

Medical students dissecting human bodies at the Women's College Hospital, Philadelphia, March 1911. (Hulton Archive/Getty Images)

careers. Some women felt defeated before they began and often made alternative career choices or settled for mediocre career success as a result.

Even when universities began to admit women freely to their science programs, women continued to be at a disadvantage at the university. At the graduate level, every student has a supervisor who guides the direction of his or her work. If the student is fortunate, the supervisor will become a career mentor who not only oversees the student's work but will also provide an introduction to the professional network so vital for professional advancement. This relationship can be plagued with difficulties in any set of circumstances, but there can be additional roadblocks for women with male supervisors. A male supervisor might feel obligated to take on a female student but may nevertheless believe women should not be, or could not be, scientists. Other men who oversee research associates in their labs have taken credit for these men and women's work. This situation has affected women more often than men because women have had less opportunity to move out from under a director who took advantage of them. Florence Sabin (1871–1953), a prominent anatomist who had worked with Herbert McLean Evans (1882–1971) at the University of California, noted to a close friend that Evans had been unfair to Katherine Scott Bishop (1889–1976), who had assisted him in the discovery of vitamin E in 1922: "She did lots of his good work, made lots of the good observations and lost out in the end. He used her and then let her go" (quoted in Rossiter 1982, 82). Apparently Evans was infamous for such behavior. Although educational institutions changed their policies toward women science students, many individual science professors did not alter their own personal views toward women in science. Supervisors might make a woman student's life very difficult during the process of her education, which might have caused her to do worse than she otherwise would have done. Such a situation might even influence her to terminate her studies. Supervisors could also deny their students an introduction to the professional network that would ease the advancement of their careers, leaving a woman instead to find her way in or drop out.

Women have tended to flock to specific institutions or fields within science and to individual supervisors known to be supportive of women's careers in science. One field that was a haven for women in the first half of the twentieth century was crystallography. Crystallography was a new and unproven field where fame and fortune were still uncertain and that men therefore tended to avoid. Fields such as this were often headed by young, male scientists who were not as likely to have been influenced by established gender norms and tended to be supportive of women. At Cambridge University, the father-and-son team of William Henry Bragg (1862–1942) and William Lawrence Bragg (1890–1971) believed that crystals diffracted X-rays and that X-rays thus could help determine the structure of crystals. W. H. Bragg constructed the X-ray spectrometer to analyze crystal structure, and W. L. Bragg determined the relationship between the diffraction pattern and the atomic spacing. After World War I, W. H. Bragg moved to University College London and then to the Royal Institution in London, and his son took a position at the University of Manchester. Both men hired a significant number of women as part of their research teams; both established friendly, uncompetitive environments that focused on the pleasure of scientific exploration; and both were well-known for their fairness in attributing work to the women who had accomplished it. In 1934, I. Ellie Knaggs's crystallography work with cyanuric triazide was referenced at a meeting of the Faraday Society without mention of the fact that it was her work. W. H. Bragg wrote a letter to Sir Richard Gregory, editor of the science journal *Nature*, dated July 3, 1934, connecting Knagg with this work in what he called "a little act of justice to the lady," as the work "is her magnum opus [and] . . . she is naturally disappointed." Actions such as Bragg's helped to make women scientists visible contributors to their fields and increased the likelihood that their work would be acknowledged.

Kathleen Yardley Lonsdale (1903–1971) worked for five years at the University College laboratory under W. H. Bragg after completing her bachelor of science degree in 1922. She had been impressed by the elder Bragg, who had inspired her in her work while allowing her to follow her own research pursuits.

The Harvard College Observatory was known in the 1890s for the many women it employed to classify stellar spectra. Among those shown here in 1892 are Henrieta S. Leavitt (third from left), Williamina P. Fleming (standing), and Annie Jump Cannon (far right). (Harvard College Observatory)

When she left his laboratory she wrote to him, "I should like to take this opportunity of thanking you again for all the help you have given me in so many ways. I feel sure that it will be difficult to find a place where I shall be as happy in my work as I was at the Davy-Faraday [the Royal Institution Laboratories]" (quoted in Rayner-Canham 1998, 73). Despite the fact that Lonsdale's work contradicted his own, Bragg praised her research and, when she was struggling to balance working on her research at home with raising her three children, arranged a grant from the institution that allowed her to hire home-help so she could return to work. He ensured that she received this grant every year from 1931 until his death in 1942. His emotional, intellectual, and financial support enabled her to continue her work, which in 1946 led her to be appointed as reader in crystallography at University College and three years later as professor of chemistry and head of the Department of Crystallography. The younger Bragg nominated her as one of the first two women elected as a fellow to the Royal Society, and in 1966 she was elected president of the International Union of Crystallography. The support of a mentor in the early years of a woman's scientific career was important for entry into and advancement within the scientific profession. Lonsdale recognized the benefits she had incurred and made a point of arguing the need to erase gender stereotyping in education and to introduce communal nurseries, kindergartens, eating houses, and domestic help to facilitate women's ability to pursue a career.

Earlier, William Bateson (1861–1924), a scientist at Cambridge University who became an important researcher in the early history of genetics, wanted to investigate the internal causes of external variation in organisms. Such work was not well supported at Cambridge at that time, and yet Bateson required a large research group to help him carry out his breeding experiments. He turned to seven women at nearby Newnham College at a time when there were vociferous disputes over women obtaining degrees at the university. These women, who made up half of Bateson's research team, were his sister Anna Bateson (1863–1928), Dorothea Frances Matilda (Dora) Pertz (1859–1939), Edith Rebecca Saunders (1865–1945), Hilda Blanche Killby (1877–1962), Florence Margaret Durham (1869–1948), Igerna Brunhild Johnson Sollas (1875–1965), and Dorothea Charlotte Edith Marryat (1880–1928).

Some of these women were already studying inheritance when they met Bateson and contributed their work to his project. He supported and encouraged their work and recognized their achievements. At the 1906 meeting of the Third International Conference on Genetics, for which Bateson was president, he made a point of acknowledging the women's research in his inaugural address:

> Had it not been for the work that has been done by my friends and pupils—first of all by my colleague, Miss Saunders, whose name has been so deservedly honoured to-night—there would have been nothing at all to justify me in speaking of the significance of the work of inheritance; but for that vast reservoir of work they have piled together, I could never have dared, without that force behind me, to have asserted that Mendelian research has been and is of the importance that we now know it must possess (quoted in Richmond 2001).

Saunders had already been awarded the "silver-gilt" Banksian medal for her work in the physiology of inheritance in plants.

These women worked with Bateson to find experimental proof for Mendelian genetics and in the process helped to establish a new scientific discipline and their own scientific credentials. Subsequently, other scientific researchers referenced these women's work in their own research. Unfortunately, Bateson's women researchers have suffered from what historian Margaret Rossiter refers to as the "Matilda Effect,'" whereas Bateson has benefited from the "Matthew Effect." Bateson provided the scientists with excellent research opportunities, which they relished, and in turn he benefited from their research contributions. Historians only recently began to include these women in the history of genetics, as their role in Bateson's work has become clear. Nevertheless, the importance of male mentors for women scientists has been unequivocal, and the Newnham College Mendelians were no exception.

Many other male scientists refused women admission to faculty clubs and associations until the 1970s. Segregation of men and women into separate lunch rooms further prevented women from socializing with their male peers, an important aspect of professional networking. At London's University College, the common room was segregated until 1969, and the professional dining club excluded women until 1979. Icie Macy Hoobler (1892–1984), a renowned nutritionist, was not permitted to eat dinner at the Michigan Faculty Club even when she was the club's after-dinner speaker.

Even so, some societies that had excluded women throughout the nineteenth century began to admit women in the early part of the twentieth century. The Royal Geographical Society admitted women beginning in 1914, although not without controversy. Many of the male members believed that the admission of women would degrade the society to a social club and negate its serious character. One member even suggested that militant suffragettes be excluded. This suggested amendment was thrown out, and the subsequent women members did not disturb the status quo but rather contributed extensively to the discipline. Women were similarly admitted to the British Chemical Society in 1920 after what historian Joan Mason refers to as a "forty years' war" had been waged against their admission (1991, 233).

Whereas Miss Bowlby of Cheltenham was admitted to full membership of the British Association for the Advancement of Science (BAAS) in 1853, it was not until 1913 that the botanist Ethel Sargant (1863–1918) became the president of a section of the BAAS (the botanical section), and 1914 when E. R. Saunders of Newnham College became one of the first members of the BAAS council. When Saunders attempted to elect Agnes Arber as president of the section the following year, so strong was the feeling that a woman should not be president two years running that Arber was asked to decline and gracefully resigned from the position of president. Fears that feminists were going to take over the BAAS and make it a "botanical gynocracy," as A. C. Seward had written to Bayley Balfour in 1921, were perhaps excessive expressions of the male members' desire to retain control over the association despite the still-small presence of women. It would take until 1966 before there was a woman president of the BAAS, when the position went to Kathleen Lonsdale. Admitting women to the society but keeping them in the lower echelons outside of positions of power in the society was another tactic used by male members to retain control over and ensure the predominantly masculine identity of these societies.

Still, the progress of admitting women to the BAAS was substantially advanced over that of the Royal Society of London, which refused admittance of women until 1943. Although the 1919 Sex Disqualification (Removal) Act disallowed barriers to the admittance of women to learned societies based on sex, members of the Royal Society simply refused to consider any woman scientist worthy of nomination until this late date. This poor record was in keeping with other national scientific societies that nominated very few women before World War II. The United States National Academy of Sciences elected three women by 1944, the Soviet Academy of Sciences elected one, and Canada elected one woman in 1946. The Paris Academy of Sciences did not include women as full members until 1979. The Academy of Exact, Physical, and Natural Science in Spain did not admit women until 1986, although other Spanish scientific societies admitted women much earlier, but their numbers remained low as they did in these other societies. Being kept out of societies, like being segregated in the lunch room, ostracized women from the community of their peers, making it difficult for women scientists to be fully recognized or to participate fully in their profession. At society meetings, members could hear papers given on the latest research and could give papers themselves. Absence from such venues made it difficult for women to keep abreast of new developments in their fields and to be recognized for their own.

Women who entered the science profession and found a place for themselves in an academic or governmental position were not necessarily set for life. Women's lives as wives and mothers could end or restrict their careers. The women who worked for the Council of Scientific and Industrial Research, an Australian scientific organization, in the 1930s and 1940s faced either termination of their employment upon marriage or tenuous and/or part-time employment, accompanied by low pay and low status; in addition, they were continually threatened by the possibility of dismissal. The advent of the tenure track in the 1930s disadvantaged women because candidates had to prove their potential within the first seven years of their appointment, and those who did not do so were asked to leave. Women were often burdened with more teaching than were men, and they had fewer chances to advance quickly, yet they were perceived to be treated fairly under this new promotion system that was applied to everyone equally. Moreover, in many instances a marriage bar was in effect for women. Women who married were expected to give up their jobs to men who needed them and to concentrate on raising their family, especially in the depression of the 1930s. Of course, such women were often welcome to become the unpaid assistants of their husbands. Other women did manage to continue their scientific work alone despite being unemployed. One such woman was Maria Goeppert Mayer (1906–1972) who won a Nobel Prize in 1963 for the nuclear shell model of atomic nuclei, which demonstrated that atom-like discrete energy levels of individual protons and neutrons explained the stability of nuclei. Moreover, women who arrived at a university institution as the "trailing spouse" would be unlikely to obtain a paid position at the same university due to often-informal nepotism rules. Nepotism rules applied generally to family members and were in place to avoid a person obtaining a job solely on the basis of his or her relationship rather than his or her credentials. In practice, these rules appear to have been applied only to wives, not to husbands or other family members.

Many women chose not to marry and bear children and to instead focus on their career. But some women managed to find creative ways to have both a career and a family. Unmarried women often formed strong friendships to satisfy their need for companionship and even adopted children on their own. Some women made egalitarian marriages in which their husbands took on some of the domestic burdens and supported them in their work, and others interrupted their career to have children and then returned to their work thereafter. Women, whether married or unmarried, recognized the disadvantages of their position whether it was the double-day for married women or the sense of isolation or guilt suffered by some unmarried women who felt they were selfishly pursuing their own interests. Yet many women attempted to balance both a home life and a work life, believing that they could not be personally satisfied with only one or the other.

Many women scientists who felt ostracized from the scientific profession suffered in silence, but others formed female subcultures either outside or alongside those of the masculine professional culture that allowed them to enjoy a sense of professional cachet. The first known women's scientific society, Natuurkundig Genootschap der Dames (Women's Society for Natural Knowledge), was in operation between 1785 and 1887 in the town of Middleburg, on the southern Dutch island of Walcheren in the province of Zeeland. Founded by forty-four elite women who wished to be educated in natural philosophy, the society's members collected scientific instruments and books and organized scientific lectures for themselves, meeting regularly at the Musaeum Medioburgenese, the home of their society. Just as this society was coming to an end, American Matilda Cox Stevenson, who was rejected for membership from the Anthropological Society of Washington in 1885, formed the Women's Anthropological Society of America with ten other women in Washington. They met twice monthly, gave and heard papers to one another, and invited distinguished guests, including Alfred Russel Wallace (1823–1913), coauthor of the theory of natural selection, to speak to them.

Women could also form their own subcultures within specific scientific disciplines. The transformation of the field of physiology in several of the elite U.S. women's colleges in the 1930s, for instance, from its emphasis on hygiene to its association with the biomedical sciences, allowed women to fully participate in a normally male-dominated discipline. Such a transformation enabled women students to access laboratory research and experimentation and to earn under-graduate and graduate degrees in physiology; in addition, it created teaching opportunities for women who obtained Ph.D.'s in the field and continued their research thereafter, forming a generation of female role models and teacher mentors for women students in the field.

Finally, women established their own scientific prizes and awards to com-pensate for the persistent lack of recognition for women scientists in a male-dom-inated profession. These award-granting organizations and institutions attempted to make women's accomplishments visible, with the intent of putting pressure on male organizations to recognize their work as well. The longest-lasting award was the Ellen Richards Prize, given beginning in 1901 by the Association to Aid Women in Science. Frustrated by what they perceived to be a lack of satisfactory candi-dates for the prize, association members upgraded the Richards Prize into a woman's Nobel Prize in 1928, switching the focus from young women scientists to those further along in their careers who had made important contributions to sci-ence. The first award after this change was given jointly to physicist Lise Meitner of the University of Berlin and to chemist Pauline Ramart-Lucas of the University of Paris. The award, however, was discontinued in 1932. The reason given was that women were by then engaging in scientific research equally with men and gaining recognition for their achievements. Astronomer Annie Jump Cannon (1863–1941) was not so sure about this matter and arranged for her own winnings from the Ellen Richards Prize to go to the council of the American Astronomical Society for a triennial award for distinguished contributions to astronomy by a woman of any nationality. The Annie Jump Cannon Prize was given out every three years until the 1970s. These awards brought the faces of women scientists before the general pub-lic. Though such awards recognized women's achievements and publicized them, they did not assist women in gaining entry into male organizations and access to male-dominated prize winnings. Certainly a few exceptional women won presti-gious awards, but the majority of women researchers remained hidden behind the men for whom they worked. Women continued to struggle to be heard, recognized, and rewarded for their scientific achievements.

Women Scientists' Contributions to War, Industry, and Public Health

Gaining access to scientific employment outside of the academic world was rid-dled with the same difficulties and barriers to women as employment within the academy. When women were hired, they were paid poorly and admitted to the jobs that held the lowest status. Nevertheless, with the increasing acceptability of women in the workplace in the twentieth century and the demands created for extra workers during two world wars, new career opportunities opened to women. Some women were relegated to "women's work" in this context, but this did not negate the usefulness of the jobs they were performing or necessarily imply that these jobs were not fulfilling to the women doing them. Moreover, in some instances, women could manipulate certain fields to their own professional and political ends. Although opportunities for professional success were rare, a significant number of women made their way into male-dominated fields and achieved recognition for their contributions.

Some job opportunities were constructed within the confines of "women's work" in science. The most obvious fields for women's work were home economics, medicine, and public health. Home economics had a number of industrial applications and thus provided job opportunities outside of teaching for women. During the 1920s, 1930s, and 1940s, the U.S. government needed women with home economics degrees for the Cooperative Extension Service as home demonstration agents. These home economists assisted women in the countryside to better feed and care for their families on small budgets. Although women were not initially responsive to such assistance, these positions provided home economists with an opportunity to put their theories into practice. During the Great Depression they helped to alleviate the very worst of the deprivation. Home economists' skills were also used in hospitals by dieticians, an occupation that had a caring aspect as well as a scientific one. Although women in this field professionalized themselves by organizing as a professional body and moving into authoritative administrative positions, they were not entirely successful in this endeavor. Doctors increasingly recognized the worth of a dietitian's skills, but these women remained subservient to the doctors, and their position was categorized as "women's work," which automatically demoted it within the hospital hierarchy.

Though some science jobs were ostensibly "women's work," women utilized their position within this context to create their own ideological frameworks in attempts to change the gendered status quo. For example, in Weimar Germany (1918–1933) the government facilitated the formation of women's health clinics staffed by women doctors and funded by the state. Although their jobs were a feminized position within the medical profession, these women organized, headed, and supported these modern institutions that offered advice on birth control and performed abortions. These women were both "independent new women" and "maternal healers" (Grossman 1993, 66). Although these public health positions had low status and offered low pay, the women doctors

Katherine Burr Blodgett (1898–1979)

Katherine Blodgett's work is a good example of the scientific achievements accomplished by women in industry in the first half of the twentieth century. Her life reflected both the growing opportunities and the continuing inequities faced by

women who wanted to do scientific work. She was born in Schenectady, New York, in 1898. She spent her early years in France, returning to New York in 1906 to attend a small private school that she graduated from in 1913. A strong student, she won a scholarship to Bryn Mawr College, where she was encouraged to become a research scientist. Blodgett's father had been a patent attorney for the General Electric Company, and in her senior year a former colleague of her father's gave her a tour of the GE laboratories and suggested she might work for the company. Upon finishing her degree at Bryn Mawr, she went on to earn her MS at the University of Chicago and in 1918 went to work for GE, becoming its first woman scientist.

(General Electric Company)

Blodgett was fortunate to find a mentor at GE, Irving Langmuir. He publicly recognized the excellent work Blodgett did as his assistant working on the properties of monomolecular films on water. Langmuir also encouraged her to go to Cambridge University between 1926 and 1928 to earn her

believed they were doing important political as well as medical work. They saw themselves as assisting women and children who might not otherwise ever consult a doctor, and their work was about more than just physical healing. They were not just treating individual bodies but attempting to alter German society's attitudes toward women's sexuality, women's work, and women's childbearing, thereby improving women's and children's lives, family life, and society. They did not argue against the gender roles of wife and mother, but rather these women doctors aimed to improve women's lot within those roles through medicine, counseling, and changes in the law. The rise of Nazi Germany after 1933 ended their crusade. Many of them were Jewish and were forced to flee Germany with their families. Those women doctors who remained found their clinics turned into "Centers for Gene and Race Care" that promoted the sterilization of the "unfit." Still, during this period German women doctors had found a niche, and some used it to advance care for women.

Alternatively, British medical women who were part of the eugenics movement in the early twentieth century, such as Mary Scharlieb (1845–1931) and

Ph.D., where she became the first woman to receive a doctorate in physics. Langmuir described Blodgett as a "gifted experimenter" with a "rare combination of theoretical and practical ability." After obtaining her doctorate degree at Cambridge, she returned to the United States to work for GE from 1926 to 1963.

Blodgett not only continued to assist Langmuir but also began to conduct and publish work of her own. During World War II she also developed ways to prevent icing of aircraft wings and helped to produce more effective methods of generating smoke screens. By experimenting with the coatings on the surface of glass, she eventually produced a nonreflective glass that allowed 99 percent of light to pass through it. In 1938 Blodgett was awarded a patent for her nonreflective glass, for which she remains known today. Her coating has since found many applications, including periscopes, range finders, and cameras.

Blodgett received the American Association of University Women's Annual Achievement Award in 1945 and the Garvan Medal, the American Chemical Society's award for excellence for American women chemists, in 1951. Yet the scientific community largely ignored Blodgett's research, which was not even mentioned in a 1953 review of the first seventy-five years of research at GE in the journal *Science*. Even so, Blodgett's work and story had popular appeal, and her name and picture appeared in *Time* magazine. Moreover, she received numerous honorary doctorates from colleges and universities in the United States. Blodgett's quiet and unassuming personality and her disregard of titles and preferments within GE certainly did not assist her in achieving all the titles and honors she perhaps deserved, but probably did allow her to work amicably with the men in her field throughout her lifetime. She died in her hometown of Schenectady in 1979.

Elizabeth Sloan Chesser (1878–1940), attempted to feminize the agenda of the Eugenics Society. They argued, for example, that birth control was not just about limiting women's opportunities to conceive for the good of the race but also could allow women to choose not to have children or to explore their sexuality. In this way women could have a voice and make an impact within a male-dominated, masculine profession.

Some women did manage to gain scientific positions in fields that were considered male-dominated. Due to the late professionalization and trade origins of the pharmacy profession, American women entered this male-dominated field more readily than other areas in science and medicine between the 1870s and 1920s. Colleges offered an education in pharmacy, and pharmacy associations readily sought members of both sexes to keep up their numbers. As a result, the American Pharmaceutical Association accepted its first woman member in 1882. Women pharmacists were more likely to be college-educated than men. Although women without a family connection to pharmacy found it difficult to find an apprenticeship position or, subsequently, a full-time job in a drugstore, women both with and without family connections became pharmacists. More positions opened up for women during World War I, and in the interwar period many women moved into hospital pharmacy. Male pharmacists found the pharmacy business enterprise more lucrative than hospital work so were content to leave this low-paid and low-status work to women. Many hospitals favored women as pharmacists as they were more likely to stay for longer periods. Unfortunately, the professionalization of pharmacy in the 1920s and 1930s marginalized women as it had done in other science professions. While at the height of their participation women had made up about 10 percent of the profession, after this period women's participation in pharmacy was reduced to between 3 and 4 percent as they were squeezed out by professionalizing male pharmacists.

One area of employment in which women were welcomed during World War I was industrial chemistry in Britain, Germany, and the United States. Women chemists had been working as researchers at universities and continued to do so, filling posts left vacant by men who had gone off to war. Others were put to work by the government in analytical laboratories, studying the purity of various explosives or analyzing iron and steel samples used to make ships and tanks; they also were assigned to synthesizing essential chemicals, especially for medications. Such work gave women opportunities for research and publication as well as important war work. Following the armistice in 1918, opportunities for women in chemistry decreased as men returned to their positions. Those who remained employed tended to be excluded from laboratory research, production, or positions of authority and instead filled positions as laboratory assistants, an increasingly feminized occupation. This pattern repeated itself during and after World War II.

Another niche for women in the United States in the 1920s and 1930s was in the burgeoning field of commercial geology, due to the discovery of oil fields in Texas. In 1919 Esther Richards Applin (1895–1972) discovered that the microfossils accompanying exploratory drilling for oil could determine the age of the subterranean formations and predict whether more oil might be found in that location. Oil companies hired women to examine samples and map the regions. Women in these positions were rarely promoted or permitted to do their own fieldwork, and many were laid off during the depression in favor of men using seismographic detection instruments. However, many women had gained a strong enough reputation to work independently as consulting micropaleontologists.

Home economists also entered the business world. Lucy Maltby (1900–1984), who worked for Corning Glass Works between 1929 and 1965, epitomized this move. Hired initially to improve the sales of Corning's Pyrex ovenware, Maltby established a test kitchen, a laboratory in which she tested kitchen equipment produced by Corning, compared it with competitors' products, and created recipes for women to use in their Pyrex dishes. She hired other women home economists to work with her, but Maltby suffered from low pay in comparison to other department heads at Corning, as did her staff. The company's image of these women was one of errand girls in their publications. In the October 1947 edition of the *Corning Glass Works Gaffer*, these "girls'" jobs were to "go over town, get a pound of butter, and make the dish fit." Such work was then considered appropriate for women, connected as it was with improving the home and other women's lives. Yet these women's work required both scientific and domestic knowledge that the company men, including engineers, physicists, chemists, designers, salesmen, and managers, relied heavily upon for the success of their products.

During World War II many women scientists were responsible for the design and construction of the atomic bomb, although the majority of publications recounting the story of the creation of the bomb do not mention women's scientific participation. At least three hundred women physicists, chemists, metallurgists, biologists, biomedical scientists, and mathematicians were involved in the Manhattan Project, the code name for the bomb-building endeavor in the United States. Women were involved in making theoretical scientific discoveries that enabled others to perceive the notion of the atomic bomb. They also worked more directly in laboratory research, experimentation, and the manufacturing of materials for the bomb, as well as the mathematical computing at Los Alamos, and later in using computers that were developed to facilitate this work.

In some situations, women could move into a male-dominated scientific profession and take it over. This happened in the context of X-ray technology. In the first two decades of the twentieth century both men and women practiced as radiologists, taking X-rays and interpreting them. Radiologists professionalized in the early 1920s, by which time a division had occurred between the radiologists, who could interpret X-rays for doctors, and radiographers, who were proficient at taking patient X-rays. As work for radiologists and radiographers moved into hospitals a clear hierarchy emerged with doctors at the top, followed by radiologists and then radiographers. Male radiographers, feeling increasingly like second-class citizens in relation to radiologists, attempted to mark themselves off from female radiographers, claiming that their own skill lay in the use of the X-ray equipment and that women's skills lay in caring for the patient being X-rayed. These men also used the Society of Radiographers, established in 1921, to stipulate that all radiographers must have a college diploma, a provision that they thought would exclude many women. However, obtaining this education did not prove to be a barrier for women even when the society increased education requirements in 1936. Often already working within the hospital and using X-ray equipment as nurses, these women took advantage of hospital training programs to achieve the diploma they required to be formally recognized. The society also refused to support equal pay

Maria Goeppert Mayer (1906–1972)

Maria Goeppert Mayer overcame the barriers to women in science to win the Nobel Prize in 1963 and achieve the position of full professor at the University of California. Born in 1906 in Katowitz, Upper Silesia, Germany, she entered a well-educated, upper-middle-class family where she was encouraged to learn. When her father moved to Göttingen University to become the professor of pediatrics, Maria was determined to study there. Although the preparatory school for girls that she had been attending

closed before she could complete the coursework, she nevertheless passed the entrance exams and entered the university in 1924. She began her studies in mathematics and soon after switched to physics. She wrote her Ph.D. thesis on the decay of excited states by the simultaneous emission of two quanta, for which she was awarded her doctorate in 1930. Having lost the supportive influence of her father, who died in 1927, Goeppert was fortunate enough to receive the encouragement, support, and mentorship of the theoretical physicist Max Born and the experimentalist James Franck. While at the university, she met Joseph Mayer, a theoretical chemist from the United States on a fellowship. They were married shortly thereafter, and she returned with him to the United States.

(The Nobel Foundation)

While Joseph held a position at Johns Hopkins, she was able to continue working with him and his colleagues, but without position or pay. Because no one at Johns

for male and female radiographers, separating the men from the women as higher paid and therefore automatically assumed to be more highly skilled.

Anne Witz calls these tactics used by male radiographers "internal demarcation" (1992, 169). She points out, though, that their attempts backfired. Male radiographers argued that they needed only technical skills, not nursing skills. However, radiologists realized the importance of the female radiographers' ability to deal with patients with broken limbs, and those under anesthesia, as well as their ability to administer injections and use the X-ray equipment. Moreover, they noted that the nurse's knowledge of anatomy assisted them in taking appropriate X-rays that allowed doctors and radiologists to more readily diagnose patients' problems. In this case, exclusionary tactics or internal demarcation did not necessarily work in keeping women out of a scientific profession. Moreover, as male radiographers were pushed out, more job opportunities existed for women.

Hopkins was working on quantum mechanics, she turned her mind to chemical physics. In 1940, Maria and Joseph, now at Columbia University, coauthored a textbook on statistical mechanics; most people assumed the work was mostly her husband's. In the 1930s she gave birth to two children, Maria Ann and Peter Conrad, and carried on her work unabated. During World War II she worked on the Manhattan Project, attempting to separate uranium-235 from uranium-238. Although her particular research did not prove fruitful, her confidence was bolstered by her paid position and by the fact that she was given a supervisory position.

After the war she joined Joseph at the University of Chicago, where many of the male scientists involved in the Manhattan Project were now located. These men were supportive of her work and presence at Chicago, and she received an associate professorship. However, due to nepotism rules she did not receive a salary. It was during this time that she made her famous discoveries concerning the nuclear shell model. She discovered the existence of the "magic numbers," that is, she realized that the most stable elements contained particular numbers of either protons or neutrons, and she went on to provide an explanation as to why this was so. She proved that certain arrangements of protons and neutrons created relatively fixed orbits. From 1961 on, both she and Joseph became professors at the University of California in La Jolla, where Maria finally received a salary commensurate with her experience and achievements. She was proud of the fact that she had become the eighth generation of professors in her family. In 1963 she won the Nobel Prize for physics for her nuclear shell model, becoming the second woman in the United States to win the prize and the first woman to win it for physics. Having suffered from health problems for several decades, she had a stroke in 1960 that left her partly paralyzed. She died of heart failure in 1972. In 1985 the American Physical Society established the Maria Goeppert Mayer Award, funded by the General Electric Foundation. The award is given annually to an American woman in the early years of her career for outstanding scientific achievement.

The majority of jobs open to women in science, inside and outside of the university, in the first half of the twentieth century were classified as "women's work." Nevertheless, in all arenas—from the university to public health, to industry and to warfare—women found employment that not only allowed them to work within their chosen profession but on occasion permitted them to work toward change within their profession and beyond.

Conclusion

At the end of World War II, the women who had lived at Los Alamos, New Mexico, where the first atomic bomb had been created, decided that a record should be kept of their experiences in that town. This record, written in 1945

Lise Meitner (1878–1968)

Lise Meitner experienced in her lifetime both the highs and lows of a professional woman scientist. Born in 1878 to a middle-class, Jewish Viennese family, Meitner had shown an early interest and facility for mathematics and science; her mother had caught her putting a towel under the crack in the door so that no one would see her light while she stayed up late reading mathematics. Fortunately her supportive family facilitated the education of Lise and her sisters. When Austrian universities opened to women in 1897, Meitner spent only two years, instead of the usual four, in intensive private tutoring preparing for the university entrance exams, which she passed.

Lise Meitner and Otto Hahn. (AIP Emilio Segre Visual Archives, Brittle Books Collection)

Meitner benefited greatly from the support of male scientific mentors. From 1902 to 1905 she studied physics at the University of Vienna under the theoretical physicist Ludwig Boltzman, who was supportive of women's education and whose lessons Meitner very much enjoyed. In 1906 she obtained her Ph.D. in physics, only the second woman to achieve the degree at the University of Vienna. Meitner wanted to pursue advanced work in physics and moved to Berlin, supported financially by her parents until 1910, to attend the lectures of Max Planck, who had done important work on radiation. Although Planck initially caused Meitner to believe he was not supportive of women's education, they nevertheless went on to become very good friends.

At the same time she began unpaid work for Otto Hahn, a chemist at the Berlin

but not published until 1988, reveals the day-to-day lives of the women at Los Alamos, "the three years of working and marrying and dying, of giving birth, of getting drunk, of laughing and crying, which culminated in that successful test at the Alamogordo bombing range." Each of the nine chapters is a memoir of one woman that tells the story of her life at Los Alamos. But there is only one woman scientist among them, and she worked in the context of "women's work" in science. She was Charlotte Serber, a scientific librarian who is noted

Institute of Experimental Physics who was researching radioactivity. In 1911 Planck hired Meitner as an assistant with a small salary to mark students' problem sets, which allowed her to continue her experimental work. Money was always a problem for Meitner, who often found herself substantially underpaid. Still, in 1919 she was appointed professor in the institute, and in 1926 she was appointed adjunct professor at the University of Berlin, becoming the first woman professor of physics in Germany. Though initially Meitner and Hahn's relationship was one of student and teacher, very quickly they formed a working partnership. Hahn said of Meitner that she would tell him "leave that to me, you haven't the faintest notion about physics!" In a switch of gender roles in science, Meitner was the thinker who asked questions and interpreted experiments, and Hahn carried out the experiments themselves. From this partnership came the most important discovery of her life's work: the discovery, in 1939, that the uranium nucleus split to form two other atoms and when it did so released a considerable amount of energy. Meitner dubbed this splitting of uranium "fission." This discovery laid the groundwork for the study of nuclear energy and the atomic bomb.

Unfortunately, despite her involvement in the discovery, both in the laboratory and at a distance through correspondence with Hahn, Meitner never received credit for her work. She had been forced to flee Nazi Germany in 1938 after being stripped of her professorship because of her Jewish heritage. Just five months later, Hahn completed the experimental work to prove that the uranium nucleus split. Hahn, distancing himself from Meitner for political reasons, received the credit for the discovery and won the 1944 Nobel Prize in chemistry. Meanwhile Meitner, exiled from her country, was fortunate to be given a position at the Nobel Institute for Physics in Stockholm but was paid less than a starting salary for an assistant. It was not until 1951 that she managed to obtain her pension from the University of Berlin, and even then her Swedish salary was deducted from the amount she received. While Meitner received numerous women's honors and awards, including the Ellen Richards Prize in 1928, and prestigious prizes such as the Max Planck Medal of the German Physics Society in 1949 and the Enrico Fermi Award in 1966, Meitner was robbed of her true place in history, partly due to historical circumstances but mostly due to the disregard of Otto Hahn for her contribution to their discovery, even after the fall of Nazi Germany in 1945. Unfortunately the $50,000 she shared three ways for the Enrico Fermi Award came too late, and she died at a nursing home in Cambridge, England, in 1968.

in *Standing By and Making Do: Women of Wartime Los Alamos* as holding "an important position in the Technical Area." Interestingly, one of the book's two editors was Charlotte Serber herself. Here was a woman who was in charge of top-secret documents as well as scientific reference books but who nevertheless was clearly viewed as doing "women's work" within the scientific community and thus whose story could be considered for inclusion among those of other women. Yet it would seem that even she did not think to include other woman scientists among the memoirs of women's experiences at Los Alamos.

Why did a group of women who consciously decided that it was important for the women of Los Alamos to tell the story of their lives in the town not include at least one woman research scientist? A reader of this collection might be apt to think that perhaps there were no such women scientists at Los Alamos, but that was not the case. Perhaps the wives, teachers, and secretaries of Los Alamos grouped women scientists with the predominantly male group of scientists, not as women. They may have believed that the women scientists, as scientists, would be remembered along with the men, whereas their own lives and experiences, including those doing "women's work" in science, were more likely to be forgotten by posterity. This would have been a reasonable supposition; however, it seems the women scientists of Los Alamos have suffered from the Matilda Effect. Their work has been subsumed by the towering personas and reputations of the men working on the Manhattan Project. Male scientists saw women scientists as women, and the other women at Los Alamos saw women scientists as scientists. As a result, these women have been given no place in history until quite recently.

Ruth Howes and Caroline Herzenberg's work *Their Day in the Sun: Women of the Manhattan Project* (1999) rectifies this gap in the historical record, revealing that women's work on the Manhattan Project was central. Elda Anderson, with a Ph.D. in atomic spectroscopy, prepared the first sample of nearly pure uranium-235 used for experimentation by the Los Alamos group. Leona Woods, completing a Ph.D. in molecular spectroscopy, worked as one of Enrico Fermi's staff, experimenting with the atomic piles; she was working on the day when a pile first sustained a chain reaction. She continued to work until two days before the birth of her baby in 1944 and then was able to continue to work afterward on the construction of the plutonium production reactors with her mother as child-minder. Metallurgist Natalie Michel Goldowski (1908–?), with a Ph.D. in physical chemistry, solved the critical problem of bonding the uranium slugs to the aluminum shells by developing a noncorroding aluminum coating. Edith Hinkley Quimby (1891–1982), with a masters in physics, collaborated in the development of the use of radioactive therapy for cancer; and Anne Perley, who had an MA in biochemistry, monitored workers at Los Alamos for exposure to radionuclides and conducted studies in the area. Howes and Herzenberg have written these women back into history, making visible the important contributions that women make to science. As physicists themselves, they had "accepted the story that the development of the atomic bomb had been an exclusively male activity. There seemed to be no reason to doubt it" (1). In educating themselves, they hope to educate others.

Changing the perception of women scientists cannot be just about numbers and the recognition of their peers. Women scientists have to be recognized by society more broadly as important contributors to the scientific endeavor. They have to be written into history not just by other scientists and by historians but by the society in which they live and work. Male scientists' poor treatment of female scientists during the professionalization of science between 1860 and 1960 has been justified by the society in which they lived. Western laws and social mores have dictated the appropriate occupational roles for men and women and have permitted and even encouraged the blacklisting of women who have attempted to penetrate into masculine professional fields not just from the profession but from the social world of women too. During the period of professionalization a woman scientist, it would seem, could never truly be a scientist because she was not a man, but a woman scientist could never truly be a woman, either, because science was a profession for men. The woman scientist was both a professional and cultural monstrosity.

Yet despite the gendered occupational and sociological barriers to women's participation in science, women have not only participated in the scientific profession but have, more often than the popular historical record might suggest, become stars in their field. Such achievement is not only an individual triumph for these particular women but a triumph for all other women who follow and for those who have benefited from their research. However, the tragedy of the snubbing and silencing of women scientists by men in the profession, and by society more generally, is the countless lost opportunities for women who had much to offer in the production of scientific knowledge. The obstacles presented to women scientists have resulted in lost or delayed scientific achievements, which have been neither to the benefit of these women nor to society.

Bibliographic Essay

The most definitive account of women's lives and work in the science profession is Margaret Rossiter's two-volume work *Women Scientists in America: Struggles and Strategies to 1940* (1982) and *Women Scientists in America: Before Affirmative Action 1940–1972* (1995). More in-depth, broad studies such as these are needed in the European context. Historians need to explore the persistent discrimination of women in the science profession that Rossiter reveals in the American context and the ways in which women adjusted, compromised, and fought back against the injustices they have experienced. Rossiter also documents the extraordinary achievements in the sciences and in professional leadership of many women. The majority of studies on the history of women and the science profession in the late nineteenth and twentieth centuries focus on individual women or specific fields in science. Biographies of individual women or groups of women lend themselves to explorations of the challenges and achievements experienced by women during the professionalization of the sciences. Ruth Howes and Caroline Herzenberg have done exhaustive research into the lives and work of the women who worked on the Manhattan Project, culminating in their recent book, *Their Day in the Sun: Women of the Manhattan Project* (1999). Marlene and Geoffrey Rayner-Canham have compiled biographical data on many women in science in their *Women in Chemistry: Their Changing Roles from Alchemical Times to the Mid-Twentieth Century* (1998) and *A Devotion to Their Science: Pioneer Women of Radioactivity* (1997). Examples of works studying the advantages and disadvantages to women of specifically feminized science fields include Sarah Stage and Virginia B. Vincenti (eds.), *Rethinking Home Economics: Women and the History of a Profession* (1997); Toby Appel, "Physiology in American Women's Colleges: The Rise and Decline of a Female Subculture," in Sally Gregory Kohlstedt (ed.), *History of Women in the Sciences: Readings from Isis* (1999); and Pamela Mack, "Strategies and Compromises: Women in Astronomy at Harvard College Observatory, 1870–1920," *Journal for the History of Astronomy* 21 (1990):65–76. Shorter but equally useful and well-researched works that explore women's struggle to enter the professional world of science include Marsha Richmond, "Women in the Early History of Genetics: William Bateson and the Newnham College Mendelians, 1900–1910," *Isis* 92 (2001): 55–90; Mary Creese, "British Women of the Nineteenth and Early Twentieth Centuries Who Contributed to Research in the Chemical Sciences," *British Journal for the History of Science* 24 (1991): 275–305; and Jeffrey Johnson, "German Women in Chemistry, 1895–1925," *Zeitschrift fur Geschichte der Naturwissenschaft, Technik und Medizin* (International Journal of History and Ethics of Natural Sciences, Technology and Medicine) 6 (1998): 1–21.

For the study of women's careers in the medical profession, see Virginia Drachman, "The Limits of Progress: The Professional Lives of Women Doctors, 1881–1926," *Bulletin of the History of Medicine* 60 (1986): 58–72; Teresa Catherine Gallager, "From Family Helpmeet to Independent Professional: Women in American Pharmacy, 1870–1940," *Pharmacy in History* 31 (1989): 60–77; Atina Grossman, "German Women Doctors from Berlin to New York: Maternity and Modernity in Weimar and in Exile," *Feminist Studies* 19 (1993): 65–88; and Greta Jones, "Women and Eugenics in Britain: The Case of Mary Scharlieb, Elizabeth Sloan Chesser, and Stella Browne," *Annals of Science* 52 (1995):481–502. Grossman and Jones both study how women in a male-dominated field can, in some instances, feminize a profession with a view to improving their own working lives and the lives of other women.

Though all of these works deal with women's personal lives to a greater or lesser extent, the only work that specifically addresses this issue is Regina Morantz-Sanchez, "The Many Faces of Intimacy: Professional Options and Personal Choices among Nineteenth- and Twentieth-Century Women Physicians," in Pnina G. Abir-Am and Dorinda Outram (eds.), *Uneasy Careers and Intimate Lives: Women in Science, 1789–1979* (1987). Her essay acknowledges the difficulties women faced in having both a satisfying professional and personal life. Other publications on women's professionalization also address or at least acknowledge these difficulties, but Morantz-Sanchez's chapter specifically focuses on the positive coping mechanisms that women physicians employed to create balanced and emotionally fulfilling lives for themselves.

Women's entry into scientific societies, an important facet of the professionalizing process, is usually told on a case-by-case basis. Again, such works are certainly useful in highlighting when, where, and how women were admitted to the profession. In the majority of cases they outline the slow and tortuous progression of women's entry into scientific societies; for example, see Elena Ausejo, "Women's Participation in Spanish Scientific Institutions, 1868–1936," *Physis* 31 (1994): 537–551; Morag Bell and Cheryl McEwan, "The Admission of Women Fellows to the Royal Geographical Society, 1892–1914: The Controversy and the Outcome," *Geographical Journal* 162 (1996): 295–312; Joan Mason, "A Forty Years' War," in *Chemistry in Britain* (March 1991):233–238; and Joan Mason, "The Admission of the First Women to the Royal Society of London," *Notes and Records of the Royal Society of London* 46 (1992):279–300. In some instances, women were completely ostracized from scientific societies, as can be seen in Eveleen Richards, "Huxley and Woman's Place in Science: The 'Woman Question' and the Control of Victorian Anthropology," in James Moore (ed.), *History, Humanity, and Evolution: Essays for John C. Greene* (1989). Yet, other articles tell the story of societies receptive to women's admission from an early period—for example, David Allen, "The Women Members of the Botanical Society of London, 1836–1856," *British Journal for the History of Science* 13 (1980):240–254. An article by A. D. Boney, "The Botanical 'Establishment' Closes Ranks: Fifteen Days in January 1921," *Linnean* 11 (1995):26–37, examines how even when women were admitted to a scientific society, they could still face marginalization and discrimination.

Research on women in the science profession can be set in the larger context of the history of professionalization more generally. Anne Witz's *Professions and Patriarchy* (1992) outlines the tactics used to keep women out of the medical profession and the tactics they used to get in, and considers gender in the context of professionalization in science and medicine. Margaret Rossiter has contributed to this genre with a useful and important article, "The (Matthew)

Matilda Effect in Science," *Social Studies of Science* 23 (1993):325–341, which similarly draws attention to the disadvantages women have faced in the profession of science on the basis of their gender. Sociological articles such as that by Shirley M. Clark and Mary Corcoran, "Perspectives on the Professional Socialization of Women Faculty: A Case of Accumulative Disadvantage?" *Journal of Higher Education* 57 (1986):20–43, provide historical insight into the barriers women have confronted in academe, using faculty interviews that reveal the formal and informal disadvantages experienced by women attempting to gain entry to the university world of science. Still, much more work needs to be done in this field building upon this excellent beginning.

Women's Advancement in Science since World War II

I read the life of Marie Curie and she was an inspiration. I thought if she
could do all that, I could also make discoveries to benefit the world.
—Elizabeth Dennis, in Ragbir Bhathal, *Profiles: Australian Women Scientists* (1999)

In 1953, James Watson (1928–) and Francis Crick (1916–2004) published a
joint paper in the journal *Nature* in which they revealed the double helical
structure of DNA. They have since been heralded as the co-discoverers of the
structure of DNA, receiving the Nobel Prize for medicine in 1962. Their now-
famous story of discovery was immortalized in Watson's personal account, *The
Double Helix* (1968). In Watson's account, he and Crick become the heroes of the
tale, working at breakneck speed against other British and U.S. scientists in the
race to be the first to discover the structure of DNA. Watson's book has also
become infamous, though, for both his sexist portrait and dishonest treatment of
one of their competitors in the race, Rosalind Franklin (1920–1958).

Franklin's life and work epitomized the challenges and achievements of
women scientists in the second half of the twentieth century. With the support of
her parents, Franklin received a university education, earning her Ph.D. in chemistry from Newnham College, Cambridge University, in 1945. She found a female
mentor, physicist Adrienne Weill, who led Franklin to the field of X-ray crystallography. Franklin subsequently worked in Paris at the Laboratoire Central des
Services Chimiques de l'Etat. Then, in 1950, she was offered a fellowship at
King's College, London. Franklin had found the laboratory environment in Paris
collegial and supportive, but on moving to King's she faced a hostile working
environment. She was angered when she found that she was not permitted, as a
woman, to share the senior common room with her male colleagues. Furthermore, she arrived at King's with the impression that she was the leader of a
research team, whereas her colleague Maurice Wilkins believed that she was
part of his research team, working for him. Her relationship with Wilkins quickly
soured, and he felt disinclined to protect Franklin's work. Wilkins gave Watson

and Crick, working at Cambridge, a revealing glance at Franklin's X-ray photograph of DNA. This incident occurred without Franklin's knowledge. As a result, Watson and Crick went on to build a model of DNA as a double helical structure. After a few years of tolerating the difficult work environment at King's, Franklin chose to leave. She took her fellowship to Birbeck College, London, never knowing that Wilkins had betrayed her confidence.

Franklin's work, it seemed, was fair game because her male coworkers found her behavior aloof and unfriendly. Franklin became a "wicked witch" in Watson's heroic retelling in *Double Helix*. He justified his treatment of Franklin on the grounds that she was a frightening specter—an intelligent woman scientist with apparently no femininity to recommend her. Referring to her condescendingly as "Rosy" (a name she never used), he wrote: "I suspect that in the beginning Maurice hoped that Rosy would calm down. Yet mere inspection suggested that she would not easily bend. By choice she did not emphasize her feminine qualities. . . . There was never any lipstick to contrast with her straight black hair, while at the age of thirty-one her dresses showed all the imagination of English blue-stocking adolescents. . . . Clearly Rosy had to go or be put in her place. The former was obviously preferable because, given her belligerent moods, it would be very difficult for Maurice to maintain a dominant position that would allow him to think unhindered about DNA" (17). Franklin, who died of ovarian cancer in 1958, could not receive the 1962 Nobel Prize with Watson, Crick, and Wilkins because the prize is only awarded to those still living. In their Nobel lectures they did not mention her.

In 1975, Anne Sayre, a friend of Franklin, decided to set the record straight in *Rosalind Franklin and DNA*. Turning Watson's story on its head, she made Franklin into a scientific martyr and a feminist hero. She described the environment at King's for women as one not of "great injustices" but of suffering "small, daily annoyances" (96). Though women had obtained admission to King's as students and as researchers, Sayre remarked that "it cannot have been entirely agreeable to the women who exercised these privileges to be kept in a kind of purdah while they did it. Quite simply, it was a great advantage in those days to be a man if one was connected with King's College. Rosalind was not a man. She was unused to purdah and often it offended her. Part of the surrounding circumstances was that, from the start, she was dealt with at King's less as a scientist than as a woman, hence inferior" (96–97). Sayre emphasized that there was no moral justification for Watson and Crick's surreptitious use of Franklin's work. Franklin was denied the opportunity to be recognized for her achievements not only by her early death but by two male scientists who did not believe that a woman scientist deserved their respect or recognition.

More recently, historian Brenda Maddox has attempted to present a balanced view of Rosalind Franklin's life and work in *Rosalind Franklin: The Dark Lady of DNA* (2002). Sayre portrays Watson and Crick as the "winners" and Franklin as the "loser." Though not denying the mistreatment of Franklin at the hands of her male colleagues, Maddox nevertheless suggests that Franklin herself was likely more disappointed by her early death, which cut her research time short. Just as Susan Quinn painted the portrait of the whole woman—Marie Curie—Brenda Maddox painted the bigger picture of Franklin's life and work. Maddox suggests that Franklin's Jewish heritage and upper-middle-class background, as well as her sex, contributed to her poor relations with her King's colleagues. She also points out that Franklin chose to remain unmarried and childless to pursue her career, but that she did indeed experience love, and close and enduring friendships with men and women. Maddox also emphasizes the success of Franklin's work after her time at King's. She was so much more than the talented crystallographer who made beautiful X-ray photographs. Franklin was successful in obtaining significant funding for her group at Birbeck College. Working in a supportive, stimulating, and collegial atmosphere there, she published a significant number of papers and became a world leader in the study of virus structure by X-ray crystallography.

Rosalind Franklin practiced science in the period between World War II and the rise of the second wave of the feminist movement. Increasing numbers of women were entering science during this period. Access to the requisite education, increased number of jobs at growing universities, and industrial and technological advancements expanded employment opportunities for women scientists. However, institutional and cultural barriers still continued to hinder women's work in science. Throughout the 1970s and 1980s, feminists began to critique science as a discipline and to call for change. Their work led to a consciousness-raising among university faculty and administrators, employers, the general public, and women scientists. Despite obstacles to working in science in the second half of the twentieth century, the numbers of successful women in science grew exponentially, as did the numbers of men, and women's contribution to scientific research grew accordingly.

Barriers to Women's Participation in Science: History Repeats Itself

Despite women's interest in and intellectual capacity to practice science, and despite legislation that should have enabled women to achieve equality with men, women in the second half of the twentieth century still faced many of the same barriers women faced one hundred years earlier. Prejudices against, and struggles

Members of the American Association of University Women (AAUW) and other proponents of the Equal Rights Amendment demonstrate in front of the White House, March 22, 1977. (UPI-Bettmann/Corbis)

by, women have been summed up nicely by Henry Etzkowitz, Carol Kemelgor, and Brian Uzzi: "The human price for the Ph.D. is higher for women than for men, and the rewards are often lower" (Etzkowitz et al. 2000, 95). Although international variations existed, the similarities in women's experiences were virtually universal. At every stage of their education, girls were reminded of their differences from men and of their lesser status in western society. Women were taught that they should not be naturally inclined toward science, and that women who are good at science are not feminine. Many women overcame these barriers and disadvantages and entered undergraduate, graduate, and post-graduate science studies, but at each level a significant number of women dropped out. This state of affairs is generally referred to as the "leaky pipeline." Though the second half of the twentieth century saw increasing numbers of women taking science degrees and obtaining employment, the numbers of women present at each career stage decreases. In Europe in 1997, an almost equal number of men and women took science degrees, yet very few women obtained top positions in science. The male culture of the university, university departments, and science laboratories, both on and off university campuses, constrained women.

There have been periods of time and particular places in the second half of the twentieth century when women were welcomed into science practice. During World War II, as had been in the case in World War I, the absence of men from universities and industry allowed women to enter fields that were inaccessible during the inter-war period. Women received a higher number of Ph.D.'s in this period than they had previously, although men still received a higher percentage of the degrees than women, and men's participation in science increased at a much higher rate. During the war, women entered university positions, albeit often as part-time replacement faculty, and benefited from increased science funding.

Although women were encouraged to leave their positions after World War II, the Cold War motivated the U.S. government to encourage women to study science as a matter of urgent national military priority. The government feared that as men were needed elsewhere, especially as the Korean War began in 1950, there would be a shortage of highly trained scientific personnel. This was an important move by the government for women, as it was not perceived as a temporary war measure but as an effort to permanently move more talented women into science. Yet women in the 1950s received a conflicting message, one that simultaneously urged them to conform to traditional roles. Many government jobs in science were connected to the military, so sexism was rife, and women were kept out of top jobs, were paid lower salaries than they would have received in industry, and were not, for example, allowed to be trained for particular jobs, such as astronauts. Still, many women scientists were pleased to work in government science because industrial managers often discouraged them from working in industry. With the end of the Cold War in 1991, and thus declining military research, positions became limited, and women again were squeezed out of jobs.

Women scientists in the post–World War II period continued to utilize many of the same tactics that women scientists before them used to access scientific careers. Some lived and worked on the margins of science, accommodating their scientific work to family life and their husbands' careers. Others worked within feminized fields of science or in women's colleges, or found other niches outside of the most obvious and prestigious environments. Despite the drawbacks, some women formed support groups, women's clubs, and women's science prizes.

The idea of the lone male scientist affected many women's belief in their capacity to practice science. There has been a strongly held image among scientists that the successful male scientist works alone to make substantial scientific breakthroughs, that his best work will be accomplished early in his career (before he is forty years old), and that in order to achieve success in science he must dedicate his life to its pursuit. Feminists argue that the belief that men work alone to achieve scientific success is a myth of the scientific establishment. In fact, in the twentieth century, science increasingly became a team-orientated profession. In the laboratory, work was done collaboratively. The successful male scientist became part of a scientific network of graduate students and academic and industrial scientists early on in his career. Likely his supervisor introduced him into the necessary scientific circles and invited him to conferences. The male science student also likely extended and possibly even created new networks from his own contacts at school and beyond. This networking was often done in a social setting during sports activities, get-togethers after hours, or study groups set up among graduate students.

Uninitiated women scientists often accepted the myth that to be a successful scientist they would have to work alone and succeed alone while men took advantage of a supportive network of colleagues. Those women who perceived the necessity of joining a team sometimes found it difficult to be accepted into such a group. Male supervisors often failed to introduce their women students into the necessary circles or to invite them to important talks or to join a research group. Australian physicist Rachel Makinson, researcher at the CSIRO Division of Textile Physics, noted that "I mostly had to work by myself," except on one occasion when the chief wanted some work done quickly; she was "never given a team." When asked if she believed that this was because she was a woman scientist in a male-dominated environment, she said, "I think so—but one can't prove these things" (quoted in Bhathal 1999, 131). Whereas women have often been successful in high school science due to the support and encouragement of a particular teacher or parent, they tend to lose this personal contact at the university level and thus often felt discouraged by their isolation. Women have either not been invited to or have felt uncomfortable in the social situations in which networking connections and decisions are made. As a result, they have been excluded from the team environment in which they would very likely thrive if only they could gain access. Instead they have often been left to struggle alone, failing to complete their studies or achieve their research goals.

The male-orientated workplace further disadvantaged women by failing to accommodate their biological life cycle. If scientists assumed that they must achieve success early in their career and dedicate their early adult life to science, then women in their twenties and thirties found themselves at a marked disadvantage. Women who gave birth and took time off to raise children were often considered a loss, or at best not believed to be as serious or committed scientists as their male colleagues who have dedicated their lives fully to science by living, sleeping, and eating around their research work. Microbiologist Nancy Millis, although herself not married and remaining childless, recognized that it is the "biological problem of children" that creates a differential between men and women. If, she speculated, "a large factory or large university [had] a really good child care place which was there from 8 in the morning to 7 at night, and you could leave your child there in the total understanding that it was well looked after, then that would make a big difference, in my view, to what women could do" (Bhathal 1999, 52). Female supervisors could be as equally guilty of disadvantaging women. One female faculty member in the research by Etzkowitz, Kemelgor, and Uzzi stated: "If a student had a baby with her, I wouldn't have her. Students who have babies here get no work done. It's not that I wouldn't take a woman with a child in the first place, but the first sign of trouble, I would just tell them to go away. If my students fail it looks bad for me" (2000, 89). The assumption that the best work is done early on in one's career and by denying oneself a life outside of the workplace meant that a woman who took time out to have children might fail to graduate, to obtain a job, to obtain tenure as quickly, or to obtain it at all. Women who took time out of their career for their children often eventually achieved tenure and the various accolades of their profession, but these came later in their profession, and the chances of achieving high status, high-level positions, awards, or comparable pay increases to those of men of the same age and with the same achievements were reduced.

Women were also constrained in their ability to move from job to job, and place to place, because of their partners' careers. Though there have been an increasing number of dual-career couples in the academic and professional worlds, women's careers have repeatedly taken second-place to their partners' careers, especially if they had children. In part this might be a financial decision, as more often than not men made more money than women, even when both were successful professionals. Such decisions may be based on child-care needs, often linked to the continued stereotypical tradition that a man's career should come first. Whereas men were able to wait until later in their careers to marry and have a family, women's biological time clocks conflict with the tenure time clock. Some women opted not to marry or have children, and others gave up science in order to fulfill these needs in their lives. In short, for women, there has been no "right time" to have children on the academic career track.

However, in the United States in 1972, Congress, encouraged by working women who had sent petitions and spoken out against inequities for women in the workplace that contravened the equal pay and sex discrimination legislation, passed the Equal Employment Opportunity Act. As a result, educational institutions were no longer exempt from equal employment opportunity laws, and women could not be discriminated against on the basis of their sex in federally assisted programs including sports, textbooks, the curriculum, and in education employment. The equal pay act was extended to cover administrative, professional, and executive employment, and the United States Commission on Civil Rights' jurisdiction was extended to include sex. The United States was not alone in taking this kind of legislative action. Most western countries adopted similar antidiscrimination laws.

Yet, even as the legislation was being enacted, the search for an increased number of professional scientists was coming to a halt as science budgets were cut and Defense Department monies could no longer be spent on university research. For the first time in many years, the number of scientists employed by the U.S. federal government decreased. A glut of doctorates appeared on the market in the 1970s, and a crisis mentality ensued that blamed, in part, the number of women who had achieved doctorates in the 1960s. By the 1980s, affirmative action was not rigorously enforced in most industrialized nations and as a result had little effect on recruitment or retention of women in science. Moreover, women were often told, after starting work in a new position, that they had attained the job to fill a gender quota, not due to their accomplishments. Naturally, women felt unwelcome, and this situation affected some women's self-confidence.

Over time, not only have women's careers been thrown by the wayside, but the status quo has persisted. Male professors set examples for their male graduate students about how to behave toward female graduate students and how male supervisors advised women. Moreover, they recreated a scientific world that often disadvantaged men who would like to balance their career and personal lives more equally. Though women often made the sacrifices necessary to gain access to and succeed in a career in science, more recently they have made different decisions. Some women science students examined the lives of the female role models, looked askance at the sacrifices they made, and decided that they did not want such a life for themselves. Even women who achieved moderate success in a science career opted not to take on the heavier responsibilities of a higher-level career position so that they could have both a career and a family life. As a result, though women's numbers in science increased, women often remained in low-level positions, unable to change the status quo for those women who followed. Thus, women's failure to progress in their careers often related to both conscious and unconscious reactions to the sexism found in much of science.

The Feminist Critique of Science

Feminists generally agree that in western society gender is a highly significant factor dictating human behavior, the nature of human relationships, and the formal and informal laws by which societies live. Beginning in the 1960s, the second wave of the feminist movement explored how patriarchal societies disadvantaged women, both to reveal and correct inequalities between the sexes. Just as the movement was never monolithic despite the spirit of cooperation between

Ruth Hubbard (1924–)

Born in Vienna, Austria, Ruth Hubbard immigrated to the United States in 1938 to escape Nazism. Her parents were both doctors and influenced Ruth in considering medicine as a career. However, she graduated from Radcliffe College in 1944 with a degree in biochemical sciences and worked in clinical laboratories until deciding to return to Radcliffe to work on her Ph.D. in biology, which she received in 1950. As a biologist, Hubbard has conducted important research on the biochemistry and photochemistry of vision in vertebrates and invertebrates. She is well published in the field. In 1967 she won the Paul Karrer Medal, along with her husband, George Wald, for their work in this area. She went to Harvard University as a research fellow in 1953, becoming a research associate, lecturer, and finally a professor in 1974. She was the first woman to be awarded tenure in biology at Harvard. Since 1990 she has been professor emerita of biology at Harvard.

(Courtesy of the National Library of Medicine)

Concurrent with her academic career, Hubbard has written on issues of women's health and on the sociology of science, editing and authoring numerous books on the subject. She argues against the notion that human beings are nothing more than their genes and works with the Council for Responsible Genetics. According to the Women in Technology International website, she is perhaps most well-known for her "brilliant and courageous challenges to colleagues who promote sociobiology." She has worked to both mentor women scientists in their careers and encourage her colleagues to think critically about science and gender issues. Her most recent works in this area include *Profitable Promises: Essays on Women, Science, and Health* (1995); *Exploding the Gene Myth* (1993), with Elijah Wald; and *The Politics of Women's Biology* (1990).

women, so too did historians and philosophers of science often disagree about the patterns of injustice against women in the science professions. Issues such as whether women were different from men on biological or cultural grounds, whether greater numbers of women in science alone could make a difference, and whether women would actually change the very practice of science because they were women, were debated. According to scientist and feminist Evelyn Fox Keller (1936–), the revealing of injustices against women scientists caused feminists in the 1970s to believe that they could "turn [their] disciplines upside down—perhaps, even change the world. . . . In other words, it was a pretty heady time" (Keller 1995, 29).

Feminists have generally depicted scientists as colluding, consciously or unconsciously, in the production and reproduction of social and cultural ideas about women's and men's place in the world. Sociobiology, the study of the evolution of animal behavior that some sociobiologists have extended to human behavior, is a case in point. Sociobiologists argue that animal behavior, like physical traits, evolved genetically. Although they typically acknowledge some role for environment to play in human development, they assign a minimal role to external forces. Instead, sociobiologists have emphasized the primacy of the genes. E. O. Wilson, a prominent sociobiologist, wrote in his 1978 work *On Human Nature*, "It pays males to be aggressive, hasty, fickle, and undiscriminating. In theory it is more profitable for females to be coy, to hold back until they can identify males with the best genes. . . . Human beings obey this biological principle faithfully." Given these stereotypes, feminists have—not surprisingly—critiqued sociobiology's attempts to explain as natural, and thus justify, distinctive and even bipolar gender roles in human society. Sociobiologists argue that females are choosy and prefer marital fidelity because these characteristics help to ensure that they reproduce their genes. Men are promiscuous and seek adulterous relationships, according to sociobiologists, even committing rape to ensure their genes are transmitted, reflecting their short-term commitment to the reproductive process.

Feminist scientists, such as Ruth Bleier in *Science and Gender: A Critique of Biology and Its Theories on Women* (1984) and Ruth Hubbard in "The Emperor Doesn't Wear Any Clothes: The Impact of Feminism on Biology" (1981), have argued that sociobiology is flawed. Feminists asserted that sociobiologists, like many scientists in the past, have distorted scientific theories to confirm the status quo of gender roles in the face of feminist contentions about them. Sociobiologists have pointed to the difference in size of male sperm and female eggs in order to support their argument that men and women have different reproductive strategies. The small size and limited energy required by men to produce sperm and inseminate a female compared to the large size and greater energy required by women to nurture eggs reflects the lesser and greater commitment that nature requires of men and women in reproduction. Sociobiologists have also examined animal behavior, evidence gathered from present-day hunter-gatherer societies, and findings in the prehistoric record and applied them to explain human behavior. They argue that if animals, prehistoric humans, and present-day hunter-gatherer societies exhibit certain gendered characteristics, these characteristics must be innate and must serve some evolutionary purpose. Feminists contend, however, that sociobiologists begin with the assumption that their white, male, capitalist viewpoint is a universal perspective. They assume, for instance, that all males are innately aggressive. They then search for evidence of aggressive male behavior in animals and in prehistoric or present-day cultures, choosing animal and human groups that exhibit such behavior and rejecting as subjects for study those that do not. They use their findings to prove that male aggression is innate as it exists in our primate ancestors and in male humans. Feminists argue that sociobiologists' logic is flawed, however, because they start with the assumption that male aggression is innate and ignore examples of primate groups that exhibit female aggression or domination.

Feminists also assert that sociobiologists impose sociological terms such as rape and harems onto animal behaviors, anthropomorphizing animal behavior and again, in a circular argument, asserting that the existence of so-called rape in nature proves that rape is a natural compulsion of all males. Sociobiologists thus seem to deny that rape in human sociological terms is a misogynistic, violent act by men against women and seem to justify the behavior. Emphasizing innate human behavior conflicts with scholarly approaches that insist even scientific knowledge is in some way socially constructed. Feminist critiques of sociobiology demonstrate that science is not value-neutral and that knowledge is gained within a particular standpoint and outcomes are influenced by social factors. Scientific theories can no longer be taken for granted as representing one true reality, but must instead be viewed as a particular reality constructed within a particular social milieu and therefore shaped by that context.

Some feminists caution, however, that the critique of science can lead down the slippery slope to cultural relativism, where objectivity loses all meaning and reality is entirely dependent on the subjectivity of the individual. Evelyn Fox Keller has argued that to avoid this end, objectivity needs to be redefined rather than rejected outright. Scientists need to be conscious that objectivity is affected by what they bring to it, such as individual idiosyncrasies, cultural baggage, and personal needs and desires. Such acknowledgment, she argues, would lead scientists to a conscious recognition of the nature and purpose of science, thus working as a corrective to the connection between objectivity, masculinity, and power and domination. Scientists could consciously redirect their objectivity, linking it instead with ideas of "conversing with nature" or interactionist, organismic, and nonhierarchical science (Keller 1992, 599–600).

That such a transformation is possible, according to historian of science Marianne van den Wijngaard, can be seen in the contested scientific theories of sex hormones and their impact on the male and female brains. In the 1950s, scientists introduced "organization theory," which stipulated that sex hormones had an impact on brain development. According to this theory, male hormones, androgens, define male sexual behavior in animals such as "mounting," climbing onto the backs of other animals. The absence of androgens resulted in the development of female sexual behavior in animals, defined as "lordosis," the arching of the back. This theory, developed from experiments with guinea pigs, was quickly extrapolated to other mating behaviors, and the research implications were extended to humans: "masculinity was, apparently, translated as presence—of the male hormones, resulting in sexual initiative, action, intelligence, and career-orientation; and femininity was translated as absence—of the male hormones, resulting in sexual receptivity, passivity, and preoccupation with nice clothes and motherhood" (van den Wijngaard 1997, 38–39). Female sexual behavior was not researched in its own context, but only in relation to the absence of male hormones or the receptivity of the female to the male. These studies reproduced stereotypical gender norms while the science produced a biological explanation for these norms. However, van den Wijngaard recovered earlier research from the 1930s that had already noted that both androgens and estrogens (female hormones) influenced brain differentiation and behavior. In these studies the female state was considered the basic condition, and the male condition the modification. A few researchers in the 1970s argued for a "conversion theory" that suggested that estrogens played a role in mediating the actions of androgens in the brain. Why were such theories largely ignored in the 1950s through to the 1980s by the majority of biologists working in this area? Van den Wijngaard argues that the cultural presence of the dichotomy of masculinity and femininity and the desire of scientists to maintain this duality caused scientists to be readily swayed by the organization theory. Though more recent scientists have called for the development of more complex theories—theories that encompass biological, behavioral, and environmental factors—the majority of biologists still support the organization theory. However, the existence of competing theories suggests the possibility that alternative ways of understanding brain differentiation can be established when scientists resist the influence of the dichotomies of male/female, nature/nurture, and mind/body that permeate western culture.

The feminist critique of science has resulted in some positive changes for women in western society. Issues of women's physical health are given more attention as researchers are encouraged to consider both male and female bodies instead of basing their research on the male as the universal human form. Women's bodies have been both under- and over-researched. Diseases most studied in women are those that are specific to women's sex or gender; other more general health concerns, such as heart disease, have been under-studied in women. Researchers have long made the incorrect assumption that data collected for men can be automatically applied to women without further investigation. However, during the 1990s efforts were made, including the passing of governmental legislation in the United States, to ensure that women are part of all kinds of health and drug studies, and that they do not suffer overtreatment in obstetrics and gynecological contexts. But even here, critics argue that women are now receiving too many funding dollars and that their higher life expectancy compared with men indicates that the health care they already receive is more than sufficient.

Nonetheless, feminists argue that spending more money on women's health does not necessarily result in best practices. Feminists have gone on to criticize the "biomedical model" of health care and have argued that a more successful "community," "social," or "ecosocial model" would benefit women greatly. Considering social and environmental hazards and supports that affect health beyond disease or the body's biochemical processes, they argue, would address the broader concerns in women's health. While acknowledging progress, feminists warn their audiences not to decrease their vigilance. One concern focuses on reproductive technologies that give women choices about whether to reproduce, noting that this technology also pressures women to produce genetically perfect children or to abort those not considered perfect.

Feminists' continued vigilance about addressing women's place in science and thus in the larger social world is underscored by the advent of the so-called science wars of the 1990s. The majority of scientists agree that there should be more women in science and that women's questions in science need to be addressed. The idea that there can be or ought to be a feminist science, or indeed that the science that is practiced today and the knowledge that is acquired by it is in any way "masculine" or tainted by the politics of gender, is an anathema to some scientists. The strongest indictment of feminist epistemology of the sciences comes from Paul Gross and Norman Levitt's *Higher Superstition: The Academic Left and Its Quarrels with Science* (1994). The book serves as an attack on the authors' colleagues who believe that the scientific method not only constructs knowledge but is itself socially constructed. Gross and Levitt believe that science methodology is self-correcting and therefore cannot be affected by the social and cultural world of its practitioners. Within the group they term the "academic left," those who share a "sense of injury, resentment, and indignation against modern science," Gross and Levitt include the feminists who, they patronizingly argue, do not know what they are talking about and have got it all wrong (5). Gross and Levitt acknowledge that the system of science has been "unfair and exclusionary" to women and that feminists have a right to "complain," but they argue that this state of affairs is "largely vestigial in the universities" (107, 110). However, they are outraged that the universities have rewarded feminists for their attempts to refashion the epistemology of science. They are dismayed by the construal that objectivity in science is an impossibility and that

Evelyn Fox Keller (1936–)

Evelyn Fox Keller works as both a scientist and as a feminist, critiquing the nature of the science discipline in which she continues to work. She is one of three children born to Russian Jewish immigrants to the United States. She discovered her interest in physics while writing her English essays on the subject to improve her grade. Although she had initially been interested in psychoanalysis, she fell in love with theoretical physics at Brandeis University and wrote her senior thesis on the physicist Richard Feynman.

(Photograph by Marleen Wynants)

After graduating in 1957 she won a scholarship from the National Science Foundation and attended graduate school at Harvard. Though she had experienced a supportive environment at Brandeis, she was shocked to find herself ostracized by professors and peers alike at Harvard for being a woman who wanted to be a theoretical physicist. A summer spent at Cold Spring Harbor, a popular research laboratory, with her biologist brother, Maurice, encouraged her to switch her attention to molecular biology. She obtained her doctorate in theoretical physics in 1963 with her thesis entitled "Photoinactivation and the Expression of Genetic Information in Bateriophage-Lambda." In 1964 she was working as a theoretical physicist teaching night school at New York University, where she met and married Joseph B. Keller, a math-

the knowledge obtained by modern science is shaded by the masculine perspective, initially birthed with the scientific method. Gross and Levitt point particularly to feminist analysis of the language and metaphors of science as an example of their claims that the feminist critique of science is weak and superficial. What difference does it make, they argue, if mathematics problems are couched in stereotypical gender situations or if descriptions of sperm and ova are characterized with stereotypical gender characteristics? They assert that this language is utilized to convey scientific fact and theories to a general audience, but that scientists do not use this emotionally and often sexually charged or gender-biased language in their scientific papers. What really matters, they argue, is that the knowledge obtained through mathematics and the sciences is correct, proven by scientific methods to which all scientists strictly adhere. They also accuse feminist critics of critiquing the softer biological sciences, yet avoiding discussion of the so-called hard sciences, such as physics. They argue that the epistemological basis of a science such as physics cannot possibly have had any ematician; they had two children within two years. During this time she began work in the field of mathematical biology.

In the early 1970s, riding on the cusp of the feminist movement, she underwent psychoanalysis and explored her inner conflict between being a woman and a scientist. She began to think about the relationship of women and science more broadly. She utilized the tools of mathematical biology to analyze data on women scientists at Stanford University and developed an epidemiological model that highlighted the extensive attrition rate of women scientists. She presented these findings at a lecture series on mathematical biology and by 1974 found herself teaching a course in women's studies at the State University of New York–Purchase. In 1977 she wrote an article about her treatment as a woman scientist in graduate school. The time was ripe for asking the sort of questions she was asking, and she became a prominent feminist critic of science, publishing one of her first articles on the subject of gender and science in 1978. She was also beginning to interview Barbara McClintock for her widely read biography of McClintock, *A Feeling for the Organism*, published in 1982.

Keller has maintained her belief in the value of the scientific endeavor and continues to publish scientific works while at the same time exploring the masculine orientation of the field and attempting to persuade others of the need for change. Her work on McClintock led her to write other groundbreaking works: *Reflections on Gender and Science* (1985) and *Secrets of Life, Secrets of Death* (1992). She is currently a professor of science studies at Massachusetts Institute of Technology and has most recently turned her attention back to her scientific work with her *Refiguring Life: Metaphors of Twentieth Century Biology* (1995) and *The Century of the Gene* (2000), which argue against the idea that genes alone are responsible for biological change.

impact on the mass of the population's social and cultural paradigms, as feminists postulate, because the majority of people know nothing about physics.

These concerns raised by the "science wars" may seem at first sight to have some legitimacy, notes historian Londa Schiebinger, especially when they are raised by women who themselves claim to be feminists. But generally speaking they all, perhaps purposely or perhaps unconsciously, miss the point. As Schiebinger has noted, "language shapes even as it articulates thought" (*Has Feminism Changed Science?* 1999, 147). Scientists cannot help but draw upon their own cultural ideas and ideals in the language they use to express their scientific ideas. Language is enmeshed with cultural stereotypes that are then transferred into our scientific way of seeing the natural world. In turn, science, supposedly objective in its determinations, reads back onto our culture scientific ideas permeated with the ideological norms of a given time and place in history. Thus, even the general population, whose knowledge of biology or physics may be slim at best, will nevertheless find their culture and society permeated by gender norms that are in turn legitimized by science. Feminists believe that an analysis of language reveals the cultural shaping of science and therefore that their attempts to create a conscious awareness in both scientific and lay audiences of the impact of culture on science, and science on culture, are of the utmost importance.

Evelyn Fox Keller drives home the importance of men and women learning from one another in *Reflections on Gender and Science* (1985) in her examination of the inability of scientists to perceive contemporary physics outside of the traditional Newtonian framework. Quantum mechanics and relativity theory bring into question any absolute and objective qualities of scientific practice and method. Yet, she argues, scientists still persist in trying to fit concepts such as wave particle duality, complementarity, and uncertainty into a classical physics framework of "objective, material reality" (145). She points out that such an attitude "militates against the acceptance of a more realistic, more mature, and more humble relation to the world in which the boundaries between subject and object are acknowledged to be never quite rigid and in which knowledge of any sort is never quite total" (148–149). Both male and female scientists need to overcome this traditional way of thinking, and if they do, according to Keller, a "more mature" science will result.

The rise of women's studies and gender studies programs at universities since the 1970s have been seen by feminists as integral to the understanding and recognition of gender bias in the sciences. Ironically, however, these programs have sometimes been responsible for segregating women's issues within a feminized discipline. Consequently, gender analysis has not always had an impact on the science disciplines within the university. In other places gender analysis of science remains an academic exercise rather than leading to research that has a concrete impact on policy. Mineke Bosch has demonstrated that in the Netherlands, the strength of women's and gender studies has resulted in a divide between historians and philosophers of women and gender in science, and scientists. In an era of cutbacks, women scientists believe that they have suffered at the expense of the well-established and flourishing women's studies departments. These programs allow universities to save face as they are forced to dismiss junior science faculty first, where the majority of women scientists are currently situated, while pointing to the numbers of women remaining in women and gender studies departments. Yet, Ilse Costas (2002) has noted that in Germany, the absence, until recently, of a strong feminist movement and institutionalized gender studies has resulted in an antifeminist faculty that has kept women almost totally out of the university and thus away from science careers in academe.

Despite the difficulties feminists have faced in their critique of science, they have nevertheless drawn attention to the historical construction of science by men. The feminist critique of science reveals how masculine science constructs women as objects of science practice and as incapable of participating in the scientific enterprise. In the feminists' view, science has been a powerful tool of patriarchal oppression and continues to be so today. Feminists have seen their critique of science as a first step toward change. Recognizing that masculine science has been constructed to produce results that suit the cultural needs and desires of men may result in a reassessment of the nature, use, and purpose of science in the western world. The majority of feminists do not want to banish science, rather they hope that scientific practice and methodology can be reshaped to benefit the whole of humanity.

Women's Success in Science

Despite the impact of the feminist movement and the feminist critique of science, women have still struggled to find a prominent place in the scientific community since World War II. Berenice Carroll (1990) has argued that, in part, women continue to remain invisible and marginalized in science because "among the various techniques of depreciation and dismissal of the work of women as intellectuals and scholars, one of the most prevalent has been denial of its 'originality'" (136). Women are rarely noted as founding "fathers" or discoverers. Carroll goes on to argue that the term *originality* is problematic, as thinkers have always drawn and built upon others' ideas and inventions to create their own, combining past creations to produce new combinations and syntheses.

As researchers put women back into the history of science by "collecting" women scientists, the question of whom to include comes to the fore. There have been attempts to escape a "great women of science" history akin to that of the great men in history/science and to break down the hierarchy of ideas, the "politics of originality," that often devalues women's work in science. Carolyn Rasmussen in *On the Edge of Discovery* (1993) provides the biographies of six Australian women in the physical sciences and notes: "These women were not all high fliers in the sense usually understood in science, which often seems to rate only a research career as worthy of public record, but all made necessary and valuable contributions to Australian science and science teaching" (108). She highlights the life and work of these women in the post–World War II era, emphasizing the enjoyment, self-fulfillment, and accomplishments they attained in science as well as the barriers they faced—placing engineer Beth Coldicutt, biometrician Mildred Barnard Prentice, physicist Jean Laby, chemist Gertrude Rubinstein, secondary school principal Flora Dickson, and chemist Eva Nelson alongside one another. In their more expansive biographical survey of eighty-

Valentina Tereshkova, Russian cosmonaut and the first woman in space, 1963 (Library of Congress)

eight women, the authors of *Journeys of Women in Science and Engineering* (Susan A. Ambrose et al. 1997) subtitle their work *No Universal Constants*, drawing attention to the diversity of women scientists' lives, scientific interests, and accomplishments. They note that they particularly want "to focus in this book on how women define success for themselves" (xvi). In choosing whom to include in her work *Nobel Prize Women in Science* (1993), Sharon Bertsch McGrayne includes not only women who actually won the Nobel Prize but also women who "played a critical role in discoveries that won a Nobel for someone else" (3). Marilyn Ogilvie and Joy Harvey's *Biographical Dictionary of Women in Science* (2000) includes 2,500 women, and again their focus was to go "beyond the now well-known or award-winning women to others who have remained obscure, even though, as we learn here, they were often the mainstays of their government bureaus, colleges, nonprofit institutions, local or regional groups, and specialties. So varied have been their accomplishments that reading about them broadens one's notion of what constitutes a contribution to science—there is a lot more to science than winning the Nobel Prize" (vii). G. Kass-Simon and Patricia Farnes would agree with this statement as they note in the introduction to their edited *Women of Science: Righting the Record* (1990) that "the inclusion of minor work is important in dispelling the mistaken notion that women's scientific work was the extraordinary exception rather than simply the rarer occurrence" (xiii).

Since World War II, women have been found in every area of science and in every conceivable employment venue. The difficulties faced by women in universities led many women scientists to seek employment outside of faculty research and teaching positions. Women scientists worked in nonprofit institutions including research institutes, hospitals, museums, gardens, zoos, scientific societies, journals, and foundations, as researchers, assistants, curators, editors, librarians, and clerical workers. Women also worked in corporations' laboratories, libraries, and customer service departments as researchers, economists, bibliographers, writers, journalists, and secretaries, and occasionally as managers and directors. The U.S. government, in an effort to set an example as an employer in the 1950s and 1960s, hired women scientists in various departments including grant administration, drug regulation, defense, health, education, welfare, atomic energy, and space flight, although few women were included in the latter two. A few women even became self-employed scientists publishing bestsellers, textbooks, and encyclopedias, working as plant doctors, routine lab testers, or even wildlife filmmakers.

The majority of women remained ghettoized in areas of "women's work," but some climbed beyond the glass ceiling and accomplished important, albeit long unacknowledged, work in their fields. Icie Macy Hoobler (1892–1984) researched the chemistry of mother's milk to understand its composition as the director of research for the Merrill-Palmer School in Detroit, Michigan, and director of research of the Children's Fund of Michigan. Belle Benchley (1882–1973) began as a bookkeeper at the San Diego Zoo in 1925 and worked there for decades, eventually becoming director emeritus for her success in building up the zoo's collection of monkeys. Mary Magill, working at the Chemical Abstracts Services Inc., from 1933 to 1971, held the challenging position of index editor for organic chemistry. Companies such as the Polaroid Corporation and IBM employed significant numbers of women scientists, although IBM initially recruited women as part of its attempt to make its computers more user-friendly. Meanwhile in government, women were involved in grant administration for such bodies as the National Institutes of Health and the National Science Foundation. Few women made it to the level of program directors until the late 1960s; the majority worked as aides and assistants. However, Margaret C. Green, a developmental biologist, was a program director from 1953 to 1955, and Helen Sawyer Hogg (1905–1993), an astronomer, was a program director from 1955 to 1956. Though women would not be trained as astronauts in the United States until the advent of equal opportunity, resulting in Sally Ride's (1951–) 1983 trip into space, in the Soviet Union, Valentina Tereshkova (1937–) had traveled into space in 1963. Women entered science when and where they could and often achieved successful and rewarding careers in various fields. Many remain considered as "lesser figures," although recent works in the history of women scientists are recognizing the importance of their work.

Women continue to face barriers to their practice of science, but the increasing acknowledgment of women scientists throughout the western world demonstrates that women attain scientific successes and the accompanying accolades for their work. A recent collection of interviews with Australian women scientists, titled *Profiles: Australian Women Scientists* (Bhathal 1999), brings together, for the first time, successful women scientists in that country discussing the joys, successes, and challenges they have experienced in their scientific careers and recognizing and celebrating the diversity of their achievements. Although the fact that the majority of women still work in the biological and medical sciences is reflected in the fact that twelve of the sixteen interviewees' careers are in these fields, nevertheless all the women interviewed demonstrate that women contribute advances in their field and can achieve positions of power and authority. Maria Skyllas-Kazacos, an industrial chemist, is the discoverer of the revolutionary vanadium redox battery. Plant biochemist Jan Anderson provided the first evidence that two light reactions were involved in the process of photosynthesis. She has recently been elected a fellow of the Royal Society of London. Ann Woolcock, a medical scientist, demonstrated that diet was an important factor in causing asthma in children. Her plan for asthma management has been copied by physicians in several other countries. Adrienne Clarke, a plant biologist, and her research team uncovered the molecular basis of self-incompatibility, that is, how plants can recognize and reject their own pollen to prevent inbreeding. She became the first and only woman chair of the Council of Scientific and Industrial Research Organization (the largest such organization in Australia) from 1991 to 1996 and was elected international president of the Plant Molecular Biology Society.

Recognizing "exceptional" women scientists though, as well as the everyday workers in science, is still important. Even when women accomplish osten-

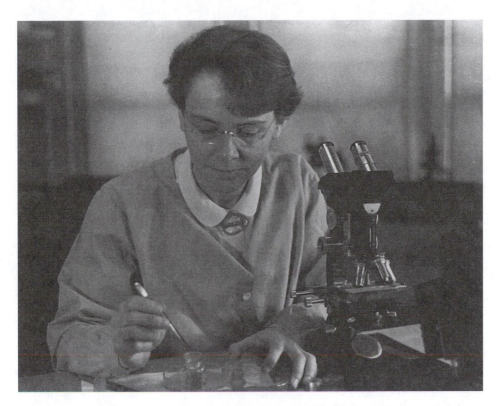

Barbara McClintock, 1947 (Courtesy of the Barbara McClintock Papers, American Philosophical Society)

sibly "original" work, their struggles do not seem to diminish accordingly, although some do eventually achieve recognition for their research. The life and work of geneticist Barbara McClintock (1904–1992) is a case in point. Born in Hartford, Connecticut, she was considered by her family to be a rather masculine child, independent and solitary. In 1919 she began attending Cornell University to study science; she obtained her bachelor of science in 1923, her master's in 1925, and her Ph.D. in science in 1927. She flourished in the academic community during these years and continued her research with fellowships at Cornell, the University of Missouri, and the California Institute of Technology.

Initially her research work was well received and her career got off to a good start, working in the field of genetics and supported by some of the most prominent male geneticists, such as T. H. Morgan and Lewis Stadler. During this time she studied characteristics on the same chromosome that are inherited together, observed the structure responsible for reforming the nucleolus after late prophase, and studied the ring chromosome. Though her colleagues recognized her talent, they were apprehensive about female scientists. However, Stadler managed to convince the faculty at the University of Missouri, where he

S. Jocelyn Bell Burnell (The Open University, courtesy AIP Emilio Segre Visual Archives)

worked, to offer her an assistant professorship in 1936, which she accepted. But the position did not work out, and she left in 1941. Nevertheless, in 1944 she was the third woman to be elected to the National Academy of Sciences, and in 1945 she became president of the Genetics Society of America.

The 1940s were McClintock's most fertile period of research. During this time she began to explore an activity in corn that has become known as transposition, the seeming ability of pieces of chromosomes to move or "jump" at will. Her scientific method during this time involved painstaking experimentation with corn plants. So detailed did her experimental work become that she professed to know each plant individually and claimed that her close association with her plants generated a respect for them that encouraged her to listen to what the experiment told her to do. Her ability to keep her mind open to what the plants, through her experiments, were telling her allowed her to respect and accept complexity in her answers and to draw conclusions that were different from what other scientists were seeing. She conducted this work at the Carnegie Institution of Washington at Cold Spring Harbor and delivered two papers on this work at the Cold Harbor Symposium in 1951 and 1956. On both occasions her audience was unable to understand her research work. They refused to believe the implications of her work, that an organism's genetic structure could respond to its individual needs. This work further ostracized her from the scientific community of her peers, and McClintock became something of a rogue scientist following her own path of research. Not until the 1970s, as the complexity of molecular biology increased, did other scientists begin to understand and support McClintock's research. In 1983, the importance of McClintock's earlier work on transposition was formally recognized when she received the Nobel Prize in physiology or medicine.

Although there have been many hurdles along the way, women scientists have come into their own as a group in the twentieth century. Women are working in science in ever-greater numbers and continue to accomplish important work in their respective fields. The fact that women scientists have to continue to struggle against gender norms, work in chilly social climates, and are continually passed over for promotions and awards highlights that still more needs to be done to enable women to be fully accepted. Nevertheless, women's scientific achievements in the twentieth century cannot be questioned. Women who enjoy science can make successful careers for themselves and will, hopefully, encourage other women to do so as well.

Conclusion

In his epilogue to *The Double Helix*, James Watson reflected that he had been unfair in his assessment of Rosalind Franklin in the early days of their acquaintance and noted the superb quality of her work both at that time and up until her death in 1958. He noted that he had realized "years too late the struggles that the intelligent woman faces to be accepted by a scientific world which often regards women as mere diversions from serious thinking" (1968, 226). Another woman scientist who, like Franklin, reflects the tensions between women's achievements and challenges in science is astronomer and physicist Jocelyn Bell Burnell (1943–). Some scientists believe she too ought to have received the Nobel Prize for physics but did not because of her status and position within the scientific community.

After studying physics at the University of Glasgow in Scotland and struggling with feelings that she did not belong in physics classes because she was a woman, she nevertheless was awarded a bachelor of physics degree with honors in 1965. She then moved on to Cambridge for graduate work in astronomy, investigating radio waves. Working as an assistant to the radioastronomer Anthony Hewish, who was building a large radio telescope, she first helped him construct the telescope and then spent hours operating and analyzing the sheets of data the telescope created. After two months Burnell began to notice some unusual readings and apprised Hewish that she had become aware of a series of regular pulses. At first she only noticed them coming from one part of the sky, although later she recognized that they came from other parts of the sky as well. Hewish believed that this was a significant enough discovery to publish in the British journal *Nature*. The article, with Burnell as the second of five authors, was published on February 9, 1968, announcing the discovery of "pulsars," likely neutron stars that have run out of nuclear fuel and have fallen in on themselves. As they collapse they spin faster and faster, producing magnetic fields that produce electric fields that in turn generate radio waves. Until Burnell's detection, neutron stars had only been considered theoretically possible. The discovery took the astronomical world by storm.

In 1973 Hewish and Burnell received the Franklin Institute of Philadelphia's Albert A. Michelson medal, and there was speculation that they would receive the Nobel Prize together. But in 1974 the Nobel Prize committee awarded the prize to Hewish alone. Debate has since ensued among scientists as to whether Burnell, an assistant to Hewish at the time and not a collaborator, should have nevertheless received the prize along with Hewish or even perhaps instead of Hewish. Burnell herself did not believe she should have won the prize, stating that the "Nobel Prizes are based on long-standing research, not on a flash-in-the-pan observation of a research student. The award to me would have debased the prize" (Bertsch McGrayne 1992, 375). She went on to marry and work part-time while raising her child and moving from place to place following her husband's career path in government. Her work in discovering pulsars gave her employment opportunities she would not otherwise have had, and she claimed that she was content with the life she had chosen. Still, a divorce, followed by her first full-time job as professor of physics at the Open University in 1991, caused her to reflect on the effects of part-time work: "The problem with part-time work is that it assumes that domestic and child care remain with women, and part-time and lower status jobs have often gone together traditionally. . . . I have some reservations about the emphasis on part-time jobs" (377). Perhaps there is some satisfaction that she was the third woman to become a professor of physics in Britain.

As historian of science Margaret Rossiter has pointed out in her second volume of *Women Scientists in America* (1995), tens of thousands of women have become scientists and engineers in the second half of the twentieth century, yet their invisibility has continued. Often hidden from view of both the general public and their male and female colleagues, they remain out of mind. While historians of science have sought to resurrect these women's work and lives and bring them to the attention of scientists and the general public alike, other feminists (many of them scientists themselves) have worked to critique the science discipline for its sexism. Feminist critics and historians of science hope that their efforts will result in change for the working lives of women scientists, bringing women like Franklin and Burnell the respect and recognition they deserve. They also aim, through their study and critique, to improve the lives of all people around the world as they encourage both male and female scientists to ask different questions about the natural world and enable them to conceptualize solutions that do not disadvantage, oppress, or ostracize.

Bibliographic Essay

Numerous books address the challenges and achievements of women scientists in the postwar period. Margaret Rossiter's *Women Scientists in America: Before Affirmative Action, 1940–1972* (1995) stands out as the most all-encompassing work. Rossiter addresses why there were not more women in science in a period in which there was an increasing demand for trained scientists. She demonstrates that there were more women in science than ever before by the 1960s but also examines how this was not reflected in women's status or prestige in the profession. Not until the affirmative action legislation of 1972 was there hope for change for women scientists. Other useful works that both celebrate women scientists' accomplishments and examine the barriers they faced throughout their careers include Sharon Bertsch McGrayne, *Nobel Prize Women in Science: Their Lives, Struggles, and Momentous Discoveries* (1992); and Susan Ambrose et al., *Journeys of Women in Science and Engineering: No Universal Constants* (1997). Other works explore women's firsthand experiences as they attempt to become scientists. These works include accounts written by the women themselves and compilations of thoughts and experiences constructed through collected interviews. They examine women's experiences in the past, both positive and negative, and address the need for change. Examples include Mary Morse, *Women Changing Science: Voices from a Field in Transition* (1995); Angela Pattatucci, *Women in Science: Meeting Career Challenges* (1998); and Ragbir Bhathal, *Profiles: Australian Women Scientists* (1999).

The feminist critique of science literature is extensive and wide-ranging. Two useful starting points are the collections of important articles that cover all areas of feminism and science: Sally Gregory Kohlstedt and Helen Longino (eds.), *Women, Gender, and Science: New Directions*, Osiris 12 (1997); and Janet A. Kourany (ed.), *The Gender of Science* (2002).

Despite the significant amount of work in this field, there are key texts that have been considered particularly important or groundbreaking by feminists working in the field. In the area of the feminist critique of science, the most useful works are Lynda Birke, *Women, Feminism, and Biology: The Feminist Challenge* (1986); Ruth Bleier, *Science and Gender: A Critique of Biology and Its*

Theories on Women (1984); Sandra Harding, *The Science Question in Feminism* (1986); Ruth Hubbard, *The Politics of Women's Biology* (1990); Evelyn Fox Keller, *Reflections on Science and Gender* (1985); Evelyn Fox Keller, *Secrets of Life, Secrets of Death: Essays on Language, Gender, and Science* (1992); and Marianne van den Wijngaard, *Reinventing the Sexes: The Biomedical Construction of Femininity and Masculinity* (1997). The focus of these works is to demonstrate how masculine science has constructed the natural world and the practice of science and scientific knowledge to reflect the gendered norms of western society, privileging men and subordinating women. The majority of these works focus upon critiques of the biological sciences.

Brian Easlea's work may also be categorized as a feminist critique of science. Easlea emphasizes the damaging nature of masculine science to both women and the world. His two major works in this field are *Science and Sexual Oppression: Patriarchy's Confrontation with Woman and Nature* (1981) and *Fathering the Unthinkable: Masculinity, Scientists, and the Nuclear Arms Race* (1983). In these works, Easlea argues that men, oppressed by their insecurity about their masculinity, use their aggressive sexuality and their science as a tool of oppression over women and nature. Men, he argues, seek creative, productive power, yet their traditional masculinity subverts their creative ability, manifesting itself in the destruction of the natural world, such as in the formation of the atomic bomb, resulting in the social, political, and sexual oppression of women and the feminine within themselves. Easlea believes that the modern world of masculine insecurity, material scarcity, and mounting world problems could be solved if men would only accept a female society. In other words, Easlea calls for a world that is nonpatriarchal, where men participate fully and equally with women in child care and in the political arena, and where an "intellectually and sensuously rich society, characterized by reciprocity, gentleness, and at times joyfulness, coupled with compassionate assistance for people experiencing unavoidable suffering" can exist (1981, xiii). Easlea, along with other feminists who critique science, seems uncertain that such a society will ever come into being.

Works by critics of feminist epistemology took the academic community by storm in the 1990s with works such as Paul R. Gross, Norman Levitt, and Martin W. Lewis, *The Flight from Science and Reason* (1996); and Paul R. Gross and Norman Levitt, *Higher Superstition: The Academic Left and Its Quarrels with Science* (1994). These works, though, have encouraged feminists to respond with a critical examination of the impact of feminism on science. For example, see Sue Rosser, *Biology and Feminism: A Dynamic Interaction* (1992); and Londa Schiebinger, *Has Feminism Changed Science?* (1999).

7

Creating a Future for Women in Science

> We have all been injured, profoundly. We require regeneration, not rebirth, and the possibilities for our reconstitution include the utopian dream of the hope for a monstrous world without gender.
> —Donna Haraway, "A Cyborg Manifesto"(1991)

Since World War II, opportunities for women appear to have multiplied exponentially. Barriers to the participation of women in science have reportedly been reduced, and stereotypes about women's capabilities have dissolved. Increased access to education via the opening of higher education to women in both Europe and the Americas continued in the postwar period as universities expanded quickly. Women demanded admission to disciplines traditionally regarded as male and were, in the 1970s and 1980s, being encouraged to enter these fields by government, industry, and universities. Riding on the euphoria of this feminist movement, many women were convinced that they were capable of participating in the areas of mathematics, engineering, medicine, and the sciences alongside men. Equal opportunity and affirmative/positive action laws, programs that encouraged girls' interest in these fields of study, and women's desire to achieve recognition for their work all supported the public perception that significant change was happening and that more and more women had entered traditionally male disciplines.

Since the 1980s, however, feminists and established women scientists have faced a conundrum. The path to these nontraditional fields of study for women are still scattered with pitfalls and barriers, but women who are established in their fields hesitate to speak in earnest to young women about the challenges they face ahead of them, believing that many women will opt for alternative careers. At the same time, though, they wonder about the fairness of encouraging and enticing women into these fields without warning them about the possible difficulties they will face and discussing with them how to deal with these problems. Established women scientists know both the joy and the anguish they faced in their scientific education, their admission to a career in science, and their attempts to climb the ladder of scientific success. Rather than tell selective stories of their successes, some women scientists and feminists suggest that they share their experiences of the hurdles they faced as women scientists and discuss how they can be overcome.

Many women who have accomplished high achievements in science have been held up for emulation. They often hold themselves up as examples of what can be done, talking to young women about how they succeeded and sharing their love for science. Meanwhile, some of these same women, in their day-to-day lives, fail as mentors for young women scientists. Believing that the only way for women to succeed in science is to practice science like men, only better, working harder, and denying themselves a life outside of their career, they drive their protégés to work as hard as they have done. Some women scientists have attempted to change this state of affairs by providing mentoring relationships that respect women's ways of pursuing careers in science and the demands of their personal lives. Either way, women interested in traditionally male careers often face the dilemma that on the one hand they have to be like men to succeed in their profession, yet on the other hand they are often castigated for not fulfilling the female roles western societies stereotypically expect of them.

Many men in science continue to see women as outsiders encroaching on a male domain. Like their historical predecessors, some men fear that women will undermine the status of the profession, waste resources by dropping out of the profession to raise their family, or change the nature of science practice. Such men use both formal and informal methods to discourage women from pursuing a career in science. Sometimes such exclusionary policies are represented as an attempt to provide fair and equal opportunities and challenges for all, an equal playing field on which men and women alike are ostensibly offered the same advantages. Of course, such equality is impossible to achieve between men and women whose social responsibilities differ significantly, and thus this "equity" often disadvantages women. At other times, institutions implement policies in a purposefully sexist manner to force women out of science. Either way, women who drop out are considered failures who could not compete on the level playing field and therefore did not deserve to be there. Certainly some men provide excellent role models and mentors for women. Moreover, some men have welcomed feminist demands for change, realizing these reforms will improve their own lives, not just the lives of women. But these positive mentoring circumstances are not available as frequently as is needed, and male scientists do not always have the power to encourage or implement change any more than do female scientists.

Continuing concerns in the 1970s and 1980s about the lack of women in science, especially in the physical sciences and engineering, has led to programs by foundations established to encourage and even entice women to study in these areas. Some researchers believe that the mere force of numbers will change the status quo. Increasingly, however, researchers have questioned whether involving larger numbers of women in science instigates change, because in fields where women's numbers are substantial, they still face the familiar prejudices and barriers to their success in science. Instead, feminists and women scientists in general are beginning to argue that schools need to teach science differently and posit that there needs to be room for, and respect for, men and women scientists who wish to balance their personal and professional lives. They argue that the nature of laboratory practice needs to change to accommodate different work styles and that women need to be enabled to break through the "glass ceiling" into influential decision-making positions. Possibly, they argue, even the nature of science itself needs to be reexamined and reassessed. Only in these ways will women ever be able to find a place for themselves that is made for them, rather than having to accommodate themselves into an ill-fitting framework that simply recognizes women's life cycle, or their ways of exploring, thinking, and learning about nature.

Critical Mass: More Women in Science?

In the 1970s, feminists believed that if enough women had the opportunity to succeed in science alongside men, and that if governments and institutions enacted policies to ensure fair admission and acceptance of women in these fields, women would eventually achieve parity with men in science. Though in the first half of the twentieth century women had been present in university science and in industry, their numbers remained small. This was not the result of women's disinterest or inability but of the unwillingness of the men in charge to make room for them. Feminists believed that the numbers of women in science had to reach a critical mass before they could effect changes in the scientific research life and workplace that would accommodate ever-increasing numbers of women. "Critical mass," or strong minority, is generally agreed to be 15 percent. According to the U.S. National Science Foundation's (NSF) survey of earned doctorates, many fields including biology, chemistry, geosciences, and even mathematics reached and exceeded the critical mass of women between 1966 and 1997. Forty percent of Ph.D.'s in biology during the 1990s were awarded to women; during the same period women took 27 percent of the Ph.D.'s in chemistry, 22 percent in geosciences, and 19 percent in mathematics. The participation of women in physics and engineering, however, still remains below

Two hundred sixty students from 140 colleges, some of whom are pictured here, attended a three-day symposium on women in science and engineering at the Massachusetts Institute of Technology in October 1964, at which sociologist Alice S. Rossi posed the question, "Women in Science: Why So Few?" (Photograph reprinted with permission of the MIT Museum)

this level—at 12 and 11 percent, respectively. The NSF website (http://www. nsf.gov) indicates that this trend of ever-increasing numbers of women continues apace, although the number of women studying and earning degrees in science is still for the most part substantially lower than that of men.

International figures reveal similar trends, indicating the rising numbers of women students in science. In the German Democratic Republic in 1987, 53 percent of higher education students in the natural sciences were women, and 27 percent of students in engineering and technical sciences were women. Twice as many women in West Germany took doctorates in engineering and science in 1986 as in 1976. In Australia in 1992, just over 40 percent of postsecondary science students were women. In France in 2000, there were equal numbers of men and women majoring in biology and chemistry up to the fourth year of university.

In some countries women scientists are also represented in the workplace. Although women are more likely to be professors of the social or biological sciences, women can be found in all areas of science. In France, 9 percent of women professors taught science in 1997. Within this 9 percent, 30 percent were in the life sciences, 10 percent were in nuclear physics, and 17 percent in chemical sciences. In Italy, 10 percent of women professors were in biomedicine, 22 percent of whom were in biology, 5 percent in chemistry and physics, and 27 percent in mathematics. The percentage of women in the natural sciences professoriat in Scandinavia in 1996 were as follows: Denmark 1 percent, Finland 8 percent, Norway 7 percent, and Sweden 5 percent. In the UK in 1996/7, 6 percent of women professors were in the biological sciences, 3 percent in mathematics, and 1 percent in physics. Yet at the lower level of lecturer, 26 percent of biologists were women, 18 percent in mathematics, and 9 percent in physics.

These generally positive figures, however, belie a more disturbing truth about women's participation in science. They do not reveal how many women did not succeed, who dropped out of science at lower educational levels due to the lack of support for women's interest in science from either teachers or the education system. The figures do not reveal how many women decided never to embark on a scientific career at all, or the number of women, compared to men, who remain in the lower echelons of science and industry, unable to break through the glass ceiling to obtain decision-making positions that would enable women to change the status quo. They do not reveal that although there may be a critical mass of women in a field, in any particular department they may still be one woman among many men in an institution where she is unable to voice her problems or concerns for fear that she will be considered unequal to the task of science and lose the respect of her peers. Perhaps most important, these numbers do not reveal the disappointment of the women who came to believe that science is not for them, who feel that their lack of achievement is a personal failing on their own part, or that despite their achievements they have never been fully accepted as equal participants in the scientific endeavor.

Ironically, in some countries the patriarchal tradition has played a positive role when the extended family has helped in caring for the young children of a working woman. Moreover, societies organized along socialist lines, such as Bulgaria, have instituted paid leave for grandmothers to care for grandchildren. Similarly, in Germany it is national policy through social security to support women who wish to have both a career and be mothers. Unlike in other countries where these supports do not exist, German career women tend to have on average the same number of children as their male counterparts. The simple acceptance of working women without the concurrent societal guilt of leaving children to be raised in day-care situations, as is the case in Israel, affords women a greater peace of mind and the ability to satisfy both work and family needs. The balancing of personal lives with careers is possible, if such balancing becomes an accepted practice within the work environment and societal framework. However, the success rates of women in science in these countries has still not been high, suggesting that such balance is only part of the solution to women's increased entry into the science profession.

Wartime opportunities, an opening up of society's views about women's paid work, easier access to education, and the advent of the second-wave feminist movement increased the numbers of women in science. Yet reaching critical mass does not ensure that women will unify to confront the status quo. Henry Etzkowitz, Carol Kemelgor, and Brian Uzzi in their ten-year study (*Athena Unbound: The Advancement of Women in Science and Technology*, 2000) found that women faculty members divide along generational lines. Senior women find themselves more comfortable with the male values to which they have grown accustomed, and junior women desiring to establish new workplace values that recognize the need to balance their personal and professional lives tend to contest the senior womens' perspective (105). Women may also divide along national or racial lines. Even when women do find support with one another, they are often concerned that if they demand attention be paid to women's issues, or they participate in women's support groups, they will be ostracized by their male peers. Though the logic of increased numbers having an impact on women's lives in science is clear, the reality is that "despite accretion of numbers, it is strategic power that really counts" (109).

Equal opportunity laws and ensuing affirmative action policies have had an important impact on women's entry into the science professions. In 1997 the European Union (EU) Treaty of Amsterdam strengthened Europe's commitment to women's equal opportunities; this was further aided by the establishment of the Women and Science Unit in 1998. European countries have been making concerted efforts to increase their numbers of women in science, competing with one another for the best results. The EU itself has set a quota of 40 percent for its women research directorates on internal committees. It also provides funds to member states to cultivate research equity, offering money to countries that establish initiates to increase the numbers of women in science, including special programs in science education and re-entry initiatives, and encourages fair salary, peer review, and hiring practices.

Recently, though, in the United States, there has been a backlash against equal opportunity laws and affirmative action policies. In the 1990s, courts declared affirmative action policies as prejudicial against white males, and there have been calls by the New Right to end affirmative action. This has not yet occurred, but affirmative action procedures were amended in 1995 by a Supreme Court decision to ensure that they would not be used as a quota system and that they not permit unqualified people to obtain positions, create reverse discrimination, or continue after their purposes have been achieved. The questioning of affirmative action has been done in a climate of shrinking funding and amidst cases of scientific fraud and hoaxes that have undermined scientists' credibility with the public. Calls to end affirmative action highlight a growing tendency to once again push women out of the professions, including science, and back into more traditional roles under the guise of economic necessity and, ironically, equal opportunity for all. This reversal may well be conducted under the assumption that obtaining a critical mass of women in science demonstrates the success of affirmative action, emphasizing that it is no longer needed. Numbers alone, however, do not denote the success of women in science. Further study reveals the continued barriers to women in science despite the advances they have achieved and underscores the necessity for change in the way science is taught and practiced to keep women in science, to encourage more to enter, and to enable them to stay and work their way up the ladder of career achievement.

Female-Friendly Science: Educating Women in the Twenty-First Century

Many studies published since the early 1970s have shown that girls are interested in math and science and that their aptitude for such studies equals that of boys in the pre-puberty years. But girls interested in math and science have to overcome the pressures of those who dissuade them from studying in these areas. Parents, teachers, and peers can reinforce stereotypical gender norms that women are naturally not good at math and science and that these subjects are not suitable pursuits for women. A mother simply comforting a daughter that she was "never good at math either" reinforces the acceptability of girls obtaining poor results in the sciences. Peers, particularly in the teenage years, can coerce girls who excel in math and science to purposely fail in order to remain part of the "in" group, or at the very least to hide their "unfeminine" interests in the company of boys. Images created by the media have pervaded the classroom, raising the specter of the male scientist and underlining the absence of women from science, reinforcing the message girls receive about who can do science. Sexism can even pervade the content of science itself. One female engineering undergraduate reported that her male professor suggested the following mnemonic to remember the resistor color codes: "Bad Boys Rape Our Young Girls But Violet Gives Willingly" (Morse 1995, 44). In numerous ways, some subtle and some blatant, women are discouraged from participating in science.

Discouragement has also been institutional. In a 1930s high school in Nottingham, England, all the thirteen year olds were divided into two groups, boys and girls. Without any discussion, the girls were taught biology in a descriptive format, and the boys were permitted to study advanced chemistry and physics

Teacher showing globe to students (Tom Stewart/Corbis)

(Pattatucci 1998, 205). More recent studies show that many girls still experience subtle, often unconscious deterrents from studying math and physical sciences. The toys of young boys tend to encourage manual dexterity, problem solving, and technical interest whereas girls' toys tend to be more nurturing. Young girls are often discouraged by the parents from playing with "boy" toys should they encounter them (likewise, boys are often dissuaded from playing with dolls or in playhouses). Teachers have paid more attention to boys in the classroom than they have to girls, allowing boys to talk more and to engage with the teacher more often. Teachers have also tended to encourage boys to question and to learn from their mistakes while expecting that girls will be quiet and compliant. Girls' scholastic ability in math and science has in many cases faded due to a lack of support and encouragement, reinforcing the belief that women are just not meant to be scientists.

Change has to start somewhere, and during the 1990s feminists called for a new approach to scientific education, one that engages young girls' interests and teaches science in a way that is most conducive to their learning styles. Girls' dislike of, and perceived inability in, science is a learned attitude encouraged by science's apparent contradiction with the preconceived parameters of femininity in western society. A different educational approach to science could change this attitude. By 1994 there had been a 50 percent reduction in the gap between boys' and girls' scores in mathematics, which authorities have attributed to increased exposure of girls to mathematics skills. This suggests that females do not lack a "math gene"; rather, they have been educationally disadvantaged, a rectifiable problem.

One important factor in female-friendly science is that students be educated in science and math in their early years and that teachers be trained to accomplish this task successfully. In addition, pedagogical theorists advise that educators need to be educated to an awareness of the unconscious messages they may send to girls about their participation in science. They also need to be taught about girls' learned fear of science and math that needs to be, and can be, overcome. Teachers need to be aware of the negative stereotypes that are pervasive about women and science and need to counteract them not only in their own attitudes toward girls' participation in science but also by talking about and confronting the reality of these stereotypes head-on in the classroom and address with students how they can be overcome. This approach may also include using gender-neutral language or providing problems to solve in math and science that contradict current stereotypes about men and women's role in society. Girls need to be given the opportunity to learn the skills that boys have acquired through gender-orientated toys. They need to be exposed to female scientists as role models and to information about women's achievements in science. Extracurricular activities, camps, and awards for girls interested in science need to be made available. Girls, no matter what their specific interests, require the positive reinforcement that teachers can provide to establish their own self-esteem and self-confidence. Teachers also need to monitor the negative behavior and attitudes of other children that may cause a student to question her (or his) scientific ability or self-worth.

Teachers need to expose all their students to a variety of learning skills to address the varying needs of students by including collaborative and competitive work, group projects, students as teacher-experts of units of material to encourage independent thinking and problem solving, discussion and lecture formats, and by introducing a variety of assessment techniques. Teachers also need to utilize a variety of teaching aids, such as textbooks and visual aids that recognize both verbally and visually women's participation in science today and throughout history. Such approaches to learning may also benefit boys; studies at the secondary school level indicate that boys do not necessarily enjoy science more than girls but are motivated by social pressure and career aspirations to stay in science when girls may more easily opt out. Finally, teachers need to apprise students of the many and expanding job opportunities that exist in the sciences at the beginning of the twenty-first century and the need for many career professionals such as lawyers, teachers, and ethicists to have an understanding of science. These changes need to take place at all levels of education from kindergarten through to graduate programs.

Perhaps most important, science and math need to be made relevant to girls' lives and interests. Studies show that girls and women are particularly interested in applied science, knowledge that goes beyond the theoretical to incorporate the real, physical, and social world outside the classroom and that addresses the ethical issues or discusses the social and political context of the science being discussed. Allowing students to engage in real-world projects that impact the community in which they live or to participate in some form of activism allows students to see how they can affect the world around them. For example, in the mid-1990s, a midwestern public high school, working with researchers from the University of Michigan, created a project called Foundations of Science in which students could investigate the water quality in a local stream. Students and teachers adopt a stream and spend a term studying "their" stream, writing reports of their analyses of findings, and putting forward suggestions for improving the water quality. Students then present these findings to the local environmental agency, and they are televised on a local station. These kinds of educational programs have resonance beyond a good grade for the individual student, boy or girl. They teach students the importance of scientific knowledge and understanding to the general public and allow students to see the interconnection between scientific knowledge and the community in which they live. This particular project and others like it have interested and encouraged girls in science.

Changes need to be made in educational institutions that reflect the growing female student body and consequent female work force. Many undergraduate and graduate students do not have a positive experience of science during their university-level education, especially those who have been inculcated in programs that nurture women in science or that provide all-women educational opportunities. Etzkowitz, Kemelgor, and Uzzi (2000) found that even after graduating with a doctorate degree, many women only then "realized that they were devoid of professional contacts and networks as they sought post-doctoral fellowships and employment" and thus were alone in a world in which they were not just trying to find work but attempting to determine a career path that would allow them to spend time with their children (98). At both the undergraduate and graduate levels, women can benefit remarkably from having access to support networks for women, study groups, mentoring programs, and the choice of attending all-women tutorials and seminars. Such programs exist at most universities on an extracurricular or ad hoc basis, but experiments in establishing formal programs that consciously address the needs of university women in science have had marked success. Two such projects are the Women in Science and

Engineering Program (WiSE) founded in 1989 at the University of Washington and the Women in Science Project (WISP) at Dartmouth College, New Hampshire, founded in 1990.

WiSE is one of the most successful and broad-reaching programs of its kind. Established to increase the recruitment and retention of women in science and engineering and to create an environment and climate conducive to women's approach to learning, the program has grown to serve 1,300 students on the University of Washington campus and 3,000 students off-campus. WiSE includes tutoring, professional and peer mentoring, advising, undergraduate and graduate retention programs, and an international exchange program. Retention rates at the undergraduate level have grown from 50 to 90 percent between 1990 and 1998. The number of degrees granted to women increased from 15 to 23 percent at the undergraduate level and from 6 to 22 percent at the graduate level. Moreover, the number of female faculty increased from six to twenty-two between 1988 and 1998. These figures are significantly higher than the national percentages.

The Women in Science Project at Dartmouth (WISP) offers as its cornerstone part-time, paid laboratory research internships for first-year women who work directly with faculty in their labs on current research work. The project hopes to harness the first-year female students' enthusiasm for science, rather than waiting until upper-year courses to provide such opportunities. This gives women the opportunity to network with faculty, gives them hands-on experience in the lab, and helps to build the confidence they need to decide to pursue graduate work in science. WISP, similar to WiSE, also includes peer mentoring as part of its program, as well as a women-tutored study room, special seminars for women, lunch conversations with women scientists, and trips to industrial sites. In addition, WISP has a student-created newsletter, distributed electronically to student participants, faculty, and administrators, that advises and educates subscribers about the program. WISP has also initiated faculty development activities to educate faculty about gender issues in science and to discuss approaches for teaching women in science. The founders believe that the equal numbers of male and female participants in the honors thesis program by 1995 was in no small part related to women's experiences gained through the opportunities supplied by WISP. They credit the program with having increased the number of women graduating with a science major to 42 percent of the total Dartmouth science graduates by 1998. They also claim that WISP has had an impact in engineering: 26 percent of engineering graduates were women in 1996, compared with just 15 percent in 1990 (the national average then was 18 percent). The success of such programs on a number of levels, including bringing women into science, appears clear.

Similar work is conducted by organizations such as the American Women in Science Network (AWIS), founded in 1971, and the Women in Engineering Programs and Advocates Network (WEPAN), founded in 1990. Both organizations exist to further opportunities in science and engineering for women at universities and in industry. AWIS is located in Washington, D.C., but has local chapters in forty-two states that include over 5,000 members. WEPAN has 600 individual members representing 200 engineering schools and Fortune 500 companies and nonprofit organizations. Through conferences, educational programs, publications, on-line information resources including job listings, the offering of awards and nominations, and mentoring programs, AWIS and WEPAN foster networking among women and allow women to establish a voice both as individuals and as a group within the organizations and beyond to policy makers and the general public.

Other positive initiatives include government-sponsored programs that encourage and fund women's progress in science through scholarships, fellowships, and training programs. For example, in the early 1990s, Canada's National Research Council announced that it was starting a new program to encourage women's entry into science and engineering. Called the National Research Council Training Program for Women in Science and Engineering, it funds promising women candidates in their undergraduate careers and assigns them work-related positions either in one of their own laboratories or in one of their industrial partners' laboratories. Moreover, increasing numbers of professional conferences on women and science also address the issue of science education and bring awareness about it.

Among faculty, the importance of change at the departmental as well as the university level is paramount. Supporting the hiring of women in science to achieve critical mass, despite the drawbacks of this approach, may be of some help to both junior and senior women faculty. Other positive initiatives include mentoring programs, providing clear rules and regulations about procedures such as program requirements for graduate students and tenure and promotion expectations for faculty, and ensuring that some women achieve positions of power so that they may understand and play a role in departmental and university decisions. But feminists and women scientists argue that the universities themselves have to make a verbal and financial commitment as well to a critical mass of women in science and to other department initiatives that support women's entry into and success in science. In addition, they argue, the universities need to make a commitment to change the institutional structure to accommodate women's needs by implementing regulations that agree to slow the tenure clock for women who want to have children and by valuing mentoring in tenure review to benefit women who often spend considerable time working with and for other women. Some research suggests that women publish less frequently than men, but that their work is more frequently cited, indicating that women take a different approach to research, working more intensely and longer on a project before they publish. Some women argue that the quality rather than the quantity of published work in science needs to be recognized in review of files for tenure and promotion.

Of course, alongside the need for changing education programs for women and science, feminists and women scientists point out that there must be a concurrent change in the working world of science. Research labs need to adapt to accommodate women's lab styles, which means creating nonthreatening, physically safe, cooperative work spaces for women where they are respected members of a research team. Activists hope to change science to reflect the importance of those approaches to science that tend to engage women, including concern for ethics and the social and political outcomes of their work. Western society needs to confer more respect upon essential scientific research positions that are currently low-paid, low-status jobs with little job security because, according to Eisenhart and Finkel in *Women's Science: Learning and Succeeding from the Margins* (1992), these are the positions in science to which women are most often drawn.

Some feminists fear that the attack of the New Right against women's rights and needs, including projects encompassed under the goal of instituting female-friendly science, may be justified by budget constraints. This trend may result in the lack of support for projects and initiatives that advance girls and women's participation in science. Such a backlash would not only cripple, once again, women's contributions to the scientific endeavor, robbing them of the ability to achieve their career goals, but would also reduce the quality of the pool of scientists by denying half the population access to careers in science. Policies like these would undercut the economic efficiency of the nations that pursue them. Feminists consistently reiterate the need for women scientists and science educators to be vigilant to ensure that the positions women have achieved in science thus far are not undermined, and that young girls' opportunities to journey into science are not cut short by the patriarchal agenda of men who do not wish to share these opportunities with them.

Can There Be a Feminist/Feminine Science?

Increasing the numbers of women in science and educating women to science are becoming commonplace discussions among academics, educators, and governments in an effort to advance women's entry into and success in science in the twenty-first century. But the emphasis remains, however, on changing girls rather than on changing science. The issue of whether women's increasing presence in science could change the very nature of science and science practice remains controversial. Feminists have argued that white, middle-class women may in fact be most comfortable reproducing masculine science because they are raised in the same social milieu and trained in the same school programs as their male counterparts. Women scientists range from believing absolutely that women do science differently to believing that they definitely do not. Yet the need to obtain access to and recreate science in their own image is important to feminists who believe, like scientist and feminist Ruth Hubbard, that "at present science is the most respectable legitimator of new realities" (Hubbard 1981, 218). Whereas women writers such as Ursula LeGuin and Marge Piercy have used science fiction to explore and contest present-day scientific practices, feminist philosophers of science increasingly have begun to speculate about the real-world possibilities of a feminist/feminine science that would change the larger world in which that scientific knowledge was produced and utilized. In *Has Feminism Changed Science?* (1999) Londa Schiebinger argues that feminists have had an impact on the culture of science, women's place in the scientific enterprise, and the science produced, although there is still room for further change. She also highlights the fact that the feminist project to change the method and content of science is still ongoing. Schiebinger highlights the necessity and importance of scientists, institutions, and governments using tools of gender analysis in science's projects, priorities, outcomes, language, images, culture, and theories, and in the definition of science itself.

Women's participation in science and medicine has led to attention to issues different from those selected by male scientists. Scholars note, for example, that scientists long ignored the discomfort many women suffer during menstruation, a circumstance that changed when women administrators allocated research grant money. Similarly, scientists have researched contraceptive techniques that interfere with the female reproductive system but only recently have begun to investigate contraceptives that interfere with the more straightforward male reproductive system. Despite significant sponsorship for research on AIDS, medical researchers tended to focus on women as transmitters of the disease to men or to fetuses, conducting very little research by the early 1990s on the effect of AIDS on women themselves. Feminists argue that feminism not only causes scientists to ask new questions but also to reject old questions and assumptions as irrelevant. As a result, the European Union is currently creating a system that will evaluate the gender content of government-sponsored research, aiming to bring the gender analysis of science into the mainstream of science policy, proposals, contracts, and research. The United States has yet to implement such a system.

Biologist/author Rachel Carson (center), with children near her home (Alfred Eisenstaedt/Time Life Pictures/Getty Images)

In a broader framework, feminists have argued that women may be more likely to question the purpose or use of particular scientific research. Feminists argue that presently the majority of scientists tend to see themselves studying the natural world to gain knowledge for knowledge's sake and do not take responsibility for the application of their knowledge. Most western countries' science budgets are heavily weighted toward military and industrial research. This science is pursued for its usefulness in maintaining the male-dominated status quo, its ability to allow humans control over nature, and for its profitability. Feminist reassessment of research is evident in challenges by feminists and others over the Human Genome Project, which is presented as both practically and ideologically problematic. They view this project as reductionist and controlling with its emphasis on biology and almost no consideration for the effects of the environment on genes. Most research focuses on locating genes to determine the position of disease-causing genes. Feminists respond that not only are many genetic diseases the result of genes interacting with the environment, but placing so many financial resources on this project ignores the fact that most human diseases are not caused by genetic defects. Major causes of death worldwide continue to be malnutrition, lack of education, lack of access to medical care and prenatal care, and the lack of available vaccines. Feminists suggest that the significant amount of funding allocated to the Human Genome Project should be rethought and redirected toward more practical and useful goals.

Feminist and scientist Sandra Harding has asked, "What would it mean to create a science for women, and one that is for women around the globe?" (Harding 1989, 705). Rosaleen Love, lecturer in the history and philosophy of science at Swinburne University of Technology, poses the question, "What would a woman hanging out the clothes on the line know about health risks, compared with the established knowledge of medicine, statistics and chemistry?" (1993, 182). Harding, along with other feminists, believes that if women had the opportunity to choose and/or direct research goals, science would consider women's needs and concerns, which would produce a science useful to humanity instead of one that enables the elite to remain dominant over others. Science that is connected to women's lived experience of care and concern for their bodies and their love for their families can certainly be productive for both human beings and the environment. In 1978, Lois Gibbs (1951–) raising her family in Love Canal, near Niagara Falls, New York, discovered that her child's school had been built on a toxic chemical dump. She used data to prove the connection between her neighbors' illnesses, including miscarriages, birth defects, and cancer, and the toxic waste that was being spread throughout their neighborhood through streambeds and swales that had existed prior to the building of the community. Eventually, with the assistance of scientists, Gibbs's research, together with extensive grassroots protests, convinced the government to purchase the homes in the Love Canal area and relocate nine hundred families. The Love Canal incident made Gibbs famous, and as a result of contacts with others in similar situations, she created the Center for Health, Environment, and Justice (1981), an organization that has assisted over 8,000 grassroots groups. Out of the Love Canal incident grew regulations for hazardous waste disposal and protection for human beings and the environment. Other women have similarly posed questions to science that arise from their lived experiences in their communities, such as the genetic manipulation of plants and animals, nuclear power plants and nuclear waste, and the placement of electricity transmission lines. All such issues raise concerns about their families' growing bodies and minds. The difficulty of nonscientists to be heard in public forums, however, continues.

Women scientists' standpoint can also reveal a fundamentally different knowledge than that of male scientists. A remarkable example of the impact of women's standpoint in science can be found in the discipline of primatology. Primatologists study primates, their physical biology, their psychology, and their social organizations and interactions. Male primatologists in the 1940s and 1950s focused on the males of male-dominant species of primates, observing females and their young very little, if at all. When increasing numbers of female primatologists came into the profession in the 1970s, they began to ask different questions, to study different species, and to observe the females and the young. Donna Haraway (1989) suggests that women primatologists did not reconstruct the science of primatology because of their sex, but because of the political climate of the period in which they entered the field (303). Jeanne Altmann's study of female baboons, for example, caused her to argue for a rethinking of primatology's methods and theories. She stressed the importance of long-term observation resulting in the observer coming to know individual baboons and understand the relationships between baboons. As a result of this new methodology, her research demonstrated that there was more to baboons than the previously emphasized conflict between males for access to females. Altmann posited that baboon society was in fact female-centered. Females provided a stable, cooperative social core while males came and went. Furthermore, while studying mothering among the baboons, Altmann discovered that parenting skills varied between females depending on their rank within the group. Her work demonstrates that biology is not destiny.

In contrast to Altmann's research that gives primacy to environment over biology, primatologist and feminist Sarah Blaffer Hrdy places biology front and center. She utilizes sociobiology, so often critiqued by feminists, to demonstrate how this science can contradict the usual generalizations made by sociobiologists about female animals. Hrdy's work shows that rather than being sexually discriminating and coy, female primates are active strategists in attaining reproductive advantage and are competitive, sexually assertive, and promiscuous. Not all feminists appreciate Hrdy's argument that maternal instincts reside in biology, as they fear that such an argument once again naturalizes women's role in society as mother. However, Hrdy argues that in her approach maternalism is aggressive, competitive, and selfish, breaking down traditional notions about maternal instincts and providing a more primary role for female primates both in nature and in society (Hrdy 199).

Feminists have also had a substantial impact on ecology. Ecofeminists argue for women's closer connection to and thus better understanding of the natural world. They believe the close connection between women and nature will automatically lead women to work toward protecting the environment not only for the sake of improving human existence but for the sake of nature itself. Vandana Shiva, Indian physicist, philosopher of science, and ecofeminist, has argued that the western world could learn how to coexist with its natural environment by learning from Third World women's interaction with and attitudes toward the natural world. In Indian cosmology, the energy and creativity of the world is conceptualized as a feminine principle that pervades everything. Indian women, as the "sylviculturalists, agriculturists and water resource managers, the traditional natural scientists," embody this feminine principal in both the ideological con-struction of nature as diverse, connected, and interrelated and in their beliefs that there is no separation between human and nature, and that all life is sacred (Shiva 1996, 283). This feminine standpoint produces sustainable agricultural practices rather than destructive ones. The bringing of western, masculinized agriculture to India through commercial plantation crops, river valley projects, and mining projects results in the loss of 1.3 million hectares of forest every year. Shiva and other ecofeminists argue that the destruction of the natural world will inevitably result in the disappearance of human life. They call urgently for a dif-ferent ideology about nature, one that will foster a harmonious interaction between human beings and the natural world.

Women scientists do not have to be self-defined feminists in order to sup-port the ideals of ecofeminists. In the United States, Rachel Carson (1907–1964), whose popular science writing emphasized the interconnectedness of all living things, became increasingly concerned about the human desire to control the natural world. In 1962 she published *Silent Spring*, a work that highlighted the damage insecticides and herbicides were doing to the environment. Her work touched a chord with the general public, thus disseminating her view of nature to a wide audience that forced the chemical industry to listen and respond to Carson's findings. Carson's larger project was to teach her readers to rethink the relationship humans had with the environment by creating a different view of nature than the well-established one of domination, control, and destruction. Carson painted a picture of a wonderful, awe-inspiring nature that is complex, ever-changing, uncontrollable, and interconnected, and sometimes a world that cannot be entirely known or understood. Human beings must develop humility before nature, she argued, not arrogance toward it.

Some feminists have insisted that women scientists would not only ask dif-ferent questions and broaden and deepen the understanding of the world with different standpoints, but that women's differences, whether founded in biology or culture, nature or nurture, can create an entirely new scientific methodology. They believe that if given the chance, women would formulate a holistic science. This approach to science, they argue, would recognize that all things are living in a material and in a spiritual sense and that human beings are a part of nature even when they investigate the natural world. They would reject objectivity, instead emphasizing women's personal experiences as a basis to determine knowledge, rejecting dualities, hierarchies, and linear concepts for holistic, com-plex, cyclical ones. However, there is a flaw in this feminist approach to science. Having accused male scientists of creating a scientific method where the mas-culine perspective dominates, feminists would be creating a method where the feminine perspective would reign supreme, simply replacing a one-sided perspective with another and possibly reinforcing the essentialist position that biology is destiny.

Feminists such as Linda Jean Shepherd and Hilary Rose have instead postulated a humanistic science over a feminist science, which incorporates into scientific practice and purpose characteristics long associated with the female gender, but arguing that both male and female scientists can bring these characteristics to science. Such a science would be more caring and would produce knowledge out of a desire to love and be loved rather than out of a desire for domination and destruction. Recognizing that science is inherently social, political, and ideological, scientists would by necessity have to ask themselves about the social, political, and ideological goals of their science. Ultimately, feminists hope that a new science will benefit everyone in society, not just the elite few.

Conclusion

There are signs of hope that the barriers to women's achievement in science may not survive the twenty-first century. Yet, in many instances, policies that promote women's participation in science are heavily prescriptive in nature. For example, the U.S. National Science Foundation has mandated that absence of women at conferences will result in funds being taken away as prima facie evidence of discrimination. Similarly, a representative of WiSE has suggested that the NSF cut grants to universities that do not maintain a minimum number of female faculty in science and engineering. Universities are no longer exempt from having lawsuits brought against them for gender discrimination, enabling women who feel they have been discriminated against in the process of attaining a job or doing their work to sue the university. The continued demand for such policies suggests that little has changed for women in science since the passing of the Equal Opportunities Employment Act and the introduction of affirmative action in the 1970s. Discrimination against women in science continues. Men and women in positions of power keep women out, and in turn, women themselves internalize the social norms against women's participation in science, creating a self-fulfilling prophecy.

Although feminists, women scientists, institutions, and governments have attempted to impose change, these changes have often resulted in a backlash against the promotion of women in science. Some people still believe that women's lack of participation is a personal choice and do not see the low enrollment figures as a problem. This attitude among administrators results in passivity, and changes are not implemented for none are considered necessary. In other instances, administrators are strongly opposed to changing the system. For example, Eileen Byrne, author of *Women in Science: The Snark Syndrome*

(1991), surveyed university faculty and administrators about their opinions of discussion papers on women's achievements in science. She received the following response from one university lecturer: "This is, of course, exactly the kind of garbage I associate the feminist movement with, and I hope you do not really expect me to waste my time reading it" (193). Women have often had to rely on those people, still mostly men, in positions of administrative power to enact change, many of whom continue to resist learning about and solving problems.

Women implementing programs for other women and girls, and the movement of some women into decision-making positions, are important processes to promote women's participation in science. But they often miss the point that men also need to be educated in the importance and necessity of these changes in order to lend their support and encouragement to changing the system. The education of men in positions of power and authority would ensure that other men no longer reproduce the status quo that often educates both boys and girls to assume that the disadvantages women suffer are self-imposed. Project STEM (Women in Science, Technology, and Mathematics) at Purdue University–Calumet has implemented as part of its program a Workplace Gender Equity Project. A one-day workshop for high school students encourages both men and women to consider careers outside of those dominated by their own gender by introducing them to men and women already filling such roles. Joan Rothschild in her *Teaching Technology from a Feminist Perspective* (1988) suggests that gender issues relating to technology should not be solely segregated in women's studies courses mostly taken by female students. Rather, teachers need to integrate these issues as a component of science and technology studies courses at the introductory level, where male as well as female students will be educated about feminist perspectives and the gender factors that inhibit women's participation in technology fields (77). Faculty development workshops on gender issues also promote awareness and educate men as well as women to their reality. The education of men and boys appears to be the next step needed toward the full participation of women in science.

Of Sue Rosser's six stages of evolution to a female-friendly science, stage six is "science redefined and reconstructed to include us all" (Rosser 1995, 17). History has shown that unless laws change, governments provide support, and individual women continually struggle to stay in science, the door to scientific participation, once ajar, seems to repeatedly swing shut on women. It is important for universities and industries to work toward hiring equal numbers of women in science. Perhaps more important according to feminists, however, schools and employers must have more respect for and give more recognition to women's work in science and their need to balance personal and professional responsibilities. Rosser believes that if men and women open science fully to the feminist perspective, the result will be "tremendous breakthroughs [that] will rival, if not surpass, those in molecular biology during the later half of the twentieth century" (Rosser 1986, 137). Until then, initiatives that support and encourage women in a world where they are otherwise taunted, castigated, and ignored despite their many achievements are most certainly as imperative as are those that support the education of men and boys about women in science and the positive role they can play. Only in this way will both men and women in western society become equally educated and interested in science to be able to make value judgments about it as well as benefit from learning to think about each other as equal contributors to the scientific endeavor. Most important, such men and women must retrain themselves in their understanding of gender and science. In this way they will be able to educate a new generation of boys and girls in the wave of scientific practice and discovery that will take place in the new millennium and will open new opportunities to everyone regardless of gender.

Bibliographic Essay

Margaret Rossiter's *Women Scientists in America: Before Affirmative Action, 1940–1972* (1995) again serves as an important work examining the proliferation of women in science during this period. She examines where women scientists have generally been employed and explores the professional lives of numerous individual women scientists. Rossiter also provides important statistical data on the fields in which women worked, the degrees they earned, and the positions and awards they held, often providing a comparison with men. For the European context, Veronica Stolte-Heiskanen's edited work *Women in Science: Token Women or Gender Equality* (1991) provides a starting point for the examination of the international state of women in science. Although this work contains only brief glimpses into the lives of women scientists in various countries ranging throughout eastern and western Europe, the essays present interesting comparisons and contrasts to the experiences of American women scientists. Many of the chapters also contain historical components that provide a point of comparison to the historical work of Rossiter. More recently, several articles in the 2002 volume of the journal *Science in Context* discuss the numbers of women in European science and the efforts that have been made over the past couple of decades to increase women's participation in science in the Netherlands, France, and Germany. The National Science Foundation website (http://www.nsf.gov) provides current statistical information on women in science in the United States.

In the area of science education, Sue Rosser's work stands out as the most expansive and in-depth study of the pedagogical approaches to teaching girls and women science and the feminist arguments for implementing them. Most directly useful are *Female-Friendly Science: Applying Women's Studies Methods and Theories to Attract Students* (1990) and *Teaching Science and Health from a Feminist Perspective: A Practical Guide* (1986). In her 1992 book *Biology and Feminism: A Dynamic Interaction*, there is a section on this subject that deals explicitly with the possible impact of feminism on the teaching of biology. Rosser has also edited a collection of articles on this subject, *Teaching the Majority: Breaking the Gender Barrier in Science, Mathematics, and Engineering* (1995). Published in the same series as Rosser's *Teaching Health and Science from a Feminist Perspective* is Joan Rothschild's *Teaching Technology from a Feminist Perspective: A Practical Guide* (1988). Other recent essays on this topic can be found in Jody Bart's edited work, *Women Succeeding in the Sciences: Theories and Practices across Disciplines* (2000), and in Lesley Parker, Lonnie Rennie, and Barry Fraser's *Gender, Science, and Mathematics: Shortening the Shadow* (1996). Also see Linda Samuels, *Girls Can Succeed in Science!: Antidotes for Science Phobia in Boys and Girls* (1999), for a practical discussion of how such science programs for girls can be implemented.

In *Athena Unbound: The Advancement of Women in Science and Technology* (2000), Henry Etzkowitz, Carol Kemelgor, and Brian Uzzi examine the sociological barriers to women's science practice and demonstrate how these barriers are translated into the university science community. They call for change at the departmental level. They also briefly examine international comparisons that further their argument of the sociological impact of gender roles and characteristics on women's participation in science. Sandra L. Hanson also explores the reasons why women do not seem to be able to enter or succeed in science to a greater extent than at present in *Lost Talent: Women in the Sciences* (1996). She explores the reasons why women who as young girls demonstrate an interest in and aptitude for science fail to pursue the subject.

Margaret A. Eisenhart and Elizabeth Finkel, working from interviews with women scientists in *Women's Science: Learning and Succeeding from the Margins* (1998), present an all-encompassing explanation for women's low status and marginal positions in science. They argue that women are attracted to scientific careers where their scientific work serves a social, political, or environmental purpose. Women find these kinds of scientific areas more rewarding and also more female-friendly. Eisenhart and Finkel argue that this friendliness is often misconstrued by women and that, in fact, these workplaces are not as friendly to women as they may appear. Nevertheless, women feel happier and more fulfilled and more rewarded by this kind of work despite the low pay and low status.

An interesting work that examines the lack of evidence upon which assumptions about women's scientific capabilities are based is Eileen Byrne,

Women in Science: The Snark Syndrome (1991). Discussing the Australian context, Byrne argues that future government policy must change if Australia is to advance scientifically and technologically. Presently, she argues, the majority of Australian women are lost to science. Moreover, the general populace cannot participate in scientific decision making due to lack of a sufficient scientific education. Paradigm shifts in sociological and scientific viewpoints need to take place to incorporate a wider percentage of the population in science.

Feminists have increasingly moved beyond critiquing science to a desire to explore how to reformulate science from their own epistemological standpoint. General works that discuss the issue of whether there can be such a thing as feminist science include Helen E. Longino, "Can There Be a Feminist Science?" *Hypatia: A Journal of Feminist Philosophy* 2, 3 (fall 1987): 51–64; Sue V. Rosser, "Feminist Scholarship in the Sciences: Where Are We Now and When Can We Expect a Theoretical Breakthrough?" *Hypatia: A Journal of Feminist Philosophy* 2, 3 (fall 1987): 5–17; and Sue V. Rosser, "Good Science: Can It Ever Be Gender Free?" *Women's Studies International Forum* 11 (1988):13–19.

Works that begin with the premise that there can be a feminist science and explore how such sciences might be and have been developed include Sandra Harding, *Whose Science? Whose Knowledge?: Thinking from Women's Lives* (1991); and Ruth Bleier (ed.), *Feminist Approaches to Science* (1986). Studies of the ways in which women could produce a feminist science include Mary Morse, *Women Changing Science: Voices from a Field in Transition* (1995); Hilary Rose, *Love, Power, and Knowledge: Towards a Feminist Transformation of the Sciences* (1994); and Linda Jean Shepherd, *Lifting the Veil: The Feminine Face of Science* (1993). Works that specifically explore existing feminist or feminine science practice include Donna Haraway, *Primate Visions: Gender, Race, and Nature in the World of Modern Science* (1989); Donna Haraway, *Simians, Cyborgs, and Women: The Reinvention of Nature* (1991); and Evelyn Fox Keller, *A Feeling for the Organism: The Life and Work of Barbara McClintock* (1983).

Chronology

1606	Francesca Fontana and her husband, Ulisse Aldrovandi, publish *On the Remains of Bloodless Animals.*
1620	Francis Bacon publishes *New Organon.*
1627	Francis Bacon publishes *New Atlantis.*
1637	René Descartes publishes his *Discourse on the Method of Rightly Conducting the Reason and Seeking of Truth in the Sciences.*
1649	Descartes publishes *Passions of the Soul.*
1655	Margaret Cavendish, Duchess of Newcastle, publishes *Philosophical and Physical Opinions.*
1660	The Royal Society is founded in England.
1666	The French Academy of Sciences is founded.
	Margaret Cavendish, Duchess of Newcastle, publishes her *Observations Upon Experimental Philosophy* and the appended *Blazing World.*
1677	Anton van Leeuwenhoek presents evidence to the Royal Society that he had observed male semen under the microscope and had seen "spermatic animalcules."
1679 and 1683	Maria Sibylla Merian publishes *Wonderful Transformation and Singular Plant-Food of Caterpillars* in two volumes.
1686	Bernard Le Bovier de Fontenelle publishes *Conversations on the Plurality of Worlds.*
1687	Isaac Newton publishes *The Principia.*
1690	Anne Conway publishes *The Principles of the Most Ancient and Modern Philosophy.*
1699	Maria Sibylla Merian begins her trip to Suriname (Dutch Guinea) to study and draw nature.

1700	The Prussian Academy of Sciences is founded.
1702	Maria Winkelmann discovers a previously unknown comet.
1705	Maria Sibylla Merian publishes *Metamorphosis of the Insects of Suriname*, which expands her earlier three-volume work on *European Insects* (1679, 1683, 1687).
1710	Maria Winkelmann is denied the position of Berlin Academy of Sciences astronomer upon her husband's death.
1711	The Academy of Sciences of the Institute of Bologna, Italy, is founded.
1725	The Imperial Academy of Sciences in Russia is founded.
1733	Laura Bassi receives a doctorate in philosophy at the University of Bologna, quickly becoming a professor of physics there and a member of the Academy of Sciences in Bologna.
1758	Carolus Linnaeus (a.k.a. Carl von Linné) publishes *Systema naturae*.
1774	William Hunter publishes *The Anatomy of the Human Gravid Uterus*.
1785–1887	The first known women's scientific society, Natuurkundig Genootschap der Dames (Women's Society for Natural Knowledge) is in operation in the town of Middleburg, on the southern Dutch island of Walcheren in the province of Zeeland.
1786–1797	Caroline Herschel discovers a total of eight comets.
1788	Johann Wolfgang von Goethe publishes *Metamorphosis of Plants*.
1789	Erasmus Darwin publishes *The Loves of the Plants*.
1796	Priscilla Wakefield publishes *An Introduction to Botany, in a Series of Familiar Letters*.
1802	William Paley publishes *Natural Theology; or, Evidences of the Existence and Attributes of the Deity, Collected from the Appearances of Nature*.
1821	Emma Willard opens the Troy Female Seminary in the United States.
1823	Catharine Beecher establishes the Hartford Female Seminary in the United States.
1827	Mary Somerville publishes her English translation of Laplace's *Mechanism of the Heavens*.

Jane Loudon publishes *The Mummy: A Tale of the Twenty-second Century.*

1828–1838   Sarah Wallis Bowdich Lee publishes *The Freshwater Fishes of Great Britain.*

1829   Almira Hart Lincoln Phelps publishes *Familiar Lectures on Botany.*

1834   Mary Somerville publishes *On the Connexion of the Physical Sciences.*

1835   Mary Somerville becomes a member of the Royal Astronomical Society.

1836   Louisa Twamley Meredith publishes *The Romance of Nature; or, The Flower Seasons Illustrated.*

1838   Jane Loudon begins publication of her eight-volume *Ladies Flower Garden.*

1839   Jane Marcet publishes *Conversations on Chemistry.*

Louisa Twamley Meredith publishes *Our Wildflowers Familiarly Described and Illustrated.*

1840   Jane Loudon publishes *Instruction in Gardening for Ladies.*

Jane Loudon publishes *Young Naturalist's Journey; or, The Travels of Agnes Merton and Her Mama.*

1842   Jane Loudon publishes *Botany for Ladies.*

1844   Louisa Twamley Meredith publishes *Notes and Sketches of New South Wales during a Residence in That Colony from 1839–1844.*

1847   Maria Mitchell discovers an unknown comet (later named after her) and wins a gold medal offered by the King of Denmark.

1848   Maria Mitchell is elected to the American Academy of Arts and Sciences.

Mary Somerville publishes *Physical Geography.*

1850   Maria Mitchell is elected to the American Association for the Advancement of Science.

1851   Jane Loudon's *Botany for Ladies* is republished under the title *Modern Botany.*

1852   Louisa Twamley Meredith publishes *My Home in Tasmania.*

Elizabeth Blackwell publishes *Laws of Life.*

1853   Sarah Wallis Bowdich Lee publishes *Anecdotes of the Habits and Instincts of Birds, Reptiles, and Fishes.*

Elizabeth Blackwell opens a dispensary in a tenement district that will later become the New York Infirmary for Women and Children.

1854   Sarah Wallis Bowdich Lee publishes *Trees, Plants, and Flowers: Their Beauties, Uses, and Influences.*

Catherine Parr Traill publishes The *Female Emigrants' Guide, and Hints on Canadian Housekeeping.*

1855–1871 Margaret Gatty publishes her series of children's books, *Parables from Nature.*

1859   Charles Darwin publishes *The Origin of Species.*

Elizabeth Agassiz publishes her teachers' guide *A First Lesson in Natural History.*

1864   The University of Zurich, known for its exceptional medical education, gives Russian student Maria Kniazhnina permission to attend lectures in anatomy and microscopy.

1865   Elizabeth Agassiz publishes *Journey in Brazil* and *Seaside Studies in Natural History.*

John Lindley publishes *Ladies' Botany; or, A Familiar Introduction to the Study of the Natural System of Botany.*

Maria Mitchell is offered the position of professor of astronomy and director of the observatory at Vassar College.

1867   Nadezhda Suslova, a Russian student attending the University of Zurich, requests permission to take the medical examination to obtain a degree and is accepted.

1868   Catherine Parr Traill publishes *Canadian Wild Flowers.*

Elizabeth Blackwell founds the Woman's Medical College in the United States.

1869   Girton College for women is founded at Cambridge University, England.

Elizabeth Garrett Anderson receives her MD from the University of Paris.

Sophia Jex-Blake and four other women gain access to medical education at the University of Edinburgh.

1870   Mary Somerville receives the Royal Geographical Society's Victoria Gold Medal.

1871   Newnham College for women is founded at Cambridge University.

Smith College for women opens in the United States.

1872   Margaret Gatty publishes *British Seaweeds.*

1873   The University of Edinburgh refuses to grant Sophia Jex-Blake and the four other women in her class their medical degrees.

Edward Clarke publishes *Sex in Education; or, A Fair Chance for the Girls.*

Mary Putnam Jacobi founds a medical dispensary for women and children at Mount Sinai Hospital in New York.

Ellen Swallow Richards receives her BS degree in chemistry from the Massachusetts Institute of Technology. She is the first woman admitted to the school and the first woman to be accepted at a scientific school.

1874   Elizabeth Garrett Anderson and Sophia Jex-Blake help found the London School of Medicine for Women.

Henry Maudsley publishes his article "Sex in Mind and Education" in *The Fortnightly Review.*

Elizabeth Garrett Anderson publishes her reply to Henry Maudsley's article "Sex in Mind and Education" in *The Fortnightly Review.*

Women are granted the right to sit for the degree exams at Cambridge University for the first time, including those in the natural sciences.

Charles Darwin publishes *Descent of Man and Selection in Relation to Sex.*

1875   Wellesley College for women opens in the United States.

Ellen Swallow Richards establishes the Women's Laboratory at MIT.

1876   The Russell Gurney Enabling Act allows medical examining bodies to test women in Britain.

1878   London University becomes the first university in Great Britain to admit women.

1878 *(cont.)*	Domestic science is made compulsory for girls in British elementary schools.
	Lady Margaret Hall for women is founded at Oxford University.
1879	Somerville College for women is founded at Oxford University.
1880	Spanish women are allowed to attend university without obtaining individual permission.
	Ellen Swallow Richards publishes *The Chemistry of Cooking and Cleaning.*
	Louisa Twamley Meredith publishes *Tasmanian Friends and Foes: Feathered, Furred, and Finned.*
1881	Astronomer Edward Pickering begins hiring women computers for the Harvard Observatory. Williamina Fleming is among those hired.
1882	The American Pharmaceutical Association accepts its first woman member.
	The all-women's Association of Collegiate Alumnae's study of 705 women graduates, entitled *Health Statistics of College Graduates*, is published.
	The Cavendish Laboratory for Physics is opened to women students at Cambridge.
	The Marianne North Gallery is opened at the Royal Botanic Gardens at Kew.
1883	Arabella Buckley publishes *Winners in Life's Race.*
1884	Ellen Swallow Richards is appointed instructor in sanitary chemistry in a new MIT laboratory dedicated to this study and, as such, becomes known as the founder of domestic science.
1884–1914	Balfour Biological Laboratory for women is in use at Cambridge for the female students at Girton and Newnham Colleges.
1885	Bryn Mawr College for women opens in the United States.
	Sophia Jex-Blake founds the Women's Hospital in Edinburgh.
	Matilda Cox Stevenson forms the Women's Anthropological Society of America with ten other women.
	Catherine Parr Traill publishes *Studies of Plant Life in Canada.*
	Agnes Mary Clerke publishes *A Popular History of Astronomy during the Nineteenth Century.*

Sophia Jex-Blake publishes *Medical Women: A Thesis and a History*.

Ellen Swallow Richards publishes *Food Materials and Their Adulterations*.

1886 Sophia Jex-Blake founds the Edinburgh School of Medicine for Women.

1889 Sofia Kovalevskaia publishes her work on the revolution of a solid body about a fixed point, known as the "Kovalevskaia top," which links complex function theory and analytic mechanics. For this work she will win the Prix Bordin, the highest award of the French Academy of Sciences.

1890 Agnes Mary Clerke publishes *The System of the Stars*.

1890–1895 The Royal Observatory in Greenwich hires women computers.

1892 Johns Hopkins Medical School in the United States begins admitting women.

1895 Agnes Mary Clerke publishes *The Herschels and Modern Astronomy*.

1896 Mary Kingsley publishes *Travels in West Africa*.

1897 Austrian universities opened to women.

Beatrix Potter's paper "On the Germination of the Spores of *Agaricineae*" is read at a meeting of the Linnean Society.

1898 Williamina Fleming's paper "Stars of the Fifth Type in the Magellanic Clouds" is read for her at a Harvard conference.

Agnes Mary Clerke publishes *Astronomy*.

1899 Mary Kingsley publishes *West African Studies*.

1899–1911 Williamina Fleming becomes the first woman ever to receive a corporation appointment at Harvard as the curator of astronomical photographs.

1900 Eleanor Anne Ormerod is awarded an honorary LLD degree by University of Edinburgh for her work in economic entomology.

1901–1932 The Ellen Richards Prize is awarded by the Association to Aid Women in Science.

1903 Agnes Mary Clerke publishes *Problems in Astrophysics*, which she coauthored.

1903 (cont.)	Agnes Mary Clerke is recognized for her work in astronomy by the Royal Astronomical Society with an honorary membership.
	Marie Curie becomes the first woman in France to receive a Ph.D. in physics. Her thesis is on radioactive substances.
	Pierre and Marie Curie are awarded the Davy Medal by the Royal Society.
	Marie Curie becomes the first woman to be awarded a Nobel Prize for physics, won jointly with Pierre Curie and Henri Becquerel for their research on radiation.
1904	The Education Act in Britain states that students are to receive seven and a half hours of science instruction weekly.
	Leeds University opens all its degree options to women.
1905	Agnes Mary Clerke publishes *Modern Cosmogonies*.
1906	The Sorbonne offers Marie Curie her husband's chair in physics after his death earlier in the year; Marie accepts.
	Williamina Fleming is made an honorary member of the Royal Astronomical Society in England.
	Lise Meitner obtains her Ph.D. degree in physics, only the second woman to achieve the degree at the University of Vienna.
1908	The American Home Economics Association and its *Journal of Home Economics* are founded, in part by Ellen Swallow Richards.
1910	Spanish women are allowed full access to university education.
	Domestic science is institutionalized as a field for women in the United States.
	Marie Curie publishes *Treatise on Radioactivity*.
1911	Marie Curie wins the Nobel Prize a second time for chemistry, for her work on radium and polonium.
	Marie Curie is refused a seat in the French Academy of Sciences.
1913	Reverend John A. Zahm publishes *Woman in Science with an Introductory Chapter on Woman's Long Struggle for Things of the Mind* under the pseudonym H. J. Mozans.
	Botanist Ethel Sargant becomes the president of the botanical section of the British Association for the Advancement of Science.
1914	The Royal Geographical Society begins admitting women.

E. R. Saunders of Newnham College becomes one of the first female members of the Council of the British Association for the Advancement of Science.

1918   Dr. Agnes Fay Morgan becomes chair of the well-established home economics department at the University of California.

1919   In Britain the Sex Disqualification (Removal) Act is passed, disallowing barriers to the admittance of women to learned societies based on sex.

1920   The British Chemical Society begins admitting women.

1922   Oxford University allows women to obtain degrees.

Marie Curie is made a member of the Academy of Medicine of Paris for the part she played in discovering radium, thus becoming the first woman to be made a member of an academy in France.

1923   Marie Curie is awarded a substantial lifetime pension by the French government for her discovery of radium.

1926   Charlotte Haldane publishes *Man's World*.

Lise Meitner is appointed adjunct professor at the University of Berlin, becoming the first woman professor of physics in Germany.

1928   Katherine Burr Blodgett becomes the first woman to receive a doctorate in physics at Cambridge University.

Barbara McClintock earns her Ph.D. from Cornell University.

1934   In response to the discontinuation of the Ellen Richards Prize, Annie Jump Cannon arranges for her own winnings from the Ellen Richards Prize to go to the council of the American Astronomical Society for a triennial award for distinguished contributions to astronomy by a woman of any nationality.

1935   Irene Curie and her husband, Frederic Joliot, receive the Nobel Prize in chemistry for their discovery of artificial radioactivity.

1938   Geologist Alice Wilson becomes the first woman elected a fellow of the Royal Society of Canada.

Katherine Burr Blodgett receives a patent for her invention of reflective glass.

1939   Lise Meitner, with Otto Hahn, discovers that the uranium nucleus splits to form two other atoms and in so doing releases a considerable amount of energy; they dub this splitting of uranium "fission."

1940	The Norwich Museum receives Margaret Fountaine's collection of twenty-two thousand butterflies.
1943	The Royal Society of London begins admitting women.
1944	Barbara McClintock becomes the third woman to be elected to the National Academy of Sciences.
	The U.S. National Academy of Sciences has elected only three women members to date.
1945	Kathleen Yardley Lonsdale becomes the first of two women elected as a fellow to the Royal Society.
	Barbara McClintock becomes president of the Genetics Society of America.
	The Soviet Academy of Sciences has elected only one woman member to date.
	Rosalind Franklin earns her Ph.D. in chemistry from Newnham College, Cambridge University.
1946	The Royal Society of Canada elects its first woman member.
1948	Cambridge University begins allowing women to obtain degrees.
1950	Ruth Hubbard earns her Ph.D. in biology from Radcliffe College.
1953	James Watson and Francis Crick publish their paper in the journal *Nature* in which they reveal the double helical structure of DNA.
1960s	Beginning of the second wave of the Feminist Movement.
1962	James Watson and Francis Crick receive the Nobel Prize for medicine.
	Rachel Carson publishes *Silent Spring*.
1963	Maria Goeppert Mayer wins the Nobel Prize for physics for the nuclear shell model of atomic nuclei, which demonstrates that atom-like discrete energy levels of individual protons and neutrons explain the stability of nuclei. She is the second American woman to win this prize and the first to win it for physics.
	Russian Valentina Tereshkova becomes the first woman to travel into space.
	Evelyn Fox Keller earns her doctorate in theoretical physics from Harvard University for her thesis entitled "Photoinactivation and the Expression of Genetic Information in Bateriophage-Lambda."

1966   Kathleen Yardley Lonsdale becomes the first woman president of the British Association for the Advancement of Science.

1967   Ruth Hubbard wins the Paul Karrer Medal, along with her husband George Wald, for their work on the biochemistry and photochemistry of vision in vertebrates and invertebrates.

1968   James Watson publishes his autobiographical work *The Double Helix.*

1970s   Ruth Hubbard and Evelyn Fox Keller become feminist critics of science.

1971   The American Women in Science Network is founded.

1972   The U.S. Congress passes the Equal Employment Opportunity Act, the Equal Rights Amendment, and the Education Amendments Acts.

1973   Anthony Hewish and Jocelyn Bell Burnell receive the Franklin Institute of Philadelphia's Albert A. Michelson medal for the discovery of pulsars.

1974   Anthony Hewish receives the Nobel Prize without his assistant Jocelyn Bell Burnell.

Ruth Hubbard becomes the first woman to be awarded tenure in biology at Harvard.

1975   Anne Sayre, a friend of Rosalind Franklin, publishes *Rosalind Franklin and DNA.*

1978   E. O. Wilson, a prominent sociobiologist, publishes *On Human Nature.*

Lois Gibbs discovers that her neighborhood in Love Canal, near Niagara Falls, New York, has been built on a toxic chemical dump and begins a grassroots activist campaign to convince the government to purchase the homes in the Love Canal area and relocate nine hundred families.

1979   The Paris Academy of Sciences begins including women as full members.

1981   Lois Gibbs creates the Center For Health, Environment, and Justice, formerly Citizens Clearinghouse for Hazardous Wastes, an organization that has since assisted over eight thousand grassroots groups.

1982	Evelyn Fox Keller publishes her biography of Barbara McClintock, *A Feeling for the Organism.*
1983	Sally Ride becomes the first American woman to travel into space.
	Barbara McClintock's work on transposition is formally recognized by the scientific community, and she receives the Nobel Prize.
1986	The Academy of Exact, Physical, and Natural Science in Spain begins to admit women.
	In Yugoslavia, 22 percent of the natural sciences faculty are women, and 13 percent of the engineering and technology faculty are women.
	The Soviet Union outstrips the United States in numbers of women in science, with almost 40 percent of scientists in universities and research institutes being women, although only 13 percent hold a doctorate degree.
1987	In the German Democratic Republic (West Germany), 53 percent of higher education students in the natural sciences are women, and 27 percent of students in engineering and technical sciences are women.
1989	The Women in Science and Engineering Program is founded at the University of Washington.
1990	The Women in Science Project (WiSP) at Dartmouth College, New Hampshire, is founded.
	Women in Engineering Programs and Advocates Network is founded.
1991	Jocelyn Bell Burnell becomes professor of physics at the Open University, the third woman to become a professor of physics in Great Britain.
1992	In Australia just over 40 percent of postsecondary science students are women.
1994	A 50 percent reduction in the gap between boys' and girls' scores in mathematics is obtained.
1995	Affirmative action procedures are amended by a U.S. Supreme Court decision to ensure that they are not used as a quota system and that they not permit unqualified people to obtain positions, create reverse discrimination, or continue after their purposes have been achieved.

1997   The Linnean Society issues Beatrix Potter an official apology on the one-hundredth anniversary of the reading of her paper at the society.

According to the U.S. National Science Foundation's survey of earned doctorates, many fields including biology, chemistry, geosciences, and even mathematics reach or exceed the critical mass (15 percent) of women between 1966 and 1997.

1998   The European Union Treaty of Amsterdam strengthens Europe's commitment to women's equal opportunities and establishes a Women and Science Unit.

2000   In France, there are equal numbers of men and women in biology and chemistry up to the fourth year of university.

Glossary

argument from design   a philosophical reasoning that seeks to provide proof of God from evidence in the physical world. If order can be found in the universe, design and thus a creator is inferred.

Cartesian   refers to René Descartes's dualistic philosophy of the separation of mind and body.

classical physics   encompasses all aspects of Newtonian physics that were later questioned or expanded by the theory of relativity and quantum theory.

conservation of energy (physical theory of)   the First Law of Thermodynamics is the principle of the conservation of energy. This principle states that although energy can be converted into work or into new types of energy, the amount of energy present in a system does not increase or decrease.

craniology   the study of the shape and size of the human head, popular in the nineteenth century, used for making comparisons between the races and the sexes.

crystallography   the scientific study of crystal formation and phenomena.

Davy Medal   a bronze medal award, established in 1877, that is awarded annually by the Royal Society for an important European or North American discovery in chemistry.

eugenics   the belief that the human race can be improved through selective breeding or genetic control.

Linnean classification system   an organization system constructed by Carl von Linné (1707–1778), a Swedish botanist, in which all living organisms are ordered by two Latin names, referred to as *binomial nomenclature*. The first name represents the *genus*, and the second refers to the *species* to which the organism belongs. The system fell out of fashion in the nineteenth century but does form the basis for the modern classification system.

Marie Curie strategy   wherein women's achievements are measured solely against the work of exceptional women in science, such as Marie Curie, forcing women to work much harder than most men to achieve equal recognition with men in the scientific community. Women scientists often fall short of such ideals, providing justification to employers for refusing scientific work to women, or relegating them to low-pay, low-status positions.

Matilda effect   a term coined by historian Margaret Rossiter to contrast with the term *Matthew Effect*. The term is named for Matilda Joslyn Gage (1826–1898), an American women's rights activist and early sociologist of knowledge, who recognized the social acceptability of men claiming women's work as their own, and fought against this injustice. Rossiter applies this term to women scientists and views it as a negative effect.

Matthew effect   applied to the science profession, this term recognizes that often scientists who achieve prestige within the profession are given credit not only for their own discoveries but for those of other, less well-known scientists too. On occasion such figures co-opt the work of others as their own, knowingly taking credit for something that was not fully, or even partially, theirs to claim. A positive spin has often been placed upon this effect for both parties involved.

Mendelian genetics   Gregor Mendel (1822–1884) discovered how genetic inheritance worked through his study of peas. Mendel's laws of inheritance state that characteristics are inherited intact through the reproductive cells of sexually reproducing organisms. An offspring receives one pair of traits from its parents. Some characteristics are dominant over others so may not be expressed in a particular offspring, but cannot be altered or blended by other traits and so can thereby still be passed on from one generation to the next even when they are not expressed.

mesmerism   Franz Anton Mesmer (1734–1815), an Austrian physician, claimed to be able to use a kind of hypnotism to cure his patients. Mesmerism became popular as a form of treatment in the early nineteenth century, especially for women.

natural philosopher   a person who studies the natural and physical world.

natural philosophy   the study of the natural and physical world.

natural theology   the examination of the natural world to provide, through reasoned argument, proof of the existence of God.

naturphilosophie   an approach to the study of nature practiced by the German poet and dramatist Johann Wolfgang von Goethe (1749–1832), who spent much of his life studying the natural world. He believed, in contrast to the more popular, mechanistic approach to science, that the scientist should be in sympathy with nature in order to study it.

Nobel Prize   Alfred Bernhard Nobel (1833–1896), known for inventing dynamite, established the Nobel Prize in 1901 for contributions to chemistry, physics, physiology or medicine, literature, economics, and peace. This prestigious prize is awarded annually.

patent   a government grant that gives the holder control over a specific invention for a limited time, to make, sell, or use for their own personal gain, excluding all others.

patriarchal society   the term given to a society that is socially, politically, economically, and religiously dominated by men and which, as a result, makes women second-class citizens either by law and/or via established cultural norms.

physiology   the study of organisms' life processes.

pitchblende   a mineral that contains uranium, radium, and polonium, all sources of radioactivity.

popularizer   in science, a writer or lecturer who makes scientific theories and information understandable and appealing to a lay audience.

positivist philosophy   a philosophical system asserting that the only reliable form of knowledge is to be established through experience, specifically through experimental investigation and observation.

Protestant Reformation   the sixteenth-century religious and political movement in Europe to reform the practice and teachings of the Roman Catholic Church, resulting in the establishment of the Protestant Church.

radioactivity   a property of certain elements (such as uranium, radium, and polonium) that causes them to spontaneously emit radiation from their nuclei.

recapitulation theory   states that an organism will pass through the evolutionary development stages of its ancestors during the embryonic stage of its life.

romanticism   a late-eighteenth- and early-nineteenth-century artistic movement that emphasized the importance of feeling and passion over reason.

Royal Society (of London)   the first official scientific body in England, established in 1660 by King Charles II. The Royal Society remains a prominent professional scientific association to the present day.

scientific revolution   seventeenth-century scientists established a new approach to the study of the natural world that included experimentation, mathematics, and correct reasoning, resulting in a number of scientific discoveries and advances, including the advent of Newtonian science.

sexual selection   Charles Darwin posited the idea of sexual selection as part of the evolutionary process. He argued that certain individuals have a greater chance of reproducing than others due to the mating advantages of secondary sexual characteristics in male animals that make them more attractive to the females of their species.

sociobiology   the study of the genetic basis of animal, including human, social behavior.

spiritualism   the belief that people who have died exist as spirits in another world with whom mediums can communicate.

theory of evolution by natural selection   a theory of the development, survival, and extinction of species over time determined by a species' ability to successfully adapt to its environment, thereby enabling it to reproduce more successfully than others.

theory of preformation   the scientific theory introduced in the late seventeenth century which argued that the embryo was fully formed in the gamete (sperm or egg).

vivisection   the act of dissecting a living animal to demonstrate the function and working of particular organs or to conduct scientific research.

Documents

Introduction

Chapter 1

Chapter 2

Chapter 3

Chapter 4

Chapter 5

Chapter 6

Conclusion

Appended to Marie Curie's biography of Pierre Curie, this short section titled "Autobiographical Notes" reveals the struggles she encountered in practicing science. In this excerpt Curie describes her period of intensive scientific work isolating radium and proving that it was a new element. She emphasizes the abysmal conditions under which she and Pierre conducted their research and the necessity of balancing paid work, scientific research, and home life but she also expresses the enthusiasm and excitement with which they worked toward their goal. This work was to win them the Nobel Prize, along with Henri Becquerel, for the discovery of radioactivity and radioactive elements.

Pierre Curie

MARIE CURIE

But first of all in our life was our scientific work. My husband gave much care to the preparation of his courses, and I gave him some assistance in this, which, at the time, helped me in my education. However, most of our time was devoted to our laboratory researches.

My husband did not then have a private laboratory. He could, to some extent, use the laboratory of the school for his own work, but found more freedom by installing himself in some unused corner of the Physics School building. I thus learned from his example that one could work happily even in very insufficient quarters. At this time my husband was occupied with researches on crystals, while I undertook an investigation of the magnetic properties of steel. This work was completed and published in 1897.

In that same year the birth of our first daughter brought a great change in our life. A few weeks later my husband's mother died and his father came to live with us. We took a small house with a garden at the border of Paris and continued to occupy this house as long as my husband lived.

It became a serious problem how to take care of our little Irene and of our home without giving up my scientific work. Such a renunciation would have been very painful to me, and my husband would not even think of it; he used to say that he had got a wife made expressly for him to share all his preoccupations. Neither of us would contemplate abandoning what was so precious to both.

Of course we had to have a servant, but I personally saw to all the details of the child's care. While I was in the laboratory, she was in the care of her grandfather, who loved her tenderly and whose own life was made brighter by her. So the close union of our family enabled me to meet my obligations. Things were particularly difficult only in case of more exceptional events, such as a child's illness, when sleepless nights interrupted the normal course of life.

It can be easily understood that there was no place in our life for worldly relations. We saw but a few friends, scientific workers, like ourselves, with whom we talked in our home or in our garden, while I did some sewing for my little girl. We also maintained affectionate relations with my husband's brother and his family. But I was separated from all my relatives, as my sister had left Paris with her husband to live in Poland.

It was under this mode of quiet living, organized according to our desires, that we achieved the great work of our lives, work begun about the end of 1897 and lasting for many years.

I had decided on a theme for my doctorate. My attention had been drawn to the interesting experiments of Henri Becquerel on the salts of the rare metal uranium. Becquerel had shown that by placing some uranium salt on a photographic plate, covered with black paper, the plate would be affected as if light had fallen on it. The effect is produced by special rays which are emitted by the uranium salt and are different from ordinary luminous rays as they can pass through black paper. Becquerel also showed that these rays can discharge an electroscope. He at first thought that the uranium rays were produced as a result of exposing the uranium salt to light, but experiment showed that salts kept for several months in the dark continued the peculiar rays.

My husband and I were much excited by this new phenomenon, and I resolved to undertake the special study of it. It seemed to me that the first thing to do was to measure the phenomenon with precision. In this I decided to use that property of the rays which enabled them to discharge an electroscope. However, instead of the usual electroscope, I used a more perfect apparatus. One of the models of the apparatus used by me for these first measurements is now in the College of Physicians and Surgeons in Philadelphia.

I was not long in obtaining interesting results. My determinations showed that the emission of the rays is an atomic property of the uranium, whatever the physical or chemical conditions of the salt were. Any substance containing uranium is as much more active in emitting rays, as it contains more of this element.

I then thought to find out if there were other substances possessing this remarkable property of uranium, and soon found that substances containing thorium behaved in a similar way, and that this behavior depended similarly on an atomic property of thorium. I was now about to undertake a detailed study of the uranium and thorium rays when I discovered a new interesting fact.

I had occasion to examine a certain number of minerals. A few of them showed activity; they were those containing either uranium or thorium. The activity of these minerals would have had nothing astonishing about it, if it had been in proportion to the quantities of uranium or thorium contained in them. But it was not so. Some of these minerals revealed an activity three or four times greater than that of uranium. I verified this surprising fact carefully, and could not doubt its truth. Speculating about the reason for this, there seemed to be but one explanation. There must be, I thought, some unknown substance, very active, in these minerals. My husband agreed with me and I urged that we search at once for this hypothetical substance, thinking that, with joined efforts, a result would be quickly obtained. Neither of us could foresee that in beginning this work we were to enter the path of a new science which we should follow for all our future.

Of course, I did not expect, even at the beginning, to find a new element in any large quantity, as the minerals had already been analyzed with some precision. At least, I thought there might be as much as one per cent of the unknown substance in the minerals. But the more we worked, the clearer we realized that the new radioactive element could exist only in quite minute proportion and that, in consequence, its activity must be very great. Would we have insisted, despite the scarcity of our means of research, if we had known the true proportion of what we were searching for, no one can tell; all that can be said now is that the constant progress of our work held us absorbed in a passionate research, while the difficulties were ever increasing. As a matter of fact, it was only after several years of most arduous labor that we finally succeeded in completely separating the new substance, now known to everybody as radium. Here is, briefly, the story of the search and discovery.

As we did not know, at the beginning, any of the chemical properties of the unknown substance, but only that it emits rays, it was by these rays that we had to search. We first undertook the analysis of a pitchblende from St. Joachimsthal. Analyzing this ore by the usual chemical methods, we added an examination of its different parts for radioactivity, by the use of our delicate electrical apparatus. This was the foundation of a new method of chemical analysis which, following our work, has been extended, with the result that a large number of radioactive elements have been discovered.

In a few weeks we could be convinced that our prevision had been right, for the activity was concentrating in a regular way. And, in a few months, we could separate from the pitchblende a substance accompanying the bismuth, much more active than uranium, and having well defined chemical properties. In July, 1898, we announced the existence of this new substance, to which I gave the name of polonium, in memory of my native country.

While engaged in this work on polonium, we had also discovered that, accompanying the barium separated from the pitchblende, there was another new element. After several months more of close work we were able to separate this second new substance, which was afterwards shown to be much more important than polonium. In December, 1898, we could announce the discovery of this new and now famous element, to which we gave the name of radium.

However, the greatest part of the material work had yet to be done. We had, to be sure, discovered the existence of the remarkable new elements, but it was chiefly by their radiant properties that these new substances were distinguished from the bismuth and barium with which they were mixed in minute quantities. We had still to separate them as pure elements. On this work we now started.

We were very poorly equipped with facilities for this purpose. It was necessary to subject large quantities of ore to careful chemical treatment. We had no money, no suitable laboratory, no personal help for our great and difficult undertaking. It was like creating something out of nothing, and if my earlier studying years had once been called by my brother-in-law the heroic period of my life, I can say without exaggeration that the period on which my husband and I now entered was truly the heroic one of our common life.

We knew by our experiments that in the treatment of pitchblende at the uranium plant of St. Joachimsthal, radium must have been left in the residues, and, with the permission of the Austrian government, which owned the plant, we succeeded in securing a certain quantity of these residues, then quite valueless,—and used them for extraction of radium. How glad I was when the sacks arrived, with the brown dust mixed with pine needles, and when the activity proved even greater than that of the primitive ore! It was a stroke of luck that the residues had not been thrown far away or disposed of in some way, but left in a heap in the pine wood near the plant. Some time later, the Austrian government, on the proposition of the Academy of Science of Vienna, let us have several tons of similar residues at a low price. With this material was prepared all the radium I had in my laboratory up to the date when I received the precious gift from the American women.

The School of Physics could give us no suitable premises, but for lack of anything better, the Director permitted us to use an abandoned shed which had been in service as a dissecting room of the School of Medicine. Its glass roof did not afford complete shelter against rain; the heat was suffocating in summer, and the bitter cold of winter was only a little lessened by the iron stove, except in its immediate vicinity. There was no question of obtaining the needed proper apparatus in common use by chemists. We simply had some old pine-wood tables with furnaces and gas burners. We had to use the adjoining yard for those of our chemical operations that involved producing irritating gases; even then the gas often filled our shed. With this equipment we entered on our exhausting work.

Yet it was in this miserable old shed that we passed the best and happiest years of our life, devoting our entire days to our work. Often I had to prepare our lunch in the shed, so as not to interrupt some particularly important operation. Sometimes I had to spend a whole day mixing a boiling mass with a heavy iron rod nearly as large as myself. I would be broken with fatigue at the day's end.

Other days, on the contrary, the work would be a most minute and delicate fractional crystallization, in the effort to concentrate the radium. I was then annoyed by the floating dust of iron and coal from which I could not protect my precious products. But I shall never be able to express the joy of the untroubled quietness of this atmosphere of research and the excitement of actual progress with the confident hope of still better results. The feeling of discouragement that sometimes came after some unsuccessful toil did not last long and gave way to renewed activity. We had happy moments devoted to a quiet discussion of our work, walking around our shed.

One of our joys was to go into our workroom at night; we then perceived on all sides the feebly luminous silhouettes of the bottles or capsules containing our products. It was really a lovely sight and one always new to us. The glowing tubes looked like faint, fairy lights.

Thus the months passed, and our efforts, hardly interrupted by short vacations, brought forth more and more complete evidence. Our faith grew ever stronger, and our work being more and more known, we found means to get new quantities of raw material and to carry on some of our crude processes in a factory, allowing me to give more time to delicate finishing treatment.

At this stage I devoted myself especially to the purification of the radium, my husband being absorbed by the study of the physical properties of the rays emitted by the new substances. It was only after treating one ton of pitchblende residues that I could get definite results. Indeed we know to-day that even in the best minerals there are not more than a few decigrammes of radium in a ton of raw material.

At last the time came when the isolated substances showed all the characters of a pure chemical body. This body, the radium, gives a characteristic spectrum, and I was able to determine for it an atomic weight much higher than that of the barium. This was achieved in 1902. I then possessed one decigramme of very pure radium chloride. It had taken me almost four years to produce the kind of evidence which chemical science demands, that radium is truly a new element. One year would probably have been enough for the same purpose, if reasonable means had been at my disposal. The demonstration that cost so much effort was the basis of the new science of radioactivity.

Source: Curie, Marie. 1963. "Autobiographical Notes." Pp. 88–93 in Marie Curie, *Pierre Curie*. New York: Dover Publications.

Margaret Cavendish had a lifelong interest in science. She eagerly read and responded in her published writings to the work of Francis Bacon and Rene Descartes who outlined the methodology for the new science. The Blazing World, which she appended to her Observations upon Experimental Philosophy, is a fictional work in which she satirized the Royal Society for ostracizing her because she was unwilling to accept the experimental method. She preferred following Descartes's work, which she believed encouraged her to rely solely on her own logic to understand the material world. Cavendish and her work highlight a woman's interest in the sciences and her inability to move into the inner circle of male scientists. In these excerpts, Cavendish pokes fun at scientists' reliance on flawed instruments to understand nature and concludes that the only way to know the material world is through the reasoning of one's own mind.

The Blazing World

MARGARET CAVENDISH

Lastly, the Empress asked the bird-men of the nature of thunder and lightning? and whether it was not caused by roves of ice falling upon each other? To which they answered, that it was not made that way, but by an encounter of cold and heat; so that an exhalation being kindled in the clouds, did dash forth lightning, and that there were so many rentings of clouds as there were sounds and cracking noises: but this opinion was contradicted by others, who affirmed that thunder was a sudden and monstrous blas, stirred up in the air, and did not always require a cloud; but the Empress not knowing what they meant by blas (for even they themselves were not able to explain the sense of this word) liked the former better; and to avoid hereafter tedious disputes, and have the truth of the phaenomenas of celestial bodies more exactly known, commanded the bear-men, which were her experimental philosophers, to observe them through such instruments as are called telescopes, which they did according to her Majesty's command; but these telescopes caused more differences and divisions amongst them, than ever they had before; for some said, they perceived that the sun stood still, and the earth did move about it, others were of opinion, that they both did move; and others said again, that the earth stood still, and the sun did move; some counted more stars than others; some discovered new stars never seen before; some fell into a great dispute with others concerning the bigness of the stars; some said the moon was another world like their terrestrial globe, and the spots therein were hills and valleys; but others would have the spots to be the terrestrial parts, and the smooth and glossy parts, the sea: at last, the Empress commanded them to go with their telescopes to the very end of the Pole that was joined to the world she came from, and try whether they could perceive any stars in it; which they did; and being returned to her Majesty, reported that they had seen three blazing-stars appear there, one after another in a short time, whereof two were bright, and one dim; but they could not agree neither in this observation; for some said it was but one star which appeared at three several times, in several places; and others would have them to be three several stars; for they thought it impossible, that those three several appearances should have been but one star, because every star did rise at a certain time, and appeared in a certain place, and did disappear in the same place: next, it is altogether improbable, said they, that one star should fly from place to place, especially at such a vast distance, without a visible motion, in so short a time, and appear in such different places, whereof two were quite opposite, and the third side-ways: lastly, if it had been but one star, said they, it would always have kept the same splendour, which it did not; for, as above mentioned, two were bright, and one was dim. After they had thus argued, the Empress began to grow angry at their telescopes, that they could give no better intelligence; for, said she, now I do plainly perceive, that your glasses are false informers, and instead of discovering the truth, delude your senses; wherefore I command you to break them, and let the bird-men trust only to their natural eyes, and examine celestial objects by the motions of their own sense and reason. The bear-men replied, that it was not the fault of their glasses, which caused such differences in their opinions, but the sensitive motions in their optic organs did not move alike, nor were their rational judgments always regular: to which the Empress answered, that if their glasses were true informers, they would rectify their irregular sense and reason; but, said she, nature has made your sense and reason more regular than art has your glasses, for they are mere deluders, and will never lead you to the knowledge of truth; wherefore I command you again to break them; for you may observe the progressive motions of celestial bodies with your natural eyes better than through artificial glasses. The bear-men being exceedingly troubled at her Majesty's dipleasure concerning their telescopes, kneeled down, and in the humblest manner petitioned that they might not be broken; for, said they, we take more delight in artificial delusions, than in natural truths. Besides, we shall want employments for our senses, and subjects for arguments; for were there nothing but truth, and no falsehood, there would be no occasion for to dispute, and by this means we should want the aim and pleasure of our endeavours in confuting and contradicting each other; neither would one man be thought wiser than another, but all would either be alike knowing and wise, or all would be fools; wherefore we most humbly beseech your Imperial Majesty to spare our glasses, which are our only delight, and as dear to us as our lives. The Empress at last consented to their request, but upon condition, that their disputes and quarrels should remain within their schools, and cause no factions or disturbances in state, or government. The bear-men, full of joy, returned their most humble thanks to the Empress; and to make her amends for the displeasure which their telescopes had occasioned, told her Majesty, that they had several other artificial optic-glasses, which they were sure would give her Majesty a great deal more satisfaction. Amongst the rest they brought forth several microscopes, by the means of which they could enlarge the shapes of little bodies, and make a louse appear as big as an elephant, and a mite as big as a whale. First of all they showed the Empress a grey drone-fly, wherein they observed that the greatest part of her face, nay, of her head, consisted of two large bunches all covered over with a multitude of small pearls or hemispheres in a trigonal order, which pearls were of two degrees, smaller and bigger; the smaller degree was lower-most, and looked towards the ground; the other was upward, and looked sideward, forward and backward: they were all so smooth and polished, that they were able to represent the image of any object, the number of them was in all 14000. After the view of this strange and miraculous creature, and their several observations upon it, the Empress asked them what they judged those little hemispheres might be? They answered, that each of them was a perfect eye, by reason they perceived that each was covered with a transparent cornea, containing a liquor within them, which resembled the watery or glassy humour of the eye. To which the Empress replied, that they might be glassy pearls, and yet not eyes, and that perhaps their microscopes did not truly inform them: but they smilingly answered her Majesty, that she did not know the virtue of those microscopes; for they did never delude, but rectify and inform their senses; nay, the world, said they, would be but blind without them, as it has been in former ages before those microscopes were invented. . . .

You have converted me, said the Duchess to the spirits, from my ambitious desire; wherefore I'll take your advice, reject and despise all the worlds without me, and create a world of my own. The Empress said, if I do make such a world, then I shall be mistress of two worlds, one within, and the other without me. That your Majesty may, said the spirits; and so left these two ladies to create two worlds within themselves: who did also part from each other, until such time as they had brought their worlds to perfection. The Duchess of Newcastle was most earnest and industrious to make her world, because she had none at present; and first she resolved to frame it according to the opinion of Thales, but she found herself so much troubled with demons, that they would not suffer her to take her own will, but forced her to obey their orders and commands; which she being unwilling to do, left off from making a world that way, and began to frame one according to Pythagoras's doctrine; but in the creation thereof, she was so puzzled with numbers, how to order and compose the several parts, that she having no skill in arithmetic was forced also to desist from the making of that world. Then she intended to create a world according to the opinion of Plato; but she found more trouble and difficulty in that, than in the two former; for the numerous Ideas having no other motion but what was derived from her mind, whence they did flow and issue out, made it a far harder business to her, to impart motion to them, than puppet-players have in giving motion to every several puppet; in so much, that her patience was not able to endure the trouble which those ideas caused her; wherefore she annihilated also that world, and was resolved to make one according to the opinion of Epicurus; which she had no sooner begun, but the infinite atoms made such a mist, that it quite blinded the perception of her mind; neither was she able to make a vacuum as a receptacle for those atoms, or a place which they might retire into; so that partly for the want of it, and of a good order and method, the confusion of those atoms produced such strange and monstrous figures, as did more affright than delight her, and caused such a chaos in her mind, as had almost dissolved it. At last, having with much ado cleansed and cleared her mind of these dusty and misty particles, she endeavoured to create a world according to Aristotle's opinion; but remembering that her mind, as most of the learned hold it, was immaterial, and that according to Aristotle's principle, out of nothing, nothing could be made; she was forced also to desist from that work, and then she fully resolved, not to take any more patterns from the ancient philosophers, but to follow the opinions of the moderns; and to that end, she endeavoured to make a world according to Descartes' opinion; but when she had made the ethereal globules, and set them a-moving by a strong and lively imagination, her mind became so dizzy with their extraordinary swift turning round, that it almost put her into a swoon; for her thoughts, by their constant tottering, did so stagger, as if they had all been drunk: wherefore she dissolved that world, and began to make another, according to Hobbes' opinion; but when all the parts of this imaginary world came to press and drive each other, they seemed like a company of wolves that worry sheep, or like so many dogs that hunt after hares; and when she found a reaction equal to those pressures, her mind was so squeezed together, that her thoughts could neither move forward nor backward, which caused such an horrible pain in her head, that although she had dissolved that world, yet she could not, without much difficulty, settle her mind, and free it from that pain which those pressures and reactions had caused in it.

At last, when the Duchess saw that no patterns would do her any good in the framing of her world; she resolved to make a world of her own invention, and this world was composed of sensitive and rational self-moving matter; indeed, it was composed only of the rational, which is the subtlest and purest degree of matter; for as the sensitive did move and act both to the perceptions and consistency of the body, so this degree of matter at the same point of time (for though the degrees are mixed, yet the several parts may move several ways at one time) did move to the creation of the imaginary world; which world after it was made, appeared so curious and full of variety, so well ordered and wisely governed, that it cannot possibly be expressed by words, nor the delight and pleasure which the Duchess took in making this world of her own.

Source: Cavendish, Margaret. 1994. "The Description of a New World, Called the Blazing World." Pp. 140–143 and 186–188 in Kate Lilley (ed.), *The Blazing World and Other Writings.* New York: Penguin Books.

After intensive study of Cartesian philosophy, Anne Conway rejected it in her Principles *and proposed instead her own "monastic vitalist" philosophy, arguing that nature was a living organism and that there was no separation between mind and body. Her work demonstrates that women were not only engaging with the new science but were also willing to contest it. In these excerpts Conway puts forth her notion of the inseparable nature of mind, spirit, and body.*

The Principles of the Most Ancient and Modern Philosophy

ANNE CONWAY

Now therefore let us examine, how every Creature is composed, and how the parts of its composition may be converted the one into the other; for that they have originally one and the same Essence, or Being.

In every visible Creature there is a Body and a Spirit, or *Principium magis Activum, & magis Passivum*, or, *more Active and more Passive Principle*, which may fitly be termed Male and Female, by reason of that Analogy a Husband hath with his Wife. For as the ordinary Generation of Men requires a Conjunction and Co-operation of Male and Female; so also all Generations and Productions whatsoever they be, require an Union, and conformable Operation of those Two Principles, to wit, Spirit and Body; but the Spirit is an Eye or Light beholding its own proper Image, and the Body is a Tenebrosity or Darkness receiving that Image, when the Spirit looks there-into, as when one sees himself in a Looking-Glass; for certainly he cannot so behold himself in the Transparent Air, nor in any Diaphanous Body, because the reflexion of an Image requires a certain opacity or darkness, which we call a Body: Yet to be a Body is not an Essential property of any Thing; as neither is it a Property of any Thing to be dark; for nothing is so dark that it cannot be made Light; yea, the Darkness it self may become Light, as the Light which is created may be turned into Darkness, as the Words of Christ do fully evince, when he saith, *If the Light which is in thee be darkness*, &c. where he means the Eye or Spirit which is in the Body, which beholdeth the Image of any Thing: Therefore as every Spirit hath need of a Body, that it may receive and reflect its Image, so also it requires a Body to retain the same; for every Body hath this retentive Nature, either more or less in it self; and by how much the perfecter a Body is, that is, more perfectly mix'd, so much the more retentive is it, and so Water is more retentive than Air, and Earth of some Things is more retentive than Water.

But the Seed of a Female Creature, by reason of its so perfect mixture; for that it is the purest Extraction of the whole Body, hath in it a noble retention: And in this Seed, as a Body, the Male Seed, which is the Image and Spirit of the Male, is received and retained, together with other Spirits which are in the Female; and therefore whatsoever Spirit is then strongest, and hath the strongest Image or *Idea* in the Seed, whether it be the Masculine or the Feminine, or any other Spirit from either of these received from without, that Spirit is predominant in the Seed, and forms the Body, as near as may be, after its own Image, and so every Creature receives his External Form. And after the same manner also, the Internal Productions of the Mind, *viz.* Thoughts are generated, which according to their Kind are true Creatures, and have a true Substance, proper to themselves, being all our Internal Children, and all of them Male and Female, that is, they have Body and Spirit; for if they had not a Body, they could not be retained, nor could we reflect on our own proper Thoughts; for every reflection is made by a certain Tenebrosity or Darkness, and this is a Body; so the Memory requires a Body, to retain the Spirit of the Thing thought on, otherwise it would vanish as the Image in a Glass, which presently vanishes, the Object being removed. And so likewise, when we remember any Body, we see his Image in us, which is a Spirit that proceeded from him, whilst we beheld him from without; which Image or Spirit is retained in Some Body, which is the Seed of our Brain, and thence is made a certain Spiritual Generation in us: And so every Spirit hath its Body, and every Body its Spirit; and as the Body, *sc.* of a Man or Beast, is nothing else but an innumerable multitude of Bodies, compacted together into one, and disposed into a certain order; so likewise the Spirit of a Man, or Beast, is a certain innumerable multitude of Spirits united together in the said Body, which have their Order and Government so, that there is one Captain, or Chief Governor, another a Lieutenant, and another hath a certain kind of Government under him, and so through the whole, as it is wont to be in an Army of Soldiers; wherefore the Creatures are called Armies, and God the God of Hosts, as the Devil which possessed the Man was called Legion, because there were many of them; so that every Man; yea, every Creature, consists of many Spirits and Bodies; (many of these Spirits which exist in Man) are called by the *Hebrews, Nizzuzoth, or Sparks.* See in *Kabbal. denud.* Tom. 2. Part 2. Tract. *de revolutionibus animarum,* Cap. 2. & *seq.* p. 256, 268, &c.) And indeed every Body is a Spirit, and nothing else, neither differs any thing from a Spirit, but in that it is more dark; therefore by how much the thicker and grosser it is become, so much the more remote is it from the degree of a Spirit, so that this distinction is only modal and gradual, not essential or substantial. . . .

From what hath been lately said, and from divers Reasons alledged, That Spirit and Body are originally in their first Substance but one and the same thing, it evidently appears that the Philosophers (so called) which have taught otherwise, whether Ancient or Modern, have generally erred and laid an ill Foundation in the very beginning, whence the whole House and superstructure is so feeble, and indeed so unprofitable, that the whole Edifice and Building must in time decay, from which absurd Foundation have arose very many gross and dangerous Errours, not only in Philosophy, but also in Divinity (so called) to the great damage of Mankind, hindrance of true Piety, and contempt of God's most Glorious Name, as will easily appear, as well from what hath been already said, as from what shall be said in this Chapter.

And none can Object, That all this Philosophy is no other than that of *des Cartes,* or *Hobbs* under a new Mask. For, First, as touching the *Cartesian* Philosophy, this saith that every Body is a mere dead Mass, not only void of all kind of Life and Sense, but utterly uncapable thereof to all Eternity; this grand Errour also is to be imputed to all those who affirm Body and Spirit to be contrary Things, and inconvertible one into another, so as to deny a Body all Life and Sense; which is quite contrary to the grounds of this our Philosophy. Wherefore it is so far from being a *Cartesian* Principle, under a new Mask, that it may be truly said it is *Anti-Cartesian,* in regard of their Fundamental Principles; although it cannot be denied that *Cartes* taught many excellent and ingenious Things concerning the Mechanical part of Natural Operations, and how all Natural Motions proceed according to Rules and Laws Mechanical, even as indeed Nature her self, *i.e.,* the Creature, hath an excellent Mechanical Skill and Wisdom in it self, (given it from God, who is the Fountain of all Wisdom,) by which it operates: But yet in Nature, and her Operations, they are far more than merely Mechanical; and the same is not a mere Organical Body, like a Clock, wherein there is not a vital Principle of Motion; but a living Body, having Life and Sense, which Body is far more sublime than a mere Mechanism, or Mechanical Motion.

Source: Conway, Anne. 1982. *The Principles of the Most Ancient and Modern Philosophy.* Pp. 188–190 and 221–222. Dordrecht, The Netherlands: Kluwer. (Reprinted with kind permission of Kluwer Academic Publishers.)

In The New Organon *Francis Bacon outlined his new methodology for a science of the material world. He believed that knowledge could be obtained only by looking objectively at the natural world, collecting information, and conducting experiments. In this excerpt, Bacon outlines why induction is the preferable mode of scientific investigation, noting that this approach allows the investigator to overcome the errors of the mind and the senses and will hopefully result in discoveries that will build a better life for human beings. Note that in this section he refers to the philosopher as male ("True sons of knowledge") and refers to nature as female ("we may find a way at length into her inner chambers").*

The New Organon

FRANCIS BACON

But the greatest change I introduce is in the form itself of induction and the judgment made thereby. For the induction of which the logicians speak, which proceeds by simple enumeration, is a puerile thing, concludes at hazard, is always liable to be upset by a contradictory instance, takes into account only what is known and ordinary, and leads to no result.

Now what the sciences stand in need of is a form of induction which shall analyze experience and take it to pieces, and by a due process of exclusion and rejection lead to an inevitable conclusion. And if that ordinary mode of judgment practiced by the logicians was so laborious, and found exercise for such great wits, how much more labor must we be prepared to bestow upon this other, which is extracted not merely out of the depths of the mind, but out of the very bowels of nature.

Nor is this all. For I also sink the foundations of the sciences deeper and firmer; and I begin the inquiry nearer the source than men have done heretofore, submitting to examination those things which the common logic takes on trust. For first, the logicians borrow the principles of each science from the science itself; secondly, they hold in reverence the first notions of the mind; and lastly, they receive as conclusive the immediate informations of the sense, when well disposed. Now upon the first point, I hold that true logic ought to enter the several provinces of science armed with a higher authority than belongs to the principles of those sciences themselves, and ought to call those putative principles to account until they are fully established. Then with regard to the first notions of the intellect, there is not one of the impressions taken by the intellect when left to go its own way, but I hold it as suspect and no way established until it has submitted to a new trial and a fresh judgment has been thereupon pronounced. And lastly, the information of the sense itself I sift and examine in many ways.

For certain it is that the senses deceive; but then at the same time they supply the means of discovering their own errors; only the errors are here, the means of discovery are to seek.

The sense fails in two ways. Sometimes it gives no information, sometimes it gives false information. For first, there are very many things which escape the sense, even when best disposed and no way obstructed, by reason either of the subtlety of the whole body or the minuteness of the parts, or distance of place, or slowness or else swiftness of motion, or familiarity of the object, or other causes. And again when the sense does apprehend a thing its apprehension is not much to be relied upon. For the testimony and information of the sense has reference always to man, not to the universe; and it is a great error to assert that the sense is the measure of things.

To meet these difficulties, I have sought on all sides diligently and faithfully to provide helps for the sense—substitutes to supply its failures, rectifications to correct its errors; and this I endeavor to accomplish not so much by instruments as by experiments. For the subtlety of experiments is far greater than that of the sense itself, even when assisted by exquisite instruments—such experiments, I mean, as are skillfully and artificially devised for the express purpose of determining the point in question. To the immediate and proper perception of the sense, therefore, I do not give much weight; but I contrive that the office of the sense shall be only to judge of the experiment, and that the experiment itself shall judge of the thing. And thus I conceive that I perform the office of a true priest of the sense (from which all knowledge in nature must be sought, unless men mean to go mad) and a not unskillful interpreter of its oracles; and that while others only profess to uphold and cultivate the sense, I do so in fact. Such then are the provisions I make for finding the genuine light of nature and kindling and bringing it to bear. And they would be sufficient of themselves if the human intellect were even and like a fair sheet of paper with no writing on it. But since the minds of men are strangely possessed and beset so that there is no true and even surface left to reflect the genuine rays of things, it is necessary to seek a remedy for this also.

Now the idols, or phantoms, by which the mind is occupied are either adventitious or innate. The adventitious come into the mind from without—namely, either from the doctrines and sects of philosophers or from perverse rules of demonstration. But the innate are inherent in the very nature of the intellect, which is far more prone to error than the sense is. For let men please themselves as they will in admiring and almost adoring the human mind, this is certain: that as an uneven mirror distorts the rays of objects according to its own figure and section, so the mind, when it receives impressions of objects through the sense, cannot be trusted to report them truly, but in forming its notions mixes up its own nature with the nature of things.

And as the first two kinds of idols are hard to eradicate, so idols of this last kind cannot be eradicated at all. All that can be done is to point them out, so that this insidious action of the mind may be marked and reproved (else as fast as old errors are destroyed new ones will spring up out of the ill complexion of the mind itself, and so we shall have but a change of errors, and not a clearance); and to lay it down once for all as a fixed and established maxim that the intellect is not qualified to judge except by means of induction, and induction in its legitimate form. This doctrine, then, of the expurgation of the intellect to qualify it for dealing with truth is comprised in three refutations: the refutation of the philosophies; the refutation of the demonstrations; and the refutation of the natural human reason. The explanation of which things, and of the true relation between the nature of things and the nature of the mind, is as the strewing and decoration of the bridal chamber of the mind and the universe, the divine goodness assisting, out of which marriage let us hope (and be this the prayer of the bridal song) there may spring helps to man, and a line and race of inventions that may in some degree subdue and overcome the necessities and miseries of humanity. . . .

Now my method, though hard to practice, is easy to explain; and it is this. I propose to establish progressive stages of certainty. The evidence of the sense, helped and guarded by a certain process of correction, I retain. But the mental operation which follows the act of sense I for the most part reject; and instead of it I open and lay out a new and certain path for the mind to proceed in, starting directly from the simple sensuous perception. The necessity of this was felt, no doubt, by those who attributed so much importance to logic, showing thereby that they were in search of helps for the understanding, and had no confidence in the native and spontaneous process of the mind. But this remedy comes too late to do any good, when the mind is already, through the daily intercourse and conversation of life, occupied with unsound doctrines and beset on all sides by vain imaginations. And therefore that art of logic, coming (as I said) too late to the rescue, and no way able to set matters right again, has had the effect of fixing errors rather than disclosing truth. There remains but one course for the recovery of a sound and healthy condition—namely, that the entire work of the understanding be commenced afresh, and the mind itself be from the very outset not left to take its own course, but guided at every step; and the business be done as if by machinery. Certainly if in things mechanical men had set to work with their naked hands, without help or force of instruments, just as in things intellectual they have set to work with little else than the naked forces of the understanding, very small would the matters have been which, even with their best efforts applied in conjunction, they could have attempted or accomplished. Now (to pause a while upon this example and look in it as in a glass) let us suppose that some vast obelisk were (for the decoration of a triumph or some such magnificence) to be removed from its place, and that men should set to work upon it with their naked hands, would not any sober spectator think them mad? And if they should then send for more people, thinking that in that way they might manage it, would he not think them all the madder? And if they then proceeded to make a selection, putting away the weaker hands, and using only the strong and vigorous, would he not think them madder than ever? And if lastly, not content with this, they resolved to call in aid the art of athletics, and required all their men to come with hands, arms, and sinews well anointed and medicated according to the rules of the art, would he not cry out that they were only taking pains to show a kind of method and discretion in their madness? Yet just so it is that men proceed in matters intellectual—with just the same kind of mad effort and useless combination of forces—when they hope great things either from the number and cooperation or from the excellency and acuteness of individual wits; yea, and when they endeavor by logic (which may be considered as a kind of athletic art) to strengthen the sinews of the understanding, and yet with all this study and endeavor it is apparent to any true judgment that they are but applying the naked intellect all the time; whereas in every great work to be done by the hand of man it is manifestly impossible, without instruments and machinery, either for the strength of each to be exerted or the strength of all to be united.

Upon these premises two things occur to me of which, that they may not be overlooked, I would have men reminded. First, it falls out fortunately as I think for the allaying of contradictions and heartburnings, that the honor and reverence due to the ancients remains untouched and undiminished, while I may carry out my designs and at the same time reap the fruit of my modesty. For if I should profess that I, going the same road as the ancients, have something better to produce, there must needs have been some comparison or rivalry between us (not to be avoided by any art of words) in respect of excellency or ability of wit; and though in this there would be nothing unlawful or new (for if there be anything misapprehended by them, or falsely laid down, why may not I, using a liberty common to all, take exception to it?) yet the contest, however just and allowable, would have been an unequal one perhaps, in respect of the measure of my own powers. As it is, however (my object being to open a new way for the understanding, a way by them untried and unknown), the case is altered: party zeal and emulation are at an end, and I appear merely as a guide to point out the road—an office of small authority, and depending more upon a kind of luck than upon any ability or excellency. And thus much relates to the persons only. The other point of which I would have men reminded relates to the matter itself.

Be it remembered then that I am far from wishing to interfere with the philosophy which now flourishes, or with any other philosophy more correct and complete than this which has been or may hereafter be propounded. For I do not object to the use of this received philosophy, or others like it, for supplying matter for disputations or ornaments for discourse—for the professor's lecture and for the business of life. Nay, more, I declare openly that for these uses the philosophy which I bring forward will not be much available. It does not lie in the way. It cannot be caught up in passage. It does not flatter the understanding by conformity with preconceived notions. Nor will it come down to the apprehension of the vulgar except by its utility and effects.

Let there be therefore (and may it be for the benefit of both) two streams and two dispensations of knowledge, and in like manner two tribes or kindreds of students in philosophy—tribes not hostile or alien to each other, but bound together by mutual services; let there in short be one method for the cultivation, another for the invention, of knowledge.

And for those who prefer the former, either from hurry or from considerations of business or for want of mental power to take in and embrace the other (which must needs be most men's case), I wish that they may succeed to their desire in what they are about, and obtain what they are pursuing. But if there be any man who, not content to rest in and use the knowledge which has already been discovered, aspires to penetrate further; to overcome, not an adversary in argument, but nature in action; to seek, not pretty and probable conjectures, but certain and demonstrable knowledge—I invite all such to join themselves, as true sons of knowledge, with me, that passing by the outer courts of nature, which numbers have trodden, we may find a way at length into her inner chambers. And to make my meaning clearer and to familiarize the thing by giving it a name, I have chosen to call one of these methods or ways *Anticipation of the Mind*, the other *Interpretation of Nature*.

Source: Bacon, Francis. 2000. "The Great Instauration" and "Author's Preface." Pp. 20–23 and 33–37 in Fulton H. Anderson (ed.), *The New Organon and Related Writings*. New York: Bobbs-Merrill Company.

Dr. Henry Maudsley was a British psychologist who, like Dr. Clarke in the United States, believed that women should not attend institutions of higher education nor be educated in a similar manner to men. Using medical science, specifically physiology, to support his assertions, Maudsley argued that women risked their physical and mental health, their reproductive capacity, and even their lives by engaging in higher education pursuits. According to Maudsley, women are physically incapable of the intellectual challenge of higher education, and their bodies and minds have been suited by nature for them to be wives and mothers and therefore their education should follow suit. Maudsley's "scientific" argument conveniently supports the nineteenth-century social and cultural status quo for women's place in society.

Sex in Mind and Education

HENRY MAUDSLEY

Explanatory Note

Maudsley, Henry (1835–1918), British Medical psychologist. He studied at University College Hospital, London and graduated M.B. and M.R.C.S. in 1856, and M.D. in 1857. In 1859 he became Medical Superintendent of the Manchester Royal Lunatic Asylum. He began a career in psychiatry and obtained his M.R.C.P. in 1861. He was joint editor of the Journal of Mental Science from 1863 to 1878. Maudsley was appointed physician to the West London Hospital in 1864 and lectured in psychiatry at St. Mary's Hospital from 1868–81. He became Professor of Medical Jurisprudence at University College, London from 1869–79 and was elected F.R.C.P. in 1869. His works include: The Physiology and Pathology of Mind *(1867),* Body and Mind *(1870),* Responsibility in Mental Disease *(1874),* Natural Causes and Supernatural Seemings *(1886) and* Organic to human—Psychological to Sociological *(1916), and an autobiography (1912).*

Those who view without prejudice, or with some sympathy the movements for improving the higher education of women, and for throwing open to them fields of activity from which they are now excluded, have a hard matter of it sometimes to prevent a feeling of reaction being aroused in their minds by the arguments of the most eager of those who advocate the reform. Carried away by their zeal into an enthusiasm which borders on or reaches fanaticism, they seem positively to ignore the fact that there are significant differences between the sexes, arguing in effect as if it were nothing more than an affair of clothes, and to be resolves, in their indignation at woman's wrongs, to refuse her the simple rights of her sex. They would do better in the end if they would begin by realising the fact that the male organization is one, and the female organization another, and that, let come what come may in the way of assimilation of female and male education and labour, it will not be possible to transform a woman into a man. To the end of the chapter she will retain her special functions, and must have a special sphere of development and activity determined by the performance of those functions.

It is quite evident that many of those who are foremost in their zeal for raising the education and social status of women, have not given proper consideration to the nature of her organization, and to the demands which its special functions make upon its strength. These are matters which it is not easy to discuss out of a medical journal; but, in view of the importance of the subject at the present stage of the question of female education, it becomes a duty to use plainer language than would otherwise be fitting in a literary journal. The gravity of the subject can hardly be exaggerated. Before sanctioning the proposal to subject woman to a system of mental training which has been framed and adapted for men, and under which they have become what they are, it is needful to consider whether this can be done without serious injury to her health and strength. It is not enough to point to exceptional instances of women who have undergone such a training, and have proved their capacities when tried by the same standard as men; without doubt there are women who can, and will, so distinguish themselves, if stimulus be applied and opportunity given; the question is, whether they may not do it at a cost which is too large a demand upon the resources of their nature. Is it well for them to contend on equal terms with men for the goal of man's ambition?

Let it be considered that the period of the real educational strain will commence about the time when, by the development of the sexual system, a great revolution takes place in the body and mind, and an extraordinary expenditure of vital energy is made, and will continue through those years after puberty when, by the establishment of periodical functions, a regularly recurring demand is made upon the resources of a constitution that is going through the final stages of its growth and development. The energy of a human body being a definite and not inexhaustible quantity, can it bear, without injury, an excessive mental drain as well as the natural physical drain which is so great at that time? Or, will the profit of the one be to the detriment of the other? It is a familiar experience that a day of hard physical work renders a man incapable of hard mental work, his available energy having been exhausted. Nor does it matter greatly by what channel the energy be expended; if it be used in one way it is not available for use in another. When Nature spends in one direction, she must economise in another direction. That the development of puberty does draw heavily upon the vital resources of the female constitution, needs not to be pointed out to those who know the nature of the important physiological changes which then take place. In persons of delicate constitution who have inherited a tendency to disease, and who have little vitality to spare, the disease is apt to break out at that time; the new drain established having deprived the constitution of the vital energy necessary to withstand the enemy that was lurking in it. The time of puberty and the years following it are therefore justly acknowledged to be a critical time for the female organization. The real meaning of the physiological changes which constitute puberty is, that the woman is thereby fitted to conceive and bear children, and undergoes the bodily and mental changes that are connected with the development of the reproductive system. At each recurring period there are all the preparations for conception, and nothing is more necessary to the preservation of female health than that these changes should take place regularly and completely. It is true that many of them are destined to be fruitless so far as their essential purpose is concerned, but it would be a great mistake to suppose that on that account they might be omitted or accomplished incompletely, without harm to the general health. They are the expressions of the full physiological activity of the organism. Hence it is that the outbreak of disease is so often heralded, or accompanied, or followed by suppression or irregularity of these functions. In all cases they make a great demand upon the physiological energy of the body; they are sensitive to its sufferings, however these be caused; and, when disordered, they aggravate the mischief that is going on.

When we thus look at the matter honestly in the face, it would seem plain that women are marked out by nature for very different offices in life from those of men, and that the healthy performance of her special functions renders it improbable she will succeed, and unwise for her to persevere, in running over the same course at the same pave with him. For such a race she is certainly weighted unfairly. Not is it a sufficient reply to this argument to allege, as is sometimes done, that there are many women who have not the opportunity of getting married, or who do not aspire to bear children; for whether they care to be mothers or not, they cannot dispense with those physiological functions of their nature that have reference to that aim, however much they might wish it, and they cannot disregard them in the labour of life without injury to their health. They cannot choose but to be women; cannot revel successfully against the tyranny of their organization, the complete development and function whereof must take place after its kind. This is not the expression of prejudice nor the false sentiment it is the plain statement of a physiological fact. Surely, then, it is unwise to pass it by; first or last it must have its due weight in the determination of the problem of women's education and mission; it is best to recognise it plainly, however we may conclude finally to deal with it.

It is sometimes said, however, that sexual difference ought not to have any place in the culture of the mind, and one hears it affirmed with an air of triumphant satisfaction that there is no sex in mental culture. This is a rash statement, which argues want of thought or insincerity of thought in those who make it. There is sex in mind as distinctly as there is sex in body; and if the mind is to receive the best culture of which its nature is capable, regard must be had to the mental qualities which correlate differences of sex. To aim, by means of education and pursuits in life, to assimilate the female to the male mind, might well be pronounced as unwise and fruitless a labour as it would be to strive to assimilate the female to the male body by means of the same kind of physical training and by the adoption of the same pursuits. Without doubt there have been some striking instances of extraordinary women who have shown great mental power, and these may fairly be quoted as evidence in support of the right of women to the best mental culture; but it is another matter when they are adduced in support of the assertion that there is no sex in mind, and that a system of female education should be laid down on the same lines, follow the same method, and have the same ends in view, as a system of education for men.

Let me pause here to reflect briefly upon the influence of sex upon mind. In its physiological sense, with which we are concerned here, mind is the sum of those functions of the brain which are commonly known as thought, feeling and will. Now the brain is one among a number of organs in the commonwealth of the body; with these organs it is in the closest physiological sympathy by definite paths of nervous communication, has special correspondence with them by internuncial nerve-fibres; so that its functions habitually feel and declare the influence of the different organs. There is an intimate consensus of functions. Though it is the highest organ of the body, the co-ordinating centre to which impressions go and from which responses are sent, the nature and functions of the inferior organs with which it lives in unity, affect essentially its nature as the organ of mental functions. It is not merely that disorder of a particular organ hinders or oppresses these functions, but it affects them in a particular way; and we have good reason to believe that this special pathological effect is a consequence of the specific physiological effect which each organ exerts naturally upon the constitution and function of mind. A disordered liver gives rise to gloomy feelings; a diseased heart to feelings of fear and apprehension; morbid irritation of the reproductive organs, to feelings of a still more special kind—these are familiar facts; but what we have to realise is, that each particular organ has, when not disordered, its specific and essential influence in the production of certain passions or feelings. From of old the influence has been recognised, as we see in the doctrine by which the different passions were located in particular organs of the body, the heart, for example, being made the seat of courage, the liver the seat of jealousy, the bowels the seat of compassion; and although we do not now hold that a passion is aroused anywhere else than in the brain, we believe nevertheless that the organs are represented in the primitive passions, and that when the passion is aroused into violent action by some outward cause, it will discharge itself upon the organ and throw its functions into commotion. In fact, as the uniformity of thought among men is due to the uniform operation of the external senses, as they think alike because they have the same number and kind of senses, so the uniformity of their fundamental passions is due probably to the uniform operation of the internal organs of the body upon the brain; they feel alike because they have the same number and kind of internal organs. If this be so, these organs come to be essential constituents of our mental life.

The most striking illustration of the kind of organic action which I am endeavouring to indicate is yielded by the influence of the reproductive organs upon the mind; a complete mental revolution being made when they come into activity. As great a change takes place in the feelings and ideas, the desires and will, as it is possible to imagine, and takes place in virtue of the development of their functions. Let it be noted then that this great and important mental change is different in the two sexes, and reflects the difference of their respective organs and functions. Before experience has opened their eyes, the dreams of a young man and maiden differ. If we give attention to the physiology of the matter, we see that it cannot be otherwise, and if we look to the facts of pathology, which would not fitly be in place here, they are found to furnish the fullest confirmation of what might have been predicted. To attribute to the influence of education the mental differences of sex which declare themselves so distinctly at puberty, would be hardly less absurd than to attribute to education the bodily differences which then declare themselves. The comb of a cock, the antlers of a stag, the mane of a lion, the beard of a man, are growths in relation to the reproductive organs which correlate mental differences of sex as marked almost as these physical differences. In the first years of life, girls and boys are much alike in mental and bodily character, the differences which are developed afterwards being hardly more than intimated, although some have thought the girl's passion for her doll evinces even at that time a forefeeling of her future functions; during the period of reproductive activity, the mental and bodily differences are declared most distinctly; and when that period is past, and man and woman decline into second childhood, they come to resemble one another more again. Furthermore, the bodily form, the voice, and the mental qualities of mutilated men approach those of women; while women whose reproductive organs remain from some cause in a state of arrested development, approach the mental and bodily habits of men.

No psychologist has yet devoted himself to make, or has succeeded in making, a complete analysis of the emotions, by resolving the complex feelings into their simple elements and tracing them back from their complex evolutions to the primitive passions in which they are rooted; this is a promising and much-needed work which remains to be done; but when it is done, it will be shown probably that they have proceeded originally from two fundamental instincts, or—if we add consciousness of nature and aim—passions, namely, that of self-preservation, with the ways and means of self-defence which it inspires and stimulates, and that of propagation, with the love of offspring and other primitive feelings that are connected with it. Could we in imagination trace mankind backwards along the path stretching through the ages, on which it has gone forward to its present height and complexity of emotion, and suppose each new emotional element to be given off at the spot where it was acquired, we should view a road along which the fragments of our high, special and complex feelings were scattered, and should reach a starting-point of the primitive instincts of self-preservation and propagation. Considering, then, the different functions of the sexes in the operation of the latter instinct, and how a different emotional nature has necessarily been grafted on the original differences in the course of ages, does it not appear that in order to assimilate the female to the male mind it would be necessary to undo the life-history of mankind from its earliest commencement? Nay, would it not be necessary to go still farther back to that earliest period of animal life upon earth before there was any distinction of sex?

If the foregoing reflections be well grounded, it is plain we ought to recognise sex in education, and to the specialties of woman's physical and mental nature. Each sex must develop after its kind; and if education in its fundamental meaning be the external cause to which evolution is the internal answer, if it be the drawing out of the internal qualities of the individual into their highest perfection by the influence of the most fitting external conditions, there must be a difference in the method of education of the two sexes answering to differences in their physical and mental natures. Whether it be only the statement of a partial truth, that 'for valour he' is formed, and 'for beauty she and sweet attractive grace,' or not, it cannot be denied that they are formed for different functions, and that the influence of those functions pervades and affects essentially their entire beings. There is sex in mind and there should be sex in education.

Let us consider, then, what an adapted education must have regard to. In the first place, a proper regard to the physical nature of women means attention given, in their training, to their peculiar functions and to their foreordained work as mothers and nurses of children. Whatever aspirations of an intellectual kind they may have, they cannot be relieved from the performance of those offices so long as it is thought necessary that mankind should continue on earth. Even if these be looked upon as somewhat mean and unworthy offices in comparison with the nobler functions of giving birth to and developing ideas; if, agreeing with

Goethe, we are disposed to hold—Es wäre doch immer hübscher wenn man die Kinder von den Baumen schüttelte; it must still be confessed that for the great majority of women they must remain the most important offices of the best period of their lives. Moreover they are work which, like all work, may be well or ill done, and which, in order to be done well, cannot be done in a perfunctory manner, as a thing by the way. It will have to be considered whether women can scorn delights, and live labourious days of intellectual exercise and production, without injury to their functions as the conceivers, mothers, and nurses of children. For it would be an ill thing, if it should so happen, that we got the advantages of a quantity of female intellectual work at the price of a puny, enfeebled, and sickly race. In this relation, it must be allowed that women do not and cannot stand on the same level as men.

In the second place, a proper regard to the mental nature of women means attention given to those qualities of mind which correlate the physical differences of her sex. Men are manifestly not so fitted mentally as women to be the educators of children during the early years of their infancy and childhood; they would be almost as much out of the place in going systematically to work to nurse babies as they would be in attempting to suckle them. On the other hand, women are manifestly endowed with qualities of mind which specially fit them to stimulate and foster the first growths of intelligence in children, while the intimate and special sympathies which a mother has with her child as a being which, though individually separate, is still almost a part of her nature, give her an influence and responsibilities which are specially her own. The earliest dawn of an infant's intelligence is its recognition of its mother as the supplier of its wants, as the person whose near presence is associated with the relief of sensations of discomfort, and with the production of feelings of comfort; while the relief and pleasure which she herself feels in yielding it warmth and nourishment strengthens, if it was not originally the foundation of, that strong love of offspring which, with unwearied patience, surrounds its wayward youth with a thousand ministering attentions. It can hardly be doubted that if the nursing of babies were given over to men for a generation or two, they would abandon the task in despair or in disgust, and conclude it to be not worth while that mankind should continue on earth. But 'can a woman forget her sucking child, that she should not have compassion on the son of her womb?' Those can hardly be in earnest who question that woman's sex is represented in mind, and that the mental qualities which spring from it qualify her specially to be the successful nurse and educator of infants and young children.

Furthermore, the female qualities of mind which correlate her sexual character adapt her, as her sex does, to be the helpmate and companion of man. It was a Eastern idea, which Plato has expressed allegorically, that a complete being had in primeval times been divided into two halves, which have ever since been seeking to unite together and to reconstitute the divided unity. It will hardly be denied that there is a great measure of truth in the fable. Man and woman do complement one another's being. This is no less true of mind than it is of body; is true of mind indeed as a consequence of its being true of body. Some may be disposed to argue that the qualities of mind which characterize women now, and have characterized them hitherto, in their relations with men, are in great measure, mainly if not entirely, the artificial results of the position of subjection and dependence which she has always occupied; but those who take this view do not appear to have considered the matter as deeply as they should; they have attributed to circumstances much of what unquestionably lies deeper than circumstances, being inherent in the fundamental character of sex. It would be a delusive hope to expect, and a mistaken labour to attempt, to eradicate by change of circumstances the qualities which distinguish the female character, and fit woman to be the helpmate and companion of man in mental and bodily union.

So much may be fairly said on general physiological grounds. We may now go on to inquire whether any ill effects have been observed from subjecting women to the same kind of training as men. The facts of experience in this country are not such as warrant a full and definite answer to the inquiry, the movement for revolutionizing the education of women being of a recent date. But in America the same method of training for the sexes in mixed classes has been largely applied; girls have gone with boys through the same curriculum of study, from primary to grammar schools, from schools to graduation in colleges, working eagerly under the stimulus of competition, and disdaining any privilege of sex. With what results? With one result certainly—that while those who are advocates of the mixed system bear favourable witness to the results upon both sexes, American physicians are beginning to raise their voices in earnest warnings and protests. It is not that girls have not ambition, nor that they fail generally to run the intellectual rave which is set before them, but it is asserted that they do it at a cost to their strength and health which entails lifelong suffering, and even incapacitates them for the adequate performance of the natural functions of their sex. Without pretending to endorse these assertions, which it would be wrong to do in the absence of sufficient experience, it is right to call attention to them, and to claim serious consideration for them; they proceed from physicians of high professional standing, who speak from their own experience, and they agree moreover with what perhaps might have been feared or predicted on physiological grounds. It may fairly be presumed that the stimulus of competition will act more powerfully on girls than on boys; not only because they are more susceptible by nature, but because it will produce more effect upon their constitutions when it is at all in excess. Their nerve-centres being in a state of greater instability, by reason of the development of their reproductive functions, they will be the more easily and the more seriously deranged. A great argument used in favour of a mixed education is that it affords adequate stimulants to girls for thorough and sustained work, which have hitherto been a want in girls' schools; that it makes them less desirous to fir themselves only for society, and content to remain longer and work harder at school. Thus it is desired that emulation should be used in order to stimulate them to compete with boys in mental exercises and aims, while it is not pretended they can or should compete with them in those outdoor exercises and pursuits which are of such great benefit in ministering to bodily health, and to success in which boys, not unwisely perhaps, attach scarcely less honour than to intellectual success. It is plain then that the stimulus of competition in studies will act more powerfully upon them, not only because of their greater constitutional susceptibility, but because it is left free to act without the compensating balance of emulation, in other fields of activity. Is it right, may well be asked, that it should be so applied? Can woman rise high in spiritual development of any kind unless she take a holy cate of the temple of her body?

A small volume, entitled *Sex in Education* which has been published recently by Dr. Edward Clarke of Boston, formerly a Professor in Harvard College, contains a somewhat startling description of the baneful effects upon female health which have been produced by an excessive educational strain. It is asserted that the number of female graduates of schools and colleges who have been permanently disabled to a greater or less degree by improper methods of study, and by a disregard to the reproductive apparatus and its functions, is so great as to excite the gravest alarm, and to demand the serious attention of the community.

If these causes should continue for the next half-century, and increase in the same ratio as they have for the last fifty years, it requires no prophet to foretell that the wives who are to be the mothers in our republic must be drawn from Transatlantic homes. The sons of the New World will have to re-act, on a magnificent scale, the old story of unwived Rome and the Sabines.

Dr. Clarke relates the clinical histories of several cases of tedious illness, in which he traced the cause unhesitatingly to a disregard of the function of the female organization. Irregularity, imperfection, arrest, or excess occurs in consequence of the demand made upon the vital powers at times when there should rightly be an intermission or remission of labour, and is followed first by pallor, lassitude, debility, sleeplessness, headache, neuralgia, and then by worse ills. The course of events is something in this wise. The girls enters upon the hard work of school or college at the age of fifteen years or thereabouts, when the functions of her sex has perhaps been fairly established; ambitious to stand high in class, she pursues her studies with diligence, perseverance, constancy, allowing herself no days of relaxation or rest out of the schooldays, paying no attention to the periodical tides of her organization, undeeding a drain 'that would make the stroke oar of the University crew falter.' For a time all seems to go will with her studies; she triumphs over male and female competitors, gains the front rank, and is stimulated to continue exertions in order to hold it. But in the long run nature, which cannot be ignored or defied with impunity, asserts its power, excessive losses occur; health fails, she becomes the victim of aches and pains, is unable to go on with her work, and compelled to seek medical advice. Restored to health by rest from work, a holiday at the sea-side, and suitable treatment, she goes back to her studies, to begin again the same course of unheeding work, until she has completed the curriculum, and leaves college a good scholar but a delicate and ailing women, whose furutre life is one of more or less suffering. For she does not easily regain the vital energy which was recklessly sacrificed in the acquirement of learning; the special functions which have relation to her future offices as woman, and the full and perfect accomplishment of which is essential to sexual completeness, have been deranged at a critical time; if she is subsequently married, she is unfit for the best discharge of maternal functions, and is apt to suffer from a variety of troublesome and serious disorders in connection with them. In some cases the brain and the nervous system testify to the exhaustive effects of undie labour, nervous and even mental disorders declaring themselves.

Such is a picture, painted by an experienced physician, of the effects of subjecting young women to the method of education which has been framed for young men. Startling as it is, there is nothing in it which may not well be true to nature. If it be an effect of excessive and ill-regulated study to produce derangement of the functions of the female organization, of which so far from there being an antecendent improbability there is a great probability, then there can be no question that all the subsequent ills mentioned are likely to follow. The important physiological change which takes place at puberty, accompanied, as it is, by so great a revolution in mind and body, and by so large an expenditure of vital energy, may easily and quickly overstep its healthy limits and pass into a pathological change, under conditions of excessive stimulation, or in persons who are constitutionally feeble and whose nerve-centres are more unstable than natural; and it is a familiar medical obercation that many nervous disorders of a minor kind, and even such serious disorders as chorea, epilepsy, insanity, are often connected with irregularities or suppression of these important functions.

In addition to the ill effects upon the bodily health which are produced directly by an excessive mental application, and a consequent development of the nervous system at the expense of the nutritive functions, it is alleged that remoter effects of an injurious character are produced upon the entire nature, mental and bodily. The arrest of development of the reproductive system discovers itself in the physical form and in the mental character. There is an imperfect development of the structure which Nature has provided in the female for nursing her offspring.

'Formerly,' writes another American physician, Dr. N. Allen, 'such an organization was generally possessed by American women, and they found but little difficulty in nursing their infants. It was only occasionally, in case of some defect in the organization, or where sickness of some kind had overtaken the mother, that it became necessary to resort to the wet-nurse, or to feeding by hand. And the English, the Scotch, the German, the Canadian, the French, and the Irish women who are living in thiscountry, generally nurse their children: the exceptions are rare. But how is it with our American women who become mothers? It has been supposed by some that all, or nearly all of them, could nurse their offspring just as well as not; that the disposition only was wanting, and that they did not care about having the trouble or confinement necessarily attending it. But this is a great mistake. This very indifference or aversion shows something wrong in the organization, as well as in the dispostion: if the physical system were all right, the mind and natural institncts would generally be right also. While there may be here and there cases of this kind, such an indispostion is not always found. It is a fact that large numbers of our women are anxious to nurse their offspring, and make the attempt: they persevere for a while—perhaps for weeks or months—and then fail. . . . There is still another class that cannot nurse at all, having neither the organs nor nourishment necessary to make a beginning.'

Why should there be such a difference between American women and those of foreign origin residing in the same locality, or between them and their grandmothers? Dr. Allen goes on to ask. The answer he finds in the undue demands made upon the brain and nervous system to the detriment of the organis of nutrition and secretion.

In consequence of the great neglect of physical exercise, and the continuous application to study, together with various other influences, large numbers of our American women have altogether an undue predominance of the nervous temperament. If only here and there an individual were found with such an organization, not much harm comparatively would result; but when a majority, or nearly a majority have it, the evil becomes one of no small magnitude.

Of the same effect writes Dr. Weir Mitchell, an eminent American physiologist.

> Worst of all, to my mind, most destructive in every way, is the American view of female education. The time taken for the more serious instruction of girls extends to the age of eighteen, and rarely over this. During these years they are undergoing such organic development as renders them remarkably sensitive. . . . To-day the American woman is, to speak plainly, physically unfit for her duties as woman, and is, perhaps, of all civilised females, the least qualified to udertake those weightier tasks which tax so heavily the nervous system of man. She is not fairly up to what Nature asks from her as wife and mother. How will she sustain herself under the pressure of those yet more exacting duties which nowadays she is eager to share with man?

Here then is no uncertain testimony as to the effects of the American system of female education: some women who are without the instinct or desire to nurse their offspring, some who hav the desire but not the capacity, and others who have neither the instinct nor the capacity. The facts will hardly be disputed, whatever may finally be the accepted interpretation of them. It will not probably be argued that an absence of the capacity and the instinct to nurse is a result of higher development, and that it should be the aim of woman, as she advances to a higher level, to allow the organs which minister to this function to waste and finally to become by disuse as rudimentary in her sex as they are in the male sex. Their development is notably in close sympathy wwith that of the organis of reproduction, an arrest thereof being often associated with some defect of the latter; so that it might perhaps fairly be qustioned whether it was right and proper, for the race's sake, that a woman who has not the wish and power to nurse should indulge in the functions of maternity. We may take note, by the way, that those in whom the organs are wasted invoke the dressmaker's aid in order to gain the appearance of them; they are not satisfied unless they wear the show of perfect womanhood. However, it may be in the plan of evolution to produce at some future period a race of sexless beings who, undistracted and unharassed by the ignoble troubles of reproduction, shall carry on the intellectual work of the world, not otherwise than as the sexless ants do the work and the fighting of the community.

Meanwhile, the consequences of an imperfectly developed reproductive system are not sexual only; they are also mental. Intellectually and morally there is a deficiency, or at any rate a modification answering to the physical deficiency; in mind, as in body, the individual fails to reach the ideal of a complete and perfect womanhood. If the aim of a true education be to make her reach *that*, it cannot certainly be a true education which operates in any degree to unsex her; for sex is fundamental, lies deeper than culture, cannot be ignores or defied with impunity. You may hide nature, but you cannot extinguish it. Consequently it does not seem impossible that if the attempt to do so be seriously and persistently made, the result may be a monstrosity—something which having ceased to be woman is yet not man—'ce quelque chose de monstrueux,' which the Comte A. De Gasparin forebodes, 'cet être répugnant, qui déjà paraît à notre horizon.'

The foregoing considerations go to show that the main reason of woman's position lies in her nature. That she has not competed with men in the active work of life was probably because not having had the power she had not the desire to do so, and because having the capacity of functions which man has not she has found her pleasure in performing them. It is not simply that man being stronger in body than she is, has held her in subjection, and debarred her from careers of action which he was resolved to keep for himself; her maternal functions must always have rendered and must continue to render, most of her activity domestic. There have been times enough in the history of the world when the freedom which she has had, and the position which she has held in the estimation of men, would have enabled her to assert her claims to other functions, had she so willed it. The most earnest advocate of her rights to be something else than what she has hitherto been would hardly argue that she has always been in the position of a slave kept in forcible subjection by the superior physical force of men. Assuredly, if she has been a slave, she has been a slave content with her bondage. But it may perhaps be said that in that lies the very pith of the matter— that she is not free, and does not care to be free; that she is a slave, and does not know or feel it. It may be alleged that she has lived for so many ages in the position of dependence to which she was originally reduced by the superior muscular strength of man, has been so thoroughly imbued with inherited habits of submission, and overawed by the influence of customs never questioned, that she has not the desire for emancipation; that thus a moral bondage has been established more effectual than an actual physical bondage. That she has now exhibited a disposition to emancipate herself, and has initiated a movement to that end, may be owing partly to the easy means of intellectual intercommunication in this age, whereby a few women scattered through the world, who felt the impulses of a higher inspiration have been enabled to co-operate in a way that would have been impossible in former times, and partly to the awakened moral sense and to the more enlightened views of men, which have led to the encouragement and assistance, instead of the suppression, of their efforts.

It would be rash to assert that there is not some measure of truth in these arguments. Let any one who thinks otherwise reflect upon the degraded condition of women in Turkey, where habit is so ingrained in their nature, and custom

so powerful over the mind, that they have neither thought nor desire to attain to
a higher state, and 'nought feel their foul disgrace:' a striking illustration how
women may be demoralised and yet not know nor feel it, and an instructive lesson for those who are anxious to form a sound judgment upon the merits of the
movement for promoting their higher education and the removal of the legal disabilities under which they labour. It is hardly possible to exaggerate the effects
of the laws and usages of a country upon the habits of thought of those who, generation after generation, have been born, and bred, and have lived under them.
Were the law which ordains that when a father dies intestate, all the real property of which he is possessed shall be inherited by his eldest son, his other children being sent empty away, enacted for the first time, there is no one, proably,
who would not be shocked by its singular injustice; yet the majority of persons
in this country are far from thinking it extraordinary or unjust, and a great many
of them would deem it a dangerous and wicked doctrine to question its justice.
Only a few weekss ago, a statesman who has held high offices in a Conservative
ministry, in an address to electors, conjured them not to party with the principle
of primogeniture, and declared that there was no change in the law which he
would so vehemently oppose as this: 'let them but follow the example of a neighbouring nation in this respect, and there was an end of their personal freedom
and liberty!' So much do the laws and usages of a country affect the feelings and
judgments of those who dwell therein. If we clearly apprehend the fact, and
allow it the weight which it deserves, it will be apparent that we must hesitate to
accept the subordinate position which women have always had as a valid argument for the justice of it, and a sufficient reason why they should continue for
ever in it.

But we may not fairly assert that it would be no less a mistake in an opposite direction to allow no weight to such an argument? Setting physiological considerations aside, it is not possible to suppose that the whole explanation of
woman's position and character is that man, having in the beginning found her
pleasing in his eyes and necessary to his enjoyment, took forcible possession of
her, and has ever since kept her in bondage, without any other justification than
the right of the strongest. Superiority of muscular strength, without superiority
of any other kind, would not have done that, any more than superiority of muscular strength has availed to give the lion or the elephant possession of the earth.
If it were not that woman's organization and functions found their fitting home
in a position different from, if not subordinate to, that of men, she would not so
long have kept that position. If she is to be judged by the same standard as men,
and to make their aims her aims, we are certainly bound to say that she labours
under an inferiority of constitution by a dispensation which there is no gainsay-
ing. This is a matter of physiology, not a matter of sentiment; it is not a mere

question of larger or smaller muscles, but of the energy and power of endurance of the nerve-force which drives the intellectual and muscular machinery; not a question of two bodies and minds that ware in equal physical conditions, but of one body and mind capable of sustained and regular hard labour, and of another body and mind which for one quarter o each month during the best years of life is more or less sick and unfit for hard work. It is in these considerations that we find the true explanation of what has been from the beginning until now, and what must doubtless continue to be, though it be in a modified form. It may be a pity for woman that she has been created women, but, being such, it is as ridiculous to consider herself inferior to man because she is not man, as it would be for man to consider himself inferior to her because he cannot perform her functions. There is one glory of the man, another glory of the woman, and the glory of the one differeth from that of the other.

Taking into adequate account the physiology of the female organization, some of the statements made by the late Mr. Mill in his book on the subjection of women strike one with positive amazement. He calls upon us to own that what is now called the nature of women is an eminently artificial thing, the result of forced repression in some directions, of unnatural stimulation in others; that their character has been entirely distorted and disguised by their relations with their masters, who have kept them in so unnatural a state; that if it were not for this there would not be any material difference, nor perhaps any difference at all, in the character and capacities which would unfold themselves; that they would do the same things as men fully as well on the whole, if education and cultivation were adapted to correcting, instead of aggravating, the infirmities incident to their temperament; and that they have been robbed of their natural development, and brought into their present unnatural state, by the brutal right of the strongest which man has used. If these allegations contain no exaggeration, if they be strictly true, then is this article an entire mistake.

Mr. Mill argues as if when he has shown it to be probable that the inequality of rights between the sexes has no other source than the law of the strongest, he had demonstrated its monstrous injustice. But is that entirely so? After all there is a right in might—the right of the strong to be strong. Men have the right to make the most of their powers, to develop them to the utmost, and to strive for, and if possible gain and hold, the position in which they shall have the freest play. It would be a wrong to the stronger if it were required to limit its exertions to the capacities of the weaker. And if it be not so limited, the result will be that the weaker must take a different position. Men will not fail to take the advantage of their strength over women: are no laws then to be made which, owning the inferiority of women's strength, shall ordain accordingly, and so protect them really from the mere brutal tyranny of might? Seeing that the greater power cannot be ignored, but in the long run must tell in individual competition, it is a fair question whether it ought not to be recognised in social adjustments and enactments, even for the necessary protection of women. Suppose that all legal distinctions were abolished, and that women were allowed free play to do what they could, as it may be right they should—to fail or succeed in every career upon which men enter; that all were conceded to them which their extremist advocates might claim for them; do they imagine that if they, being in a majority, combined to pass laws which were unwelcome to men, the latter would quietly submit? Is it proposed that men should fight for them in war, and that they, counting a majority of votes, should determine upon war? Or would they no longer claim a privilege of sex in regard to the defence of the country by arms? If all barriers of distinction of sex raised by human agency were thrown down, as not being warranted by the distinctions of sex which Nature has so plainly marked, it may be presumed that the great majority of women would continue to discharge the functions of maternity, and to have the mental qualities which correlate these functions; and if laws were made by them, and their male supporters of a feminine habit of mind, in the interests of babies, as might happen, can it be supposed that, as the world goes, there would not soon be a revolution in the State by men, which would end in taking all power from women and reducing them to a stern subjection? Legislation would not be of much value unless there were power behind to make it respected, and in such case laws might be made without the power to enforce them, or for the very purpose of coercing the power which could alone enforce them.

So long as the differences of physical power and organization between men and women are what they are, it does not seem possible that they should have the same type of mental development. But while we see great reason to dissent from the opinions, and to distrust the enthusiasm, of those who would set before women the same aims as men, to be pursued by the same methods, it must be admitted that they are entitled to have all the mental culture and all the freedom necessary to the fullest development of their natures. The aim of female education should manifestly be the perfect development, not of manhood but of womanhood, by the methods most conducive thereto: so may women reach as high a grade of development as men, though it be of a different type. A system of education which is framed to fit them to be nothing more than the superintendents of a household and the ornaments of a drawing-room, is one which does not do justice to their nature, and cannot be seriously defended. Assuredly those of them who have not the opportunity of getting married suffer not a little, in mind and body, from a method of education which tends to develop the emotional at the expense of the intellectual nature, and by their exclusion from appropriate fields of practical activity. It by no means follows, however, that it would be right to model an improved system exactly upon that which has commended itself as the best for men. Inasmuch as the majority of women will continue to get married and to discharge the functions of mothers, the education of girls certainly ought not to be such as would in any way clash with their organization, injure their health, and unfit them for these functions. In this matter the small minority of women who have other aims and pant for other careers, cannot be accepted as the spokeswomen of their sex. Experience may be left to teach them, as it will not fail to do, whether they are right or wrong in the ends which they pursue and in the means by which they pursue them: if they are right, they will have deserved well the success which will reward their faith and works; if they are wrong, the error will avenge itself upon them and upon their children, if they should ever have any. In the worst event they will not have been without their use as failures; for they will have furnished experiments to aid us in arriving at correct judgments concerning the capacities of women and their right functions in the universe. Meanwhile, so far as our present lights reach, it would seem that a system of education adapted to women should have regard to the peculiarities of their constitution, to the special functions in life for which they are destined, and to the range and kind of practical activity, mental and bodily, to which they would seem to be foreordained by their sexual organization of body and mind.

Source: Maudsley, Henry. 1874. "Sex in Mind and Education." *Fortnightly Review* 15: 466–483.

Elizabeth Garrett Anderson, a pioneer in women's education and the practice of medicine in Britain, wrote a response to Dr. Maudsley's article "Sex in Mind and Education," contesting his findings. She challenges Maudlsey's assertions that women are physically and mentally incapable of higher education. She notes that women's incapacity is often the result of social and cultural barriers not innate physiological causes and that given the opportunity and support, women would be likely to succeed in medical education.

Sex in Mind and Education: A Reply

ELIZABETH GARRETT ANDERSON

Explanatory Note

Anderson, Elizabeth Garrett (1836–1917). British pioneer woman doctor, nee Garrett. She began to study medicine in 1860 and in 1865 was granted a licence to practice by the Society of Apothecaries and then obtained a medical degree in Paris in 1870. She held hospital posts in London, 1866–1903, and

married in 1871, combining marriage and motherhood with a medical career,
a rare achievement. She became President of the London School of Medicine for
Women, and the only woman member of the British Medical Association from
1872–1893. She was a member of the first London School Board in 1870 and
in 1908 became Mayor of Aldeburgh, the first woman Mayor in England.

The April number of the *Fortnightly Review* contains an article under the
heading of Sex in Mind and Education, by Dr. Maudsley. It is a reproduction of
a lecture on the same subject by Dr. Clarke, formerly a Professor at Harvard
College, United States, with such additions and comments as the circumstances
of English life seemed to Dr. Maudsley to warrant. Dr. Clarke's lecture was originally addressed to an audience of women, and, whether we agree with its conclusions or not, we cannot blame him that with his view of the evil effects of
continuous mental work on women's health he thought himself justified in
speaking with a degree of frankness that would have been out of place under
other circumstances. In applying Dr. Clarke's arguments to the question of the
education of women in England, Dr. Maudsley pleads the importance of the subject as an excuse for placing medical and physiological views before the readers of a literary periodical. It is possible he was right in thinking the excuse adequate, though we cannot but suggest that there is grave reason for doubting
whether such a subject can be fully and with propriety discussed except in a
professional journal. As, however, the usual reserve has been broken through,
it would be out of place for those who approve the changes against which Dr.
Maudsley's argument is directed to be silent in obedience to those considerations which he has disregarded. We will therefore venture to speak as plainly
and directly as he has spoken.

Dr. Maudsley's paper consists mainly of a protest against the assimilation
of the higher education of men and women, and against the admission of women
to new careers; and this protest is founded upon a consideration of the physiological peculiarities of women. It derives much of its importance from the
assumption that what is now being tried in England has already been tried in
America, and that it has there produced the results which Dr. Maudsley thinks
are inevitable. When, however, we turn to Dr. Clarke's book (from which the
American evidence quoted by Dr. Maudsley is taken) we find that the American
system is, in many important features, and especially in those most strongly condemned by Dr. Clarke and the other witnesses, widely different from that now
being advocated in England. Even if what Dr. Maudsley urges could be admitted
as a correct inference from physiological fact, the result of a very different experiment in America could not fairly be used to support his argument. We shall show
later on in what the difference between the American and English systems consists, but it is necessary, first of all, to warn readers of Dr. Maudsley's article that

his use of American evidence is misleading and is not confirmed by reference to Dr. Clarke's book.

One other preliminary statement needs to be made before entering upon the consideration of Dr. Maudsley's argument. In the beginning of his paper he brings a serious charge against those who are advocating the changes he disapproves. He says, and in one form or another he repeats the charge again and again, that their aim is to change women into men, or, as he puts it, 'to assimilate the female to the male mind.' Much pains is taken to convince them that this is impossible; they are assured that 'women cannot choose but to be women; cannot rebel successfully against the tyranny of their organization.' and much more to the same effect. To meet such charges is difficult on account of their vagueness. We can but ask with unfeigned surprise what ground Dr. Maudsley can conceive that he has for them? We ask, what body of persons associated together in England for the purpose of promoting the education of women has made any statement, in any form or degree, implying such aims? That no injudicious advocate has ever made a remark which might bear such an interpretation we are not prepared to assert, but we can confidently challenge the production of any manifesto possessing a fair claim to authority in which anything of the kind is said or implied. The single aim of those anxious to promote a higher and more serious education for women is to make the best they can of the materials at their disposal, and if they fail, it assuredly will not be from thinking that the masculine type of excellence includes all that can be desired in humanity.

The position Dr. Maudsley has undertaken to defend is this, that the attempt now being made in various directions to assimilate the mental training of men and women is opposed to the teachings of physiology, and more especially, that women's health is likely to be seriously injured if they are allowed or encouraged to pursue a system of education laid down on the same lines, following the same method, and having the same ends in view, as a system of education for men.

He bases his opinion on the fact that just at the age when the real educational strain begins, girls are going through an important phase of physiological development, and that much of the health of their after-life depends upon the changes proper to this age being effected without check and in a normal and healthy manner. Moreover, the periodical recurrence of the function thus started, is attended, Dr. Maudsley thinks, with so great a withdrawal of nervous and physical force, that all through life it is useless for women to attempt, with these physiological drawbacks, to pursue careers side by side with men.

We have here two distinct assertions to weigh and verify: 1st, that the physiological functions started in girls between the ages of fourteen and sixteen are likely to be interfered with or interrupted by pursuing the same course of study as boys, and by being subjected to the same examinations; and 2nd, that even when these functions are in good working order and the woman has arrived at maturity, the facts of her organization interfere periodically to such an extent with steady and serious labour of mind or body that she can never hope to compete successfully with men in any career requiring sustained energy. Both with girls and women, however, it is the assimilation of their education and the equality of their aim with those of boys and men which, in Dr. Maudsley's eyes, call for special condemnation. And in each case he grounds his objection on the fact that physiologically important differences are found in the two sexes. He says, 'It would seem plain that as women are marked out by nature for very different offices from those of men, the healthy performance of her special functions renders it improbable she will succeed, and unwise for her to persevere in running over the same course at the same pace with him.' But surely this argument contains a *non sequitur.* The question depends upon the nature of the course and the quickness of the pace, and upon the fitness of both for women; not at all on the amount of likeness or unlikeness between men and women. So far as education is concerned it is conceivable, and indeed probable that, were they ten times as unlike as they are, many things would be equally good for both. If girls were less like boys than the anthropomorphic apes, nothing but experience would prove that they would not benefit by having the best methods and the best tests applied to their mental training. And if the course of study which Dr. Maudsley is criticising be one as likely to strengthen the best powers of the mind as good food is to strengthen the body, if it tend to develop habits as valuable to women as to men, and if the pace is moderate, there would seem to be no good reason why the special physiological functions of women should prevent them from running it, any more than these same functions prevent them from eating beef and bread with as much benefit as men. The question is not settled by proving that both in mind and body girls are different from boys.

The educational methods followed by boys being admitted to be better than those hitherto applied to girls, it is necessary to show that these better methods would in some way interfere with the special functions of girls. This Dr. Maudsley has not done. He has not attempted to show how the adoption of a common standard of examination for boys and girls, allowing to each a considerable range in the choice of subjects, is likely to interfere more with a girls's health than passing an inferior examination for girls only. Either would hurt her if unwisely pressed, if the stimulus of competition were unduly keen, or if in the desire for mental development the requirements of her physical nature were overlooked.

What we want to know is what exactly these requirements are, and especially how much consideration girls and women ought to show to the fact of the periodic and varying functions of their organization. In considering this point, we ought not to overlook the antecedent improbability of any organ or set of organs requiring exceptional attention; the rule certainly being that, when people are well, their physiological processes go on more smoothly without attention than with it. Are women an exception to this rule? When this is settled we shall be in a position to speculate upon how far in education or in after-life they will be able to work side by side with men without overtaxing their powers.

And first, with regard to adults. Is it true, or is it a great exaggeration, to say that the physiological difference between men and women seriously interferes with the chances of success a woman would otherwise possess? We believe it to be very far indeed from the truth. When we are told that in the labour of life women cannot disregard their special physiological functions without danger to health, it is difficult to understand what is meant, considering that in adult life healthy women do as a rule disregard them almost completely. It is, we are convinced, a great exaggeration to imply that women of average health are periodically incapacitated from serious work by the facts of their organization. Among poor women, where all the available strength is spent upon manual labour, the daily work goes on without intermission, and, as a rule, without ill effects. For example, do domestic servants, either as young girls or in mature life, show by experience that a marked change in the amount of work expected from them must be made at these times unless their health is to be injured? It is well known that they do not.

With regard to mental work it is within the experience of many women that that which Dr. Maudsley speaks of as an occasion of weakness, if not of temporary prostration, is either not felt to be such or is even recognised as an aid, the nervous and mental power being in many cases greater at those times than at any other. This is confirmed by what is observed when this function is prematurely checked, or comes naturally to an end. In either case its absence usually gives rise to a condition of nervous weakness unknown while the regularity of the function was maintained. It is surely unreasonable to assume that the same function in persons of good health can be a cause of weakness when present, and also when absent. If its performance made women weak and ill, its absence would be a gain, which it is not. Probably the true view of the matter is this. From various causes the demand made upon the nutritive processes is less in women than in men, while these processes are not proportionately less active; nutrition is thus continually a little in excess of what is wanted by the individual, and there is a margin ready for the demand made in childbearing. Till this demand arises it is no loss, but quite the reverse, to get rid of the surplus nutritive material, and getting rid of it involves, when the process is normal, no loss of vigour to the woman.

As to the exact amount of care needed at the time when this function is active and regular, individual women no doubt vary very much, but experience justifies a confident opinion that the cases in which it seriously interferes with active work of mind or body are exceedingly rare; and that in the case of most women of good health, the natural recurrence of this function is not recognised as causing anything more than very temporary *malaise*, and frequently not even that.

The case is, we admit, very different during early womanhood, when rapid growth and the development of new functions have taxed the nutritive powers more than they are destined to be taxed in mature life. At this age a temporary sense of weakness is doubtless much more common than it is later in life, and where it exists wise guardians and teachers are in the habit of making allowance for it, and of encouraging a certain amount of idleness. This is, we believe, as much the rule in the best English schools as it is in private schoolrooms and homes. No one wishes to dispute the necessity for care of this kind; but in our experience teachers, as a rule, need a warning on the point even less than parents. Fathers especially are apt to be thoughtless in expecting their girls to be equally ready at all times to ride or take long walks, and so far as individual experience may be used as a guide, we venture to think that far more harm is done to young women in ways of this kind while they are at home, than when they are protected by the quiet routine of school life. While, too, we would not deny that very great pressure of mental work at this age is to be deprecated, we believe that practically the risk of injury from undue or exceptional physical fatigue at an inopportune moment is much greater. Riding, long standing, lifting heavy weights,—*e.g.*, young brothers and sisters—dancing, and rapid or fatiguing walks, are, we believe, the chief sources of risk to delicate girls at these times, and of them all riding is probably much the most serious. The assertion that, as a rule, girls are unable to go on with an ordinary amount of quiet exercise or mental work during these periods, seems to us to be entirely contradicted by experience. Exceptional cases require special care, and under the arrangements of school life in England, whatever may be the case in America, they get it. But does it follow from this, that there is any ground for suspecting or fearing that the demands made by the special functions of womanhood during the time of development are really more in danger of being overlooked, or inadequately considered, under the new system of education than they were under the old? Dr. Maudsley seems to think there is, but he brings no evidence in support of his opinion. He is apparently not aware that most important improvements in physical training are being introduced alongside with other reforms. The time given to education is being prolonged, and the pressure in the early years of womanhood, when continuous work is less likely to be well borne, is being lightened; girls are no longer kept standing an hour or more at a time, or sitting without support for their backs; school hours and school terms are shortened; and, above all, physical exercise is no longer limited to the daily monotonous walk which was thought all-sufficient in old-fashioned schools and homes. In spite of these undeniable facts, Dr. Maudsley charges the reformers with having neglected the physical requirements of girls, in order to stimulate their mental activity. 'It is quite evident,' he says, 'that many of those who are foremost in their zeal for raising the education of women, have not given proper consideration to the nature of her organization.' In another place he blames them for having neglected physical training and exercise. To those in a position to know the facts, such a charge as this seems peculiarly misplaced and unjust. It is no doubt true, that twenty years ago the physical training of girls was deplorably neglected, and that it still is so in homes and schools of the old-fashioned type. But the same people who during recent years have been trying to improve the mental training of girls, have continually been protesting in favour also of physical development, and to a great extent their protests have been successful. The schoolmistresses who asked that girls might share in the Oxford and Cambridge Local Examinations, were the first also to introduce gymnastics, active games, daily baths, and many other hygienic reforms sorely needed in girls' schools. The London Association of Schoolmistresses, which was formed expressly 'to promote the higher education of women,' took so comprehensive a view of what this included, that the first paper issued by them was one upon 'Physical Exercises and Recreation', in which it was laid down as a maxim that 'good results are not obtained by sacrificing any one part of our nature to another. If study takes up so much time that there is not enough left for play, there must be too much study going on. The lessons must be too many or too long, and ought to be curtailed.' What more could Dr. Maudsley himself say? The same body also applied itself in the very infancy of its existence to the study of School Hygiene, and took particular trouble to ascertain in what way the health and vigour of the girls in their schools could be improved. So far from their deserving censure, all who are interested in advancing the education of girls feel themselves deeply indebted to these ladies for the zeal and self-sacrifice they have shown in adopting, at great cost often to themselves, hygienic reforms not expected from them, and often indeed in advance of the sense of the parents of their pupils.

But it may still be urged, that admitting the advantage to girls of assimilating their play-ground hours to those of boys, of substituting outdoor games for worsted work or crouching over the fire with a story-book, yet that when it comes to school work the case is different, and that to make girls work as hard as boys do, and especially to allow them to work for the same examinations, would-be to press unfairly upon their powers. In answer to this, we must take note of some facts about boys.

It must not be overlooked, that the difficulties which attend the period of rapid functional development are not confined to women, though they are expressed differently in the two sexes. Analogous changes take place in the constitution and organization of young men, and the period of immature manhood is frequently one of weakness, and one during which any severe strain upon the mental and nervous powers is productive of more mischief than it is in later life. It is possible that the physiological demand thus made is lighter than that made upon young women at the corresponding age, but on the other hand it is certain that, in many other ways unknown to women, young men still tax their strength, *e.g.*, by drinking, smoking, unduly severe physical exercise, and frequently by late hours and dissipation generally. Whether, regard is being had to all these varying influences, young men are much less hindered than young women in intellectual work by the demands made upon their physical and nervous strength during the period of development, it is probably impossible to determine. All that we wish to show is that the difficulties which attend development are not entirely confined to women, and that in point of fact great allowance ought to be made, and has already been made, for them in deciding what may reasonably be expected in the way of intellectual attainment from young men. It is not much to the point to prove that men could work harder than women, if the work demanded from either is very far from overtaxing the powers of even the weaker of the two. If we had no opportunity of measuring intellectual athletes Dr. Maudsley's warnings would lead us to suppose them to be, the question, 'Is it well for women to contend on equal terms with men for the goal of man's ambition?' might be as full of solemnity to us as it is to Dr. Maudsley. As it is, it sounds almost ironical. Hitherto most of the women who have 'contended with men for the goal of man's ambition' have had no chance of being any the worse for being allowed to do so on equal terms. They have had all the benefit of being heavily handicapped. Over and above their assumed physical and mental inferiority, they have had to start in the race without a great part of the training men have enjoyed, or they have gained what training they have been able to obtain in an atmosphere of hostility, to remain in which has taxed their strength and endurance far more than any amount of mental work could tax it. Would, for instance, the ladies who for five years have been trying to get a medical education at Edinburgh find their task increased, or immeasurably lightened, by being allowed to contend 'on equal terms with men' for that goal? The intellectual work required from other medical students is nothing compared with what it has been made to them by obliging them to spend time and energy in contesting every step of their course, and yet in spite of this heavy additional burden they have not at present shown any signs of enfeebled health or of inadequate mental power. To all who know what it is to pursue intellectual work under such conditions as these, Dr. Maudsley's pity for the more fortunate women who may pursue it in peace and on equal terms with men sounds superfluous. But Dr. Maudsley would probably say that, in speaking of the pace at which young men at the Universities work as being dangerously rapid for average women, he was not referring to anything less ambitious that the competition for honours. No one denied that in some cases this is severe; many men knock up under it, and it would doubtless tax the strength of women. But it must be borne in mind that that element in the competition which incites men to the greatest effort, and increases the strain to its upmost, is one which, for the present at least, would not operate upon women. Pecuniary rewards, large enough to affect a man's whole after-life, are given for distinction in these examinations; and it is the eager desire for a Fellowship which raises the pressure of competition to so high and, as many think, to so unwholesome a point. As there are at present no Fellowships for women, this incentive does not operate upon them.

It must always be remembered, too, that University work does not come at the age when Dr. Maudsley and Dr. Clarke think it is likely to be too exciting. No one is proposing that girls of seventeen and eighteen should be allowed to try for a place in the Cambridge Honours' Lists. What is proposed is that after a girlhood of healthful work and healthful play, when her development is complete and her constitution settled, the student, at the age of eighteen or nineteen, should begin the college course, and should be prepared to end it at twenty-two or twenty-three. As we shall see later on, this is a very different plan from that pursued in America, and censured by Dr. Clarke.

In estimating the possible consequences of extending the time spent in education, and even those of increasing somewhat the pressure put upon girls under eighteen, it should be borne in mind that even if the risk of overwork, pure and simple, work unmixed with worry, is more serious than we are disposed to think it, it is not the only, nor even the most pressing, danger during the period of active physiological development. The newly developed functions of womanhood awaken instincts which are more apt at this age to make themselves unduly prominent than to be hidden or forgotten. Even were the dangers of continuous mental work as great as Dr. Maudsley thinks they are, the dangers of a life adapted to develop only the specially and consciously feminine side of the girl's nature would be much greater. From the purely physiological point of view, it is difficult to believe that study much more serious than that usually pursued by young men would do a girl's health as much harm as a life directly calculated to over-stimulate the emotional and sexual instincts, and to weaken the guiding and controlling forces which these instincts so imperatively need. The stimulus found in novel-reading, in the theatre and ball-room, the excitement which attends a premature entry into society, the competition of vanity and frivolity, these involve far more real dangers to the health of young women than the competition for knowledge, or for scientific or literary honours, ever has done, or is ever likely to do. And even if, in the absence of real culture, dissipation be avoided, there is another danger still more difficult to escape, of which the evil physical results are scarcely less grave, and this is dulness. It is not easy for those whose lives are full to over-flowing of the interests which accumulate as life matures, to realise how insup-portably dull the life of a young woman just out of the schoolroom is apt to be, nor the powerful influence for evil this dulness has upon her health and morals. There is no tonic in the pharmacopoeia to be compared with happiness, and hap-piness worth calling such is not known where the days drag along filled with make-believe occupations and dreary sham amusements.

The cases that Dr. Clarke brings forward in support of his opinion against continuous mental work during the period of development could be outnum-bered many times over even in our own limited experience, by those in which the break-down of nervous and physical health seems at any rate to be dis-tinctly traceable to want of adequate mental interest and occupation in the years immediately succeeding school life. Thousands of young women, strong and blooming at eighteen, become gradually languid and feeble under the depressing influence of dulness, not only in the special functions of woman-hood, but in the entire cycle of the processes of nutrition and innervation, till in a few years they are morbid and self-absorbed, or even hysterical. If they had had upon leaving school some solid intellectual work which demanded real thought and excited genuine interest, and if this interest had been helped by the stimulus of an examination, in which distinction would have been a legiti-mate source of pride, the number of such cases would probably be indefinitely smaller than it is now. It may doubtless be objected that even if this plan were pursued, and young women were allowed and expected to continue at tolera-bly hard mental work till they were twenty-one or twenty-two, it would only be postponing the evil day, and that when they left college they would dislike idle-ness as much, and be as much injured by it as when they left school. This is true; but by this time they would have more internal resources against idleness and dulness, and they would have reached an age in which some share in prac-tical work and responsibility—the lasting refuge from dulness—is more easily obtained than it is in girlhood. Moreover, by entering society at a somewhat less immature age, a young women is more able to take an intelligent part in it; is prepared to get more real pleasure from the companionship it affords, and, suffering less from *ennui*, she is less apt to make a hasty and foolish marriage. From the physiological point of view this last advantage is no small or doubt-ful one. Any change in the arrangements of young women's lives which tends to discourage very early marriages will probably do more for their health and for the health of their children than any other change would do. But it is hope-less to expect girls, who are at heart very very dull, to wait till they are physiologically fit for the wear and tear consequent upon marriage if they see their way to it at eighteen or nineteen. There is always a hope that the unknown may be less dull than the known, and in the mean time the mere mention of a change gives life a fillip. It is also hopeless to expect them to be even reasonably critical in their choice. Coleridge says. 'If Ferdinand hadn't come, Miranda *must* have married Caliban;' and many a Miranda finds her fate by not being free to wait a little longer for her Ferdinand.

But Dr. Maudsley supports his argument by references to American experience. He says in effect, 'That which the English educational reformers advocate has been tried in America and has failed; the women there go through the same educational course as the men, and the result is that they are nervous, specially prone to the various ailments peculiar to their sex, not good at bearing children, and unable to nurse them.' These are grave charges, and we can scarcely wonder at Dr. Maudsley's thinking 'it is right to call attention to them.' But it is also right to see if they are true. One fact certainly seems to be plain, and that is, that American women are frequently nervous, and do too often break down in the particular ways described in the quotation, though, if we may judge at all from those whom we have an opportunity of seeing in Europe, it may be hoped that the race is not quite in such a bad plight as Dr. Maudsley's quotation would lead us to fear. But granting that the facts are stated correctly, the doubtful point is, what causes this condition of things? Dr. Clarke says that, among other causes, it is due to an education which is taken at too early an age. But against this we have to notice the testimony of many independent witnesses to the effect that the evils complained of are seen to a much greater extent among the fashionable and ideal American women—those guiltless of ever having passed an examination—than they are among those who have gone through the course of study complained of. Then, again, it is notorious that the American type in both sexes is 'nervous'. The men show it as distinctly, if not even more distinctly than the women; and not those men only who have any claim to be considered about the average in intellect or culture. If Dr. Clarke's explanation of the existence of this type in women is correct, what is its explanation in men?

Dr. Clarke himself gives us some valuable hints as to possible causes, other than study. He says: 'We live in a zone of perpetual pie and doughnut;' 'our girls revel in these unassimilable abominations.' He also justly blames the dress of American women, 'its stiff corsets and its heavy skirts;' but somewhat inconsequently, as it seems to us, he says, 'these cannot be supposed to affect directly the woman's special functions.' If one thing more than another is likely to do a woman harm in these directions, we should say it is heavy skirts; and it certainly shakes our faith in Dr. Clarke's acumen to find him attributing less direct influence to them than to mental occupation. Our own notion would be that till American girls wear light dresses and thick boots, and spend as much time out of doors as their brothers, no one knows how many examinations they could pass not only without injury but with positive benefit to their health and spirits. We find, however, no mention made by Dr. Clarke of the influence of the stove-heated rooms in which American women live, nor of the indoor lives they lead. These two things only would, we believe, suffice to explain the general and special delicacy of which he complains, and the inferiority in point of health of American to English women.

But the truth is, that the system against which Dr. Clarke protests, and to which his arguments are directed, is, in some of the very points upon which he most insists, essentially different from that which is now being gradually introduced in England. Dr. Maudsley has, with what we must call some unfairness, applied what was written against one plan, to another which is unlike it in almost every important point. Whether the system in America deserves all that Dr. Clarke says against it, Americans must determine. We are not in a position at this distance to weigh conflicting evidence, or to determine which out of many causes is the most potent in producing the ill-health he deplores. But we can speak of the conditions under which English girls work, and we are able to say distinctly that on many vital points they are just those which Dr. Clarke and the other American doctors urge as desirable.

For instance, the stress of educational effort comes in America before eighteen. Graduation takes place at that age. At our own college for women at Girton, girls under eighteen are not admitted, and the final examinations take place three or more years later. Dr. Weir Mitchell's evidence on this point, as quoted by Dr. Clarke, is very emphatic. He says: 'Worst of all is the American view of female education. The time taken for the more serious instruction of girls extends to the age of eighteen, and rarely over this.' There is nothing that the English advocates of a change of system have striven more heartily to effect, than an extension of the time given to education; and what they have urged is in complete agreement with the opinions of Dr. Clarke and Dr. Mitchell. Then, again, Dr. Clarke distinguishes very clearly between girls learning the same subjects as boys, and sharing the same final examinations (which he does not disapprove), and identical co-education, where they are subjected to exactly the same rules and daily system, and where emulation between the two is constantly at work. He says (p.135): 'It is one thing to put up a goal a long way off—five or six months, or three or four years distant—and to tell girls and boys, each in their own way, to strive for it; and quite a different thing to put up the same goal, at the same distance, and oblige each sex to run their race for it side by side on the same road, in daily competition with each other, and with equal expenditure of force at all times. Identical co-education is racing in the latter way.' Now, there is no organized movement in England for identical co-education in this sense. What is advocated is just what Dr. Clarke approves, viz. Setting up the same goal, and allowing young men and young women to reach it each in their own way, and without the stimulus of daily rivalry. The public recitations, and the long hours of standing they involve, so much blamed by Dr. Clarke, are unknown in England, except in schools of the most old-fashioned and unenlightened type. The number of hours per day spent in mental work seems also to be much greater than that which is usual or even allowed in the best English schools. Eight or ten hours is said to be the usual time given to study in the American schools. In England, six hours is the time suggested by the Schoolmistresses' Association, and this is to include time given to music and needlework. Naturally, there is not time in America for physical exercise or outdoor games.

Dr. Maudsley appends to the physiological argument others which do not press for immediate attention. They are already familiar to all who are interested in noticing what can be said in support of the policy of restriction, whether as applied to Negroes, agricultural labourers, or women. They remind us more of an Ashantee fight than of a philosophical essay; so abundant is the powder used in their discharge, and so miscellaneous and obsolete are the projectiles. Happily, too, like the Ashantee slugs, though they wound, they are not very deadly. However, even Dr. Maudsley seems to relent when he comes to the end of the subject, and he goes so far as to allow that if the women whose policy he has been opposing fail, they will still be useful as failures, and that therefore they may go on their way, not too much discouraged by his disapproval. We will venture to draw another conclusion from the discussion, and it is this; that those who wish to give a fair hearing to all that is urged in support of a higher education for women must examine the evidence for themselves, not saying to themselves loosely that medical men seem to be afraid of this higher education, or that it seems to have been tried in America, and to have failed. Let them inform themselves thoroughly of what is proposed, and of the difference between the new system and the old; and if the result be, that, by improvement in the training and education of women, as much may be hoped for their physical as for their mental development, let them, in the interests not of women only, but of the children who claim from their mothers so much more than mere existence and nurture, give to those who are labouring at this difficult work, not languid approval, but sustained and energetic support.

And to those who share Dr. Maudsley's fears, we may say, that though under any system there will be some failures, physiological and moral, neither of which will be confined to one sex, yet that experience shows that no system will live from which failure in either of these directions as a rule results. Nature in the long run protects herself from our mistakes: and when we are in doubt, we may be guided by the general principles of equity and common sense, while waiting for the light of a larger experience.

Source: Anderson, Elizabeth Garrett. 1874. "Sex in Mind and Education: A Reply." *Fortnightly Review* 15: 582–594.

Margaret Gatty was a well-respected and highly successful nineteenth-century popularizer of science. She is most well known for her series of books for children, Parables from Nature, *in which she uses natural history to instruct her readers how to live moral and religious lives. In "A Lesson of Faith" she instructs children about the life cycle and transformation of a caterpillar into a butterfly. She utilizes this transformation, unforeseen to the caterpillar and yet inevitable, to teach her readers that human beings must have faith in the life to come after death.*

A Lesson of Faith

MARGARET GATTY

> "If a man die, shall he live *again?* All the days of my appointed
> time will I wait, till my change come."—JOB xiv. 14.

"Let me hire you as a nurse for my poor children," said a Butterfly to a quiet Caterpillar, who was strolling along a cabbage-leaf in her odd lumbering way. "See these little eggs," continued the Butterfly; "I don't know how long it will be before they come to life, and I feel very sick and poorly, and if I should die, who will take care of my baby butterflies when I am gone? Will *you*, kind, mild, green Caterpillar? But you must mind what you give them to eat, Caterpillar!—they cannot, of course, live on *your* rough food. You must give them early dew, and honey from the flowers; and you must let them fly about only a little way at first; for, of course, one can't expect them to use their wings properly all at once. Dear me! it is a sad pity you cannot fly yourself. But I have no time to look for another nurse now, so you will do your best, I hope. Dear! dear! I cannot think what made me come and lay my eggs on a cabbage-leaf! What a place for young butterflies to be born upon! Still you will be kind, will you not, to the poor little ones? Here, take this gold-dust from my wings as a reward. Oh, how dizzy I am! Caterpillar! you will remember about the food—"

And with these words the Butterfly closed her eyes and died; and the green Caterpillar, who had not had the opportunity of even saying Yes or No to the request, was left standing alone by the side of the Butterfly's eggs.

"A pretty nurse she has chosen, indeed, poor lady!" exclaimed she, "and a pretty business I have in hand! Why, her senses must have left her or she never would have asked a poor crawling creature like me to bring up her dainty little ones! Much they'll mind me, truly, when they feel the gay wings on their backs, and can fly away out of my sight whenever they choose! Ah! how silly some people are, in spite of their painted clothes and the gold-dust on their wings!"

However, the poor Butterfly was dead, and there lay the eggs on the cabbage-leaf; and the green Caterpillar had a kind heart, so she resolved to do her best. But she got no sleep that night, she was so very anxious. She made her back quite ache with walking all night long round her young charges, for fear any harm should happen to them; and in the morning says she to herself—

"Two heads are better than one. I will consult some wise animal upon the matter, and get advice. How should a poor crawling creature like me know what to do without asking my betters?"

But still there was a difficulty—whom should the Caterpillar consult? There was the shaggy Dog who sometimes came into the garden. But he was so rough!—he would most likely whisk all the eggs off the cabbage-leaf with one brush of his tail, if she called him near to talk to her, and then she should never forgive herself. There was the Tom Cat, to be sure, who would sometimes sit at the foot of the apple-tree, basking himself and warming his fur in the sunshine; but he was so selfish and indifferent!—there was no hope of his giving himself the trouble to think about butterflies' eggs. "I wonder which is the wisest of all the animals I know," sighed the Caterpillar, in great distress; and then she thought, and thought, till at last she thought of the Lark; and she fancied that because he went up so high, and nobody knew where he went to, he must be very clever, and know a great deal; for to go up very high (which *she* could never do) was the Caterpillar's idea of perfect glory.

Now, in the neighbouring corn-field there lived a Lark, and the Caterpillar sent a message to him, to beg him to come and talk to her; and when he came she told him all her difficulties, and asked him what she was to do, to feed and rear the little creatures so different from herself.

"Perhaps you will be able to inquire and hear something about it next time you go up high," observed the Caterpillar timidly.

The Lark said, "Perhaps he should;" but he did not satisfy her curiosity any further. Soon afterwards, however, he went singing upwards into the bright, blue sky. By degrees his voice died away in the distance, till the green Caterpillar could not hear a sound. It is nothing to say she could not see him; for, poor thing! she never could see far at any time, and had a difficulty in looking upwards at all, even when she reared herself up most carefully, which she did now; but it was of no use, so she dropped upon her legs again, and resumed her walk round the Butterfly's eggs, nibbling a bit of the cabbage-leaf now and then as she moved along.

"What a time the Lark has been gone!" she cried, at last. "I wonder where he is just now! I would give all my legs to know! He must have flown up higher than usual this time, I do think! How I should like to know where it is that he goes to, and what he hears in that curious blue sky! He always sings in going up and coming down, but he never lets any secret out. He is very, very close!"

And the green Caterpillar took another turn round the Butterfly's eggs.

At last the Lark's voice began to be heard again. The Caterpillar almost jumped for joy and it was not long before she saw her friend descend with hushed note to the cabbage bed.

"News, news, glorious news, friend Caterpillar!" sang the Lark; "but the worst of it is, you won't believe me!"

"I believe everything I am told," observed the Caterpillar hastily.

"Well, then, first of all, I will tell you what these little creatures are to eat"— and the Lark nodded his beak towards the eggs. "What do you think it is to be? Guess!"

"Dew, and the honey out of flowers, I am afraid," sighed the Caterpillar.

"No such thing, old lady! Something simpler than that. Something that *you* can get at quite easily."

"I can get at nothing quite easily but cabbage-leaves," murmured the Caterpillar, in distress.

"Excellent! my good friend," cried the Lark exultingly; "you have found it out. You are to feed them with cabbage-leaves."

"*Never!* " said the Caterpillar indignantly. "It was their dying mother's last request that I should do no such thing."

"Their dying mother knew nothing about the matter," persisted the Lark; "but why do you ask me, and then disbelieve what I say? You have neither faith nor trust."

"Oh, I believe everything I am told," said the Caterpillar.

"Nay, but you do not," replied the Lark; "you won't believe me even about the food, and yet that is but a beginning of what I have to tell you. Why, Caterpillar, what do you think those little eggs will turn out to be?"

"Butterflies, to be sure," said the Caterpillar.

"*Caterpillars!* " sang the Lark; "and you'll find it out in time;" and the Lark flew away, for he did not want to stay and contest the point with his friend.

"I thought the Lark had been wise and kind," observed the mild green Caterpillar, once more beginning to walk round the eggs, "but I find that he is foolish and saucy instead. Perhaps he went up *too* high this time. Ah, it's a pity when people who soar so high are silly and rude nevertheless! Dear! I still wonder whom he sees, and what he does up yonder."

"I would tell you, if you would believe me," sang the Lark, descending once more.

"I believe everything I am told," reiterated the Caterpillar, with as grave a face as if it were a fact.

"Then I'll tell you something else," cried the Lark; "for the best of my news remains behind. *You will one day be a Butterfly yourself.*"

"Wretched bird!" exclaimed the Caterpillar, "you jest with my inferiority–now you are cruel as well as foolish. Go away! I will ask your advice no more."

"I told you you would not believe me," cried the Lark, nettled in his turn.

"I believe everything that I am told," persisted the Caterpillar; "that is"—and she hesitated,—"everything that it is *reasonable* to believe. But to tell me that butterflies' eggs are caterpillars, and that caterpillars leave off crawling and get wings, and become butterflies!—Lark! you are too wise to believe such nonsense yourself, for you know it is impossible."

"I know no such thing," said the Lark, warmly. "Whether I hover over the corn-fields of earth, or go up into the depths of the sky, I see so many wonderful things, I know no reason why there should not be more. Oh, Caterpillar! it is because you crawl, because you never get beyond your cabbage-leaf, that you call *any* thing *impossible*."

"Nonsense!" shouted the Caterpillar. "I know what's possible, and what's not possible, according to my experience and capacity, as well as you do. Look at my long green body and these endless legs, and then talk to me about having wings and a painted feathery coat! Fool!—"

"And fool you! you would-be-wise Caterpillar!" cried the indignant Lark. "Fool, to attempt to reason about what you cannot understand! Do you not hear how my song swells with rejoicing as I soar upwards to the mysterious wonder-world above? Oh, Caterpillar! what comes to you from thence, receive, as *I* do, upon trust."

"That is what you call—"

"*Faith*," interrupted the Lark.

At that moment she felt something at her side. She looked round—eight or ten little green caterpillars were moving about, and had already made a show of a hole in the cabbage-leaf. They had broken from the Butterfly's eggs!

Shame and amazement filled our green friend's heart, but joy soon followed; for, as the first wonder was possible, the second might be so too. "Teach me your lesson, Lark!" she would say; and the Lark sang to her of the wonders of the earth below, and of the heaven above. And the Caterpillar talked all the rest of her life to her relations of the time when she should be a Butterfly.

But none of them believed her. She nevertheless had learnt the Lark's lesson of faith, and when she was going into her chrysalis grave, she said—"I shall be a Butterfly some day!"

But her relations thought her head was wandering, and they said, "Poor thing!"

And when she was a Butterfly, and was going to die again, she said—

"I have known many wonders—I have faith—I can trust even now for what shall come next!"

Source: Gatty, Margaret. 1876. "A Lesson in Faith." Pp. 1–13 in *Parables from Nature*. London: George Bell and Sons.

Isabella Bishop was one of a number of woman travelers who became immersed in geography and natural history and published her travel experiences and reflections on the natural world. She traveled throughout Canada and the United States, Australia, New Zealand, Hawaii, the Malay peninsula, Japan, Egypt, Tibet, Persia, Kurdistan, Armenia, Korea, and China. Her knowledge of new areas and precise descriptions were considered useful to geographers and naturalists. This excerpt from her autobiographical A Lady's Life in the Rocky Mountains *details part of her journey through the Rocky Mountains. The chapter reflects her pleasure and interest in the natural world, the variety of people she met as she traveled, and the struggles and dangers she faced as a woman traveling through arduous terrain.*

A Lady's Life in the Rocky Mountains

ISABELLA LUCY BIRD BISHOP

Letter VII

PERSONALITY OF LONG'S PEAK—"MOUNTAIN JIM"—LAKE OF THE LILIES—A SILENT FOREST—THE CAMPING GROUND—"RING"—A LADY'S BOWER—DAWN AND SUNRISE—A GLORIOUS VIEW—LINKS OF DIAMONDS—THE ASCENT OF THE PEAK—THE DOG'S LIFT—SUFFERING FROM THIRST—THE DESCENT—THE BIVOUAC.

ESTES PARK, COLORADO, *October.*

As this account of the ascent of Long's Peak could not be written at the time, I am much disinclined to write it, especially as no sort of description within my powers could enable another to realize the glorious sublimity, the majestic solitude, and the unspeakable awfulness and fascination of the scenes in which I spent Monday, Tuesday, and Wednesday.

Long's Peak, 14,700 feet high, blocks up one end of Estes Park, and dwarfs all the surrounding mountains. From it on this side rise, snow-born, the bright St. Vrain, and the Big and Little Thompson. By sunlight or moonlight its splintered grey crest is the one object which, in spite of wapiti and bighorn, skunk and grizzly, unfailingly arrests the eyes. From it come all storms of snow and wind, and the forked lightnings play round its head like a glory. It is one of the noblest of mountains, but in one's imagination it grows to be much more than a mountain. It becomes invested with a personality. In its caverns and abysses one comes to fancy that it generates and chains the strong winds, to let them loose in its fury. The thunder becomes its voice, and the lightnings do it homage. Other summits blush under the morning kiss of the sun, and turn pale the next moment; but it detains the first sunlight and holds it round its head for an hour at least, till it pleases to change from rosy red to deep blue; and the sunset, as if spell-bound, lingers latest on its crest. The soft winds which hardly rustle the pine needles down here are raging rudely up there round its motionless summit. The mark of fire is upon it; and though it has passed into a grim repose, it tells of fire and upheaval as truly, though not as eloquently, as the living volcanoes of Hawaii. Here under its shadow one learns how naturally nature worship, and the propitiation of the forces of nature, arose in minds which had no better light.

Long's Peak, "the American Matterhorn," as some call it, was ascended five years ago for the first time. I thought I should like to attempt it, but up to Monday, when Evans left for Denver, cold water was thrown upon the project. It was too late in the season, the winds were likely to be strong, etc.; but just before leaving, Evans said that the weather was looking more settled, and if I did not get farther than the timber line it would be worth going. Soon after he left, "Mountain Jim" came in, and he would go up as guide, and the two youths who rode here with me from Longmount and I caught at the proposal. Mrs. Edwards at once baked bread for three days, steaks were cut from the steer which hangs up conveniently, and tea, sugar, and butter were benevolently added. Our picnic was not to be a luxurious or "well-found" one, for, in order to avoid the expense of a pack mule, we limited our luggage to what our saddle horses could carry. Behind my saddle I carried three pair of camping blankets and a quilt, which reached to my shoulders. My own boots were so much worn that it was painful to walk, even about the park, in them, so Evans had lent me a pair of his hunting boots, which hung to the horn of my saddle. The horses of the two young men were equally loaded, for we had to prepare for many degrees of frost. "Jim" was a shocking figure; he had on an old pair of high boots, with a baggy pair of old trousers made of deer hide, held on by an old scarf tucked into them; a leather shirt, with three or four ragged unbuttoned waistcoats over it; an old smashed wideawake, from under which his tawny, neglected ringlets hung; and with his one eye, his one long spur, his knife in his belt, his revolver in his waistcoat pocket, his saddle covered with an old beaver skin, from which the paws hung down; his camping blankets behind him, his rifle laid across the saddle in front of him, and his axe, canteen, and other gear hanging to the horn, he was as awful-looking a ruffian as one could see. By way of contrast he rode a small Arab mare, of exquisite beauty, skittish, high spirited, gentle, but altogether too light for him, and he fretted her incessantly to make her display herself.

Heavily loaded as all our horses were, "Jim" started over the half-mile of level grass at a hard gallop, and then throwing his mare on her haunches, pulled up alongside of me, and with a grace of manner which soon made me forget his appearance, entered into a conversation which lasted for more than three hours, in spite of the manifold checks of fording streams, single file, abrupt ascents and descents, and other incidents of mountain travel. The ride was one series of glories and surprises, of "park" and glade, of lake and stream, of mountains on mountains, culminating in the rent pinnacles of Long's Peak, which looked yet grander and ghastlier as we crossed an attendant mountain 11,000 feet high. The slanting sun added fresh beauty every hour. There were dark pines against a lemon sky, grey peaks reddening and etherealizing, gorges of deep and infinite blue, floods of golden glory pouring through canyons of enormous depth, an atmosphere of absolute purity, an occasional foreground of cotton-wood and aspen flaunting in red and gold to intensify the blue gloom of the pines, the trickle and murmur of streams fringed with icicles, the strange *sough* of gusts moving among the pine tops—sights and sounds not of the lower earth, but of the solitary, beast-haunted, frozen upper altitudes. From the dry, buff grass of Estes Park we turned off up a trail on the side of a pine-hung gorge, up a steep pine-clothed hill, down to a small valley, rich in fine, sun-cured hay about eighteen inches high, and enclosed by high mountains whose deepest hollow contains a lily-covered lake, fitly named "The Lake of the Lilies." Ah, how magical its beauty was, as it slept in silence, while *there* the dark pines were mirrored motionless in its pale gold, and *here* the great white lily cups and dark green leaves rested on amethyst-colored water!

From this we ascended into the purple gloom of great pine forests which clothe the skirts of the mountains up to a height of about 11,000 feet, and from their chill and solitary depths we had glimpses of golden atmosphere and rose-lit summits, not of "the land very far off," but of the land nearer now in all its grandeur, gaining in sublimity by nearness—glimpses, too, through a broken vista of purple gorges, of the illimitable Plains lying idealized in the late sunlight, their baked, brown expanse transfigured into the likeness of a sunset sea rolling infinitely in waves of misty gold.

We rode upwards through the gloom on a steep trail blazed through the forest, all my intellect concentrated on avoiding being dragged off my horse by impending branches, or having the blankets badly torn, as those of my companions were, by sharp dead limbs, between which there was hardly room to pass— the horses breathless, and requiring to stop every few yards, though their riders, except myself, were afoot. The gloom of the dense, ancient, silent forest is to me awe inspiring. On such an evening it is soundless, except for the branches creaking in the soft wind, the frequent snap of decayed timber, and a murmur in the pine tops as of a not distant waterfall, all tending to produce *eeriness* and a sadness "hardly akin to pain." There no lumberer's axe has ever rung. The trees die when they have attained their prime, and stand there, dead and bare, till the fierce mountain winds lay them prostrate. The pines grew smaller and more sparse as we ascended, and the last stragglers wore a tortured, warring look. The timber line was passed, but yet a little higher a slope of mountain meadow dipped to the south-west towards a bright stream trickling under ice and icicles, and there a grove of the beautiful silver spruce marked our camping ground. The trees were in miniature, but so exquisitely arranged that one might well ask what artist's hand had planted them, scattering them here, clumping them there, and training their slim spires towards heaven. Hereafter, when I call up memories of the glorious, the view from this camping ground will come up. Looking east, gorges opened to the distant Plains, then fading into purple grey. Mountains with pine-clothed skirts rose in ranges, or, solitary, uplifted their grey summits, while close behind, but nearly 3,000 feet above us, towered the bald white crest of Long's Peak, its huge precipices red with the light of a sun long lost to our eyes. Close to us, in the caverned side of the Peak, was snow that, owing to its position, is eternal. Soon the afterglow came on, and before it faded a big half-moon hung out of the heavens, shining through the silver blue foliage of the pines on the frigid background of snow, and turning the whole into fairyland. The "photo" which accompanies this letter is by a courageous Denver artist who attempted the ascent just before I arrived, but, after camping out at the timber line for a week, was foiled by the perpetual storms, and was driven down again, leaving some very valuable apparatus about 3,000 feet from the summit.

Unsaddling and picketing the horses securely, making the beds of pine shoots, and dragging up logs for fuel, warmed us all. "Jim" built up a great fire, and before long we were all sitting around it at supper. It didn't matter much that we had to drink our tea out of the battered meat tins in which it was boiled, and eat strips of beef reeking with pine smoke without plates or forks.

"Treat Jim as a gentleman and you'll find him one," I had been told; and though his manner was certainly bolder and freer than that of gentlemen generally, no imaginary fault could be found. He was very agreeable as a man of culture as well as a child of nature; the desperado was altogether out of sight. He was very courteous and even kind to me, which was fortunate, as the young men had little idea of showing even ordinary civilities. That night I made the acquaintance of his dog "Ring," said to be the best hunting dog in Colorado, with the body and legs of a collie, but a head approaching that of a mastiff, a noble face with a wistful human expression, and the most truthful eyes I ever saw in an animal. His master loves him if he loves anything, but in his savage moods ill-treats him. "Ring's" devotion never swerves, and his truthful eyes are rarely taken off his master's face. He is almost human in his intelligence, and, unless he is told to do so, he never takes notice of any one but "Jim." In a tone as if speaking to a human being, his master, pointing to me, said, "Ring, go to that lady, and don't leave her again to-night." "Ring" at once came to me, looked into my face, laid his head on my shoulder, and then lay down beside me with his head on my lap, but never taking his eyes from "Jim's" face.

The long shadows of the pines lay upon the frosted grass, an aurora leaped fitfully, and the moonlight, though intensely bright, was pale beside the red, leaping flames of our pine logs and their red glow on our gear, ourselves, and Ring's truthful face. One of the young men sang a Latin student's song and two Negro melodies; the other "Sweet Spirit, hear my Prayer." "Jim" sang one of Moore's melodies in a singular falsetto, and all together sang, "The Star-spangled Banner" and "The Red, White, and Blue." Then "Jim" recited a very clever poem of his own composition, and told some fearful Indian stories. A group of small silver spruces away from the fire was my sleeping place. The artist who had been up there had so woven and interlaced their lower branches as to form a bower, affording at once shelter from the wind and a most agreeable privacy. It was thickly strewn with young pine shoots, and these, when covered with a blanket, with an inverted saddle for a pillow, made a luxurious bed. The mercury at 9 P.M. was 12° below the freezing point. "Jim," after a last look at the horses, made a huge fire, and stretched himself out beside it, but "Ring" lay at my back to keep me warm. I could not sleep, but the night passed rapidly. I was anxious about the ascent, for gusts of ominous sound swept through the pines at intervals. Then wild animals howled, and "Ring" was perturbed in spirit about them. Then it was strange to see the notorious desperado, a red-handed man, sleeping as quietly as innocence sleeps. But, above all, it was exciting to lie there, with no better shelter than a bower of pines, on a mountain 11,000 feet high, in the very heart of the Rocky Range, under twelve degrees of frost, hearing sounds of wolves, with shivering stars looking through the fragrant canopy, with arrowy pines for bed-posts, and for a night lamp the red flames of a camp-fire.

Day dawned long before the sun rose, pure and lemon colored. The rest were looking after the horses, when one of the students came running to tell me that I must come farther down the slope, for "Jim" said he had never seen such a sunrise. From the chill, grey Peak above, from the everlasting snows, from the silvered pines, down through mountain ranges with their depths of Tyrian purple, we looked to where the Plains lay cold, in blue-grey, like a morning sea against a far horizon. Suddenly, as a dazzling streak at first, but enlarging rapidly into a dazzling sphere, the sun wheeled above the grey line, a light and glory as when it was first created. "Jim" involuntarily and reverently uncovered his head, and exclaimed, "I believe there is a God!" I felt as if, Parsee-like, I must worship. The grey of the Plains changed to purple, the sky was all one rose-red flush, on which vermilion cloud-streaks rested; the ghastly peaks gleamed like rubies, the earth and heavens were new created. Surely "the Most High dwelleth not in temples made with hands!" For a full hour those Plains simulated the ocean, down to whose limitless expanse of purple, cliff, rocks, and promontories swept down.

By seven we had finished breakfast, and passed into the ghastlier solitudes above, I riding as far as what, rightly or wrongly, are called the "Lava Beds," an expanse of large and small boulders, with snow in their crevices. It was very cold; some water which we crossed was frozen hard enough to bear the horse. "Jim" had advised me against taking any wraps, and my thin Hawaiian riding dress, only fit for the tropics, was penetrated by the keen air. The rarefied atmosphere soon began to oppress our breathing, and I found that Evans's boots were so large that I had no foothold. Fortunately, before the real difficulty of the ascent began, we found, under a rock, a pair of small overshoes, probably left by the Hayden exploring expedition, which just lasted for the day. As we were leaping from rock to rock, "Jim" said, "I was thinking in the night about your traveling alone, and wondering where you carried your Derringer, for I could see no signs of it." On my telling him that I traveled unarmed, he could hardly believe it, and adjured me to get a revolver at once.

On arriving at the "Notch" (a literal gate of rock), we found ourselves absolutely on the knifelike ridge or backbone of Long's Peak, only a few feet wide, covered with colossal boulders and fragments, and on the other side shelving in one precipitous, snow-patched sweep of 3,000 feet to a picturesque hollow, containing a lake of pure green water. Other lakes, hidden among dense pine woods, were farther off, while close above us rose the Peak, which, for about 500 feet, is a smooth, gaunt, inaccessible-looking pile of granite. Passing through the "Notch," we looked along the nearly inaccessible side of the Peak, composed of boulders and *débris* of all shapes and sizes, through which appeared broad, smooth ribs of reddish-colored granite, looking as if they upheld the towering rock mass above. I usually dislike bird's-eye and panoramic views, but, though from a mountain, this was not one. Serrated ridges, not much lower than that on which we stood, rose, one beyond another, far as that pure atmosphere could carry the vision, broken into awful chasms deep with ice and snow, rising into pinnacles piercing the heavenly blue with their cold, barren grey, on, on for ever, till the most distant range upbore unsullied snow alone. There were fair lakes mirroring the dark pine woods, canyons dark and blue-black with unbroken expanses of pines, snow-slashed pinnacles, wintry heights frowning upon lovely parks, watered and wooded, lying in the lap of summer; North Park floating off into the blue distance, Middle Park closed till another season, the sunny slopes of Estes Park, and winding down among the mountains the snowy ridge of the Divide, whose bright waters seek both the Atlantic and Pacific Oceans. There, far below, links of diamonds showed where the Grand River takes its rise to seek the mysterious Colorado, with its still unsolved enigma, and lose itself in the waters of the Pacific; and nearer the snow-born Thompson bursts forth from the ice to begin its journey to the Gulf of Mexico. Nature, rioting in her grandest mood, exclaimed with voices of grandeur, solitude, sublimity, beauty, and infinity, "Lord, what is man, that Thou art mindful of him? or the son of man, that Thou visitest him?" Never-to-be-forgotten glories they were, burnt in upon my memory by six succeeding hours of terror.

You know I have no head and no ankles, and never ought to dream of mountaineering; and had I known that the ascent was a real mountaineering feat I should not have felt the slightest ambition to perform it. As it is, I am only humiliated by my success, for "Jim" dragged me up, like a bale of goods, by sheer force of muscle. At the "Notch" the real business of the ascent began. Two thousand feet of solid rock towered above us, four thousand feet of broken rock shelved precipitously below; smooth granite ribs, with barely foothold, stood out here and there; melted snow refrozen several times, presented a more serious obstacle; many of the rocks were loose, and tumbled down when touched. To me it was a time of extreme terror. I was roped to "Jim," but it was of no use; my feet were paralyzed and slipped on the bare rock, and he said it was useless to try to go that way, and we retraced our steps. I wanted to return to the "Notch," knowing that my incompetence would detain the party, and one of the young men said almost plainly that a woman was a dangerous encumbrance, but the trapper replied shortly that if it were not to take a lady up he would not go up at all. He went on the explore, and reported that further progress on the correct line of ascent was blocked by ice; and then for two hours we descended, lowering ourselves by our hands from rock to rock along a boulder-strewn sweep of 4,000 feet, patched with ice and snow, and perilous from rolling stones. My fatigue, giddiness, and pain from bruised ankles, and arms half pulled out of their sockets, were so great that I should never have gone half-way had not "Jim," *nolens volens*, dragged me along with a patience and skill, and withal a determination that I should ascend the Peak, which never failed. After descending about 2,000 feet to avoid the ice, we got into a deep ravine with inaccessible sides, partly filled with ice and snow and partly with large and small fragments of rock, which were constantly giving away, rendering the footing very insecure. That part to me was two hours of painful and unwilling submission to the inevitable; of trembling, slipping, straining, of smooth ice appearing when it was least expected, and of weak entreaties to be left behind while the others went on. "Jim" always said that there was no danger, that there was only a short bad bit ahead, and that I should go up even if he carried me!

Slipping, faltering, gasping from the exhausting toil in the rarefied air, with throbbing hearts and panting lungs, we reached the top of the gorge and squeezed ourselves between two gigantic fragments of rock by a passage called the "Dog's Lift," when I climbed on the shoulders of one man and then was hauled up. This introduced us by an abrupt turn round the south-west angle of the Peak to a narrow shelf of considerable length, rugged, uneven, and so over-hung by the cliff in some places that it is necessary to crouch to pass at all. Above, the Peak looks nearly vertical for 400 feet; and below, the most tremen-dous precipice I have ever seen descends in one unbroken fall. This is usually considered the most dangerous part of the ascent, but it does not seem so to me, for such foothold as there is is secure, and one fancies that it is possible to hold on with the hands. But there, and on the final, and, to my thinking, the worst part of the climb, one slip, and a breathing, thinking, human being would lie 3,000 feet below, a shapeless, bloody heap! "Ring" refused to traverse the Ledge, and remained at the "Lift" howling piteously.

From thence the view is more magnificent even than that from the "Notch." At the foot of the precipice below us lay a lovely lake, wood embosomed, from or near which the bright St. Vrain and other streams take their rise. I thought how their clear cold waters, growing turbid in the affluent flats, would heat under the tropic sun, and eventually form part of that great ocean river which renders our far-off islands habitable by impinging on their shores. Snowy ranges, one behind the other, extended to the distant horizon, folding in their wintry embrace the beauties of Middle Park. Pike's Peak, more than one hundred miles off, lifted that vast but shapeless summit which is the landmark of southern Colorado. There were snow patches, snow slashes, snow abysses, snow forlorn and soiled look-ing, snow pure and dazzling, snow glistening above the purple robe of pine worn by all the mountains; while away to the east, in limitless breadth, stretched the green-grey of the endless Plains. Giants everywhere reared their splintered crests. From thence, with a single sweep, the eye takes in a distance of 300 miles—that distance to the west, north, and south being made up of mountains ten, eleven, twelve, and thirteen thousand feet in height, dominated by Long's Peak, Gray's Peak, and Pike's Peak, all nearly the height of Mont Blanc! On the

Plains we traced the rivers by their fringe of cotton-woods to the distant Platte, and between us and them lay glories of mountain, canyon, and lake, sleeping in depths of blue and purple most ravishing to the eye.

As we crept from the ledge round a horn of rock I beheld what made me perfectly sick and dizzy to look at—the terminal Peak itself—a smooth, cracked face or wall of pink granite, as nearly perpendicular as anything could well be up which it was possible to climb, well deserving the name of the "American Matterhorn."

Scaling, not climbing, is the correct term for this last ascent. It took one hour to accomplish 500 feet, pausing for breath every minute or two. The only foothold was in narrow cracks or on minute projections on the granite. To get a toe in these cracks, or here and there on a scarcely obvious projection, while crawling on hands and knees, all the while tortured with thirst and gasping and struggling for breath, this was the climb; but at last the Peak was won. A grand, well-defined mountain top it is, a nearly level acre of boulders, with precipitous sides all round, the one we came up being the only accessible one.

It was not possible to remain long. One of the young men was seriously alarmed by bleeding from the lungs, and the intense dryness of the day and the rarefication of the air, at a height of nearly 15,000 feet, made respiration very painful. There is always water on the Peak, but it was frozen as hard as a rock, and the sucking of ice and snow increases thirst. We all suffered severely from the want of water, and the gasping for breath made our mouths and tongues so dry that articulation was difficult, and the speech of all unnatural.

From the summit were seen in unrivalled combination all the views which had rejoiced our eyes during the ascent. It was something at last to stand upon the stormrent crown of this lonely sentinel of the Rocky Range, on one of the mightiest of the vertebrae of the backbone of the North American continent, and to see the waters start for both oceans. Uplifted above love and hate and storms of passion, calm amidst the eternal silences, fanned by zephyrs and bathed in living blue, peace rested for that one bright day on the Peak, as if it were some region

> *Where falls not rain, or hail, or any snow,*
> *Or ever wind blows loudly.*

We placed our names, with the date of ascent, in a tin within a crevice, and descended to the Ledge, sitting on the smooth granite, getting our feet into cracks and against projections, and letting ourselves down by our hands, "Jim" going before me, so that I might steady my feet against his powerful shoulders. I was no longer giddy, and faced the precipice of 3,500 feet without a shiver. Repassing the Ledge and Lift, we accomplished the descent through 1,500 feet of ice and snow, with many falls and bruises, but no worse mishap, and there separated, the young men taking the steepest but most direct way to the "Notch," with the intention of getting ready for the march home, and "Jim" and I taking what he thought the safer route for me—a descent over boulders for 2,000 feet, and then a tremendous ascent to the "Notch." I had various falls, and once hung by my frock, which caught on a rock, and "Jim" severed it with his hunting knife, upon which I fell into a crevice full of soft snow. We were driven lower down the mountains than he had intended by impassable tracts of ice, and the ascent was tremendous. For the last 200 feet the boulders were of enormous size, and the steepness fearful. Sometimes I drew myself up on hands and knees, sometimes crawled; sometimes "Jim" pulled me up by my arms or a lariat, and sometimes I stood on his shoulders, or he made steps for me of his feet and hands, but at six we stood on the "Notch" in the splendor of the sinking sun, all color deepening, all peaks glorifying, all shadows purpling, all peril past.

"Jim" had parted with his *brusquerie* when we parted from the students, and was gentle and considerate beyond anything, though I knew that he must be grievously disappointed, both in my courage and strength. Water was an object of earnest desire. My tongue rattled in my mouth, and I could hardly articulate. It is good for one's sympathies to have for once a severe experience of thirst. Truly, there was

> *Water, water, everywhere,*
> *But not a drop to drink.*

Three times its apparent gleam deceived even the mountaineer's practised eye, but we found only a foot of "glare ice." At last, in a deep hole, he succeeded in breaking the ice, and by putting one's arm far down one could scoop up a little water in one's hand, but it was tormentingly insufficient. With great difficulty and much assistance I recrossed the "Lava Beds," was carried to the horse and lifted upon him, and when we reached the camping ground I was lifted off him, and laid on the ground wrapped up in blankets, a humiliating termination of a great exploit. The horses were saddled, and the young men were all ready to start, but "Jim" quietly said, "Now, gentlemen, I want a good night's rest, and we shan't stir from here to-night." I believe they were really glad to have it so, as one of them was quite "finished." I retired to my arbor, wrapped myself in a roll of blankets, and was soon asleep.

When I woke, the moon was high shining through the silvery branches, whitening the bald Peak above, and glittering on the great abyss of snow behind, and pine logs were blazing like a bonfire in the cold still air. My feet were so icy cold that I could not sleep again, and getting some blankets to sit in, and making a roll of them for my back, I sat for two hours by the camp-fire. It was weird and gloriously beautiful. The students were asleep not far off in their blankets with their feet towards the fire. "Ring" lay on one side of me with his fine head on my arm, and his master sat smoking, with the fire lighting up the handsome side of his face, and except for the tones of our voices, and an occasional crackle and splutter as a pine knot blazed up, there was no sound on the mountain side. The beloved stars of my far-off home were overhead, the Plough and Pole Star, with their steady light; the glittering Pleiades, looking larger than I ever saw them, and "Orion's studded belt" shining gloriously. Once only some wild animals prowled near the camp, when "Ring," with one bound, disappeared from my side; and the horses, which were picketed by the stream, broke their lariats, stampeded, and came rushing wildly towards the fire, and it was fully half an hour before they were caught and quiet was restored. "Jim," or Mr. Nugent, as I always scrupulously called him, told stories of his early youth, and of a great sorrow which had led him to embark on a lawless and desperate life. His voice trembled, and tears rolled down his cheek. Was it semi-conscious acting, I wondered, or was his dark soul really stirred to its depths by the silence, the beauty, and the memories of youth?

We reached Estes Park at noon of the following day. A more successful ascent of the Peak was never made, and I would not now exchange my memories of its perfect beauty and extraordinary sublimity for any other experience of mountaineering in any part of the world. Yesterday snow fell on the summit, and it will be inaccessible for eight months to come.

I. L. B.

[1] Let no practical mountaineer be allured by my description into the ascent of Long's Peak. Truly terrible as it was to me, to a member of the Alpine Club it would not be a feat worth performing.

Source: Bishop, Isabella Lucy Bird. 1960. *A Lady's Life in the Rocky Mountains*. Pp. 83–101. Norman: University of Oklahoma.

Mary Somerville was perhaps one of the most well known women scientists of her day. Her grasp of mathematical knowledge in particular enabled her to write popular explanations of scientific work that were highly respected by the scientific community. In these excerpts from her Personal Recollections *the reader obtains an insight into how women studied and wrote science within a domestic setting. The second excerpt covers the period in which Somerville was writing her* Mechanism of the Heavens *and demonstrates the esteem in which she was held by male scientists and the passion and interest with which she engaged in her work.*

Personal Recollections, from Early Life to Old Age of Mary Somerville

MARY SOMERVILLE

[After three years of married life, my mother returned to her father's house in Burntisland, a widow, with two little boys. The youngest died in childhood. The eldest was Woronzow Greig, barrister-at-law, late Clerk of the Peace for Surrey. He died suddenly in 1865, to the unspeakable sorrow of his family, and the regret of all who knew him.

I was much out of health after my husband's death, and chiefly occupied with my children, especially with the one I was nursing; but as I did not go into society, I rose early, and, having plenty of time, I resumed my mathematical studies. By this time I had studied plane and spherical trigonometry, conic sections, and Fergusson's "Astronomy." I think it was immediately after my return to Scotland that I attempted to read Newton's "Principia." I found it extremely difficult, and certainly did not understand it till I returned to it some time after, when I studied that wonderful work with great assiduity, and wrote numerous notes and observations on it. I obtained a loan of what I believe was called the Jesuit's edition, which helped me. At this period mathematical science was at a low ebb in Britain; reverence for Newton had prevented men from adopting the "Calculus," which had enabled foreign mathematicians to carry astronomical and mechanical science to the highest perfection. Professors Ivory and de Morgan had adopted the "Calculus"; but several years elapsed before Mr. Herschel and Mr. Babbage were joint-editors with Professor Peacock in publishing an abridged translation of La Croix's "Treatise on the Differential and Integral Calculus." I became acquainted with Mr. Wallace, who was, if I am not mistaken, mathematical teacher of the Military College at Marlow, and editor of a mathematical journal published there, I had solved some of the problems contained in it and sent them to him, which led to a correspondence, as Mr. Wallace sent me his own solutions in return. Mine were sometimes right and sometimes wrong, and it occasionally happened that we solved the same problem by different methods. At last I succeeded in solving a prize problem! It was a diophantine problem, and I was awarded a silver medal cast on purpose with my name, which pleased me exceedingly.

Mr. Wallace was elected Professor of Mathematics in the University of Edinburgh, and was very kind to me. When I told him that I earnestly desired to go through a regular course of mathematical and astronomical science, even including the highest branches, he gave me a list of the requisite books, which were in French, and consisted of Francœur's pure "Mathematics," and his "Elements of Mechanics," La Croix's "Algebra," and his large work on the "Differential and Integral Calculus," together with his work on "Finite Differences and Series," Biot's "Analytical Geometry and Astronomy," Poisson's "Treatise on Mechanics," La Grange's "Theory of Analytical Functions," Euler's "Algebra," Euler's "Isoperimetrical Problems" (in Latin), Clairault's "Figure of the Earth," Monge's "Application of Analysis to Geometry," Callet's "Logarithms," La Place's "Mécanique Céleste," and his "Analytical Theory of Probabilities," &c., &c., &c.*

I was thirty-three years of age when I bought this excellent little library. I could hardly believe that I possessed such a treasure when I looked back on the day that I first saw the mysterious word "Algebra," and the long course of years in which I had persevered almost without hope. It taught me never to despair. I had now the means, and pursued my studies with increased assiduity; concealment was no longer possible, nor was it attempted. I was considered eccentric and foolish, and my conduct was highly disapproved of by many, especially by some members of my own family, as will be seen hereafter. They expected me to entertain and keep a gay house for them, and in that they were disappointed. As I was quite independent, I did not care for their criticism. A great part of the day I was occupied with my children; in the evening I worked, played piquet with my father, or played on the piano, sometimes with violin accompaniment. . . .

CHAPTER XI.

LETTER FROM LORD BROUGHAM—WRITES "MECHANISM OF THE HEAVENS"—ANECDOTE OF THE ROMAN IMPROVISATRICE—LETTERS FROM SIR JOHN HERSCHEL AND PROFESSOR WHEWELL—ELECTED HON. MEMBER OF THE ROYAL ASTRONOMICAL SOCIETY—NOTICE IN THE ACADÉMIE DES SCIENCES, AND LETTER FROM M. BIOT—PENSION—LETTER FROM SIR ROBERT PEEL—BEGINS TO WRITE ON THE CONNECTION OF THE PHYSICAL SCIENCES—VISIT TO CAMBRIDGE—LETTERS FROM PROFESSOR SEDGWICK AND LAPLACE.

After my mother's return home my father received the following letter from

Lord Brougham, which very importantly influenced the further course of my mother's life. It is dated March 27th, 1827 :—

LETTER FROM LORD BROUGHAM TO DR. SOMERVILLE.

"MY DEAR SIR,

I fear you will think me very daring, for the design I have formed against Mrs. Somerville, and still more for making you my advocate with her; through whom I have every hope of prevailing. There will be sent to you a prospectus, rules, and a preliminary treatise of our Society for Diffusing Useful Knowledge, and I assure you I speak without any flattery when I say that of the two subjects which I find it most difficult to see the chance of executing, there is one, which— unless Mrs. Somerville will undertake—none else can, and it must be left undone, though about the most interesting of the whole, I mean an account of the Mécanique Céleste; the other is an account of the Principia, which I have some hopes of at Cambridge. The kind of thing wanted is such a description of that divine work as will both explain to the unlearned the sort of thing it is—the plan, the vast merit, the wonderful truths unfolded or methodized—and the calculus by which all this is accomplished, and will also give a somewhat deeper insight to the uninitiated. Two treatises would do this. No one without trying it can conceive how far we may carry ignorant readers into an understanding of the depths of science, and our treatises have about 100 to 800 pages of space each, so that one might give the more popular view, and another the analytical abstracts and illustrations. In England there are now not twenty people who know this great work, except by name; and not a hundred who know it even by name. My firm belief is that Mrs. Somerville could add two cyphers to each of those figures. Will you be my counsel in this suit? Of course our names are concealed, and no one of our council but myself needs to know it.

Yours ever most truly,

H. BROUGHAM.

[My mother in alluding to the above says:—

This letter surprised me beyond expression. I thought Lord Brougham must have been mistaken with regard to my acquirements, and naturally concluded that my self-acquired knowledge was so far inferior to that of the men who had been educated in our universities that it would be the height of presumption to attempt to write on such a subject, or indeed on any other. A few days after this Lord Brougham came to Chelsea himself, and Somerville joined with him in urging me at least to make the attempt. I said, "Lord Brougham, you must be aware that the work in question never can be popularized, since the student must at least know something of the differential and integral calculi, and as a preliminary step I should have to prove various problems in physical mechanics and astronomy. Besides, La Place never gives diagrams or figures, because they are not necessary to persons versed in the calculus, but they would be indispensable in a work such as you wish me to write. I am afraid I am incapable of such a task: but as you both wish it so much, I shall do my very best upon condition of secrecy, and that if I fail the manuscript shall be put into the fire." Thus suddenly and unexpectedly the whole character and course of my future life was changed.

I rose early and made such arrangements with regard to my children and family affairs that I had time to write afterwards; not, however, without many interruptions. A man can always command his time under the plea of business, a woman is not allowed any such excuse. At Chelsea I was always supposed to be at home, and as my friends and acquaintances came so far out of their way on purpose to see me, it would have been unkind and ungenerous not to receive them. Nevertheless, I was sometimes annoyed when in the midst of a difficult problem some one would enter and say, "I have come to spend a few hours with you." However, I learnt by habit to leave a subject and resume it again at once, like putting a mark into a book I might be reading ; this was the more necessary as there was no fire-place in my little room, and I had to write in the drawing-room in winter. Frequently I hid my papers as soon as the bell announced a visitor, lest anyone should discover my secret.

[My mother had a singular power of abstraction. When occupied with some difficult problem, or even a train of thought which deeply interested her, she lost all consciousness of what went on around her, and became so entirely absorbed that any amount of talking, or even practising scales and *solfeggi*, went on without in the least disturbing her. Sometimes a song or a strain of melody would recall her to a sense of the present, for she was passionately fond of music. A curious instance of this peculiarity of hers occurred at Rome, when a large party were assembled to listen to a celebrated improvisatrice. My mother was placed in the front row, close to the poetess, who, for several stanzas, adhered strictly to the subject which had been given to her. What it was I do not recollect, except that it had no connection with what followed. All at once, as if by a sudden inspiration, the lady turned her eyes full upon my mother, and with true Italian vehemence and in the full musical accents of Rome, poured forth stanza after stanza of the most eloquent panegyric upon her talents and virtues, extolling them and her to the skies. Throughout the whole of this scene, which lasted a considerable time, my mother remained calm and unmoved, never changing countenance, which surprised not only the persons present but ourselves, as we well knew how much she disliked any display or being brought forward in public. The truth was, that after listening for a while to the improvising, a thought struck her connected with some subject she was engaged in writing upon at the time and so entirely absorbed her that she heard not a word of all that had been declaimed in her praise, and was not a little surprised and confused when she was complimented on it. I call this, advisedly, a power of hers, for although it occasionally led her into strange positions, such as the one above mentioned, it rendered her entirely independent of outward circumstances, nor did she require to isolate herself from the family circle in order to pursue her studies. I have already mentioned that when we were very young she taught us herself for a few hours daily; when our lessons were over we always remained in the room with her, learning grammar, arithmetic, or some such plague of childhood. Any one who has plunged into the mazes of the higher branches of mathematics or other abstruse science, would probably feel no slight degree of irritation on being interrupted at a critical moment when the solution was almost within his grasp, by some childish question about tense or gender, or how much seven times seven made. My mother was never impatient, but explained our little difficulties quickly and kindly, and returned calmly to her own profound thoughts. Yet on occasion she could show both irritation and impatience—when we were stupid or inattentive, neither of which she could stand. With her clear mind she darted at the solution, sometimes forgetting that we had to toil after her laboriously step by step. I well remember her slender white hand pointing impatiently to the book or slate—"Don't you see it? there is no difficulty in it, it is quite clear." Things were so clear to her! I must here add some other recollections by my mother of this very interesting portion of her life.

I was a considerable time employed in writing this book, but I by no means gave up society, which would neither have suited Somerville nor me. We dined out, went to evening parties, and occasionally to the theatre. As soon as my work was finished I sent the manuscript to Lord Brougham, requesting that it might be thoroughly examined, criticised and destroyed according to promise if a failure. I was very nervous while it was under examination, and was equally surprised and gratified that Sir John Herschel, our greatest astronomer, and perfectly versed in the calculus, should have found so few errors. The letter he wrote on this occasion made me so happy and proud that I have preserved it.

LETTER FROM SIR JOHN HERSCHEL TO MRS. SOMERVILLE.

DEAR MRS. SOMERVILLE,

I have read your manuscript with the greatest pleasure, and will not hesitate to add, (because I am sure you will believe it sincere,) with the highest admiration. Go on thus, and you will leave a memorial of no common kind to posterity; and, what you will value far more than fame, you will have accomplished a most useful work. What a pity that La Place has not lived to see this illustration of his great work! You will only, I fear, give too strong a stimulus to the study of abstract science by this performance.

I have marked as somewhat obscure a part of the illustration of the principle of virtual velocities. . . .. Will you look at this point again? I have made a trifling remark in page 6, but it is a mere matter of metaphysical nicety, and perhaps hardly worth pencilling your beautiful manuscript for.

Ever yours most truly,

J. HERSCHEL.

[In publishing the following letter, I do not consider that I am infringing on the rule I have followed in obedience to my mother's wishes, that is, to abstain from giving publicity to all letters which are of a private and confidential character. This one entirely concerns her scientific writings, and is interesting as showing the confidence which existed between Sir John Herschel and herself. This great philosopher was my mother's truest and best friend, one whose opinion she valued above all others, whose genius and consummate talents she admired, and whose beautiful character she loved with an intensity which is better shown by some extracts from her letters to be given presently than by anything I can say. This deep regard on her part he returned with the most chivalrous respect and admiration. In any doubt or difficulty it was his advice she sought, his criticism she submitted to; both were always frankly given without the slightest fear of giving offence, for Sir John Herschel well knew the spirit with which any remarks of his would be received.

* These books and all the other mathematical works belonging to my mother at the time of her death have been presented to the College for Women, at Girton, Cambridge.

Source: Somerville, Mary. 1876. *Personal Recollections, from Early Life to Old Age of Mary Somerville.* Pp. 77–80 and 161–168. New York: AMS Press.

Lydia Becker, a leader in the early women's suffrage movement in Britain, argues in this article that if women are given educational opportunities in science equal to those of men, their attainments in science would also be equal. Becker's own love of science (she was the author of Botany for Novices *published in 1864 and the founder of the Manchester Ladies Literary Society in 1865 whose emphasis was scientific) and her inadequate childhood education prompted her to work toward the inclusion of women and girls in scientific education. She notes that women's lack of scientific achievement to date was due to the lack of educational opportunities to women, not to any innate biological aversion or incapacity to the study and practice of science. Moreover, she sees science as an important component of a general education for women, whether they become scientists or not. A science education trains women's minds and stimulates their intellect allowing them to be equal and intelligent contributors to society whatever their role in life.*

On the Study of Science by Women

LYDIA BECKER

In speaking of the study of science by women, I desire, at the outset, to guard against the supposition that I consider such study to present any exceptional peculiarity to distinguish it from the study of science by men. Male and female students, in any branch of science, must go through the same training, and have their qualifications and capacities tested by precisely the same rules; neither is there anything in these studies which is naturally more attractive or advantageous to persons of one sex than of the other.

Nevertheless, the fact is indisputable that at the present time the students of science among men greatly outnumber those among women. Some persons attribute this circumstance to an inherent specific distinction in the minds of the two sexes of man. They assume the existence of a natural distaste or incapacity for scientific pursuits among women, and they consider it neither possible nor desirable to encourage them in the successful prosecution of such studies.

Others perceive in existing social and conventional arrangements, which exclude women from those opportunities of cultivating their intellectual faculties which are freely enjoyed by men, a perfectly sufficient explanation of the difference in the numbers and the proficiency of persons of each sex engaged in scientific pursuits.

The last is, I think, the true solution of the question, "Why are there fewer scientific women than scientific men?" The assumed difference in the minds of the two sexes is purely hypothetical; the practical difference in the training and advantages given to each is a fact as indisputable as the one which it explains.

I do not deny the existence of distinct types or orders of mind among mankind—all I deny is the coincidence of any one of these types with the physical distinction of sex.

If we take an assemblage of persons of both sexes, and test the differences of thought, opinion, or capacity existing among them, by putting before them any proposition on which opposite views can be held, I believe it would be impossible to find one which would range all the men on one side, and all the women on the other. If it were true that there is a specific difference, however slight, between the minds of men and women, it would be possible to find such a proposition, if we took one which corresponded to this distinction. When a naturalist seeks to group a number of individuals into a distinct class, he fixes on some character or set of characters common to them all, and distinguishing them from other individuals. When he finds such a group distinctly defined, he calls it a species. But when he finds two individuals differing very widely from each other, yet so connected by intermediate forms that he can pass from one extreme to the other without a violent break any where in the series, he considers them to be of one and the same kind. If we apply this principle as an illustration of the variety in human intellects, taking the conventional masculine type of mind as one end of the scale, and the conventional feminine type as the other, we shall find them connected by numerous intermediate varieties, distributed indiscriminately among male and female persons; that what is called a masculine mind is frequently found united to a feminine body, and sometimes the reverse, and that there is no necessary nor even presumptive connection between the sex of a human being and the type of intellect and character he possesses.

The equality of men and women, as regards intellect, resembles the equality of men among themselves, or women among themselves. No two are alike, no two are equal, but all start fair, and all have an equal right to advance as far as they can. Like a crowd of men and women on a level floor, all stand on the same plane, but some overtop the others. If we measure them by physical stature, there will be a considerable disparity between the sexes, and it will take an unusually tall woman to reach the height of the men. If we measure them by mental stature we shall find a different result. A woman who is somewhat taller than the masses of her sisters will be found to overtop the majority of the men.

The existence of a difference in the intellectual powers of the sexes is a question fertile in endless disputations, which can only be satisfactorily set at rest by the test of observation and experiment. Where-ever this test has been impartially applied, by studies and examinations conducted without reference to the sex of the student, the honours have been fairly divided between men and women, and no line of demarcation has made itself apparent between the character of the subjects chosen, or the degree of proficiency attained. The extremely limited area in which this test has been applied renders it, as yet, hardly safe to draw a general conclusion from the results, though these have hitherto pointed all one way ; but the existence of equality or disparity between the intellectual endowments of the sexes can only be established by the result of studies pursued under a common method, under the stimulus of similar incentives, and tested by the application of a common standard.

Most of the inducements for pursuing scientific studies are common to men and women. But there are some considerations which render such pursuits of greater value to women than to men. Prevalent opinions and customs impose on women so much more monotonous and colourless lives, and deprive them of so much of the natural and healthy excitement enjoyed by the other sex in its freer intercourse with the world, that the necessity for some pursuit which shall afford scope for the activity of their minds is even more pressing in their case than in that of men. In default of mental food and exercise, the minds of women get starved out. Numbers end by falling a prey to morbid religious excitement; while others, after vain struggles against their destiny, sink at last into a weary kind of resigned apathy, and men say they are content. But no one can measure the pain that has been endured ere the yearnings for a wider and freer existence subside into deadened calm. Many women might be saved from the evil of the life of intellectual vacuity, to which their present position renders them so peculiarly liable, if they had a thorough training in some branch of science, and the opportunity of carrying it on as a serious pursuit, in concert with others having similar tastes. Many a passing moment would then be made bright with a flash of thought which would otherwise have stolen away unmarked into the irrevocable past.

Men, who have been in the habit of enjoying the advantages attending systematic study and of the liberty of thought and speech not yet attained by women, do not need to be reminded of the benefits they derive from them. But women, who have never had the opportunity of finding out by experience the value of these conditions of mental life, do not always appreciate the magnitude of the loss they endure. If they did, I think they would not be content with their enforced exclusion from the pale of scientific society.

One of the greatest benefits which intellectual pursuits bring in their train is that of affording a peaceful neutral ground in which the mind can take refuge from the petty cares and annoyances of life, or even find diversion from more serious troubles. Like prudent investors, who keep a part of their capital in the funds, those who place the sources of a portion of their income of enjoyment in some pursuit wholly unconnected with their personal affairs, will find they have an interest which is perfectly safe amid the chances and changes of life. I do not for a moment maintain that intellectual pursuits can afford consolation in sorrow—for that we must look elsewhere; but they are undoubtedly capable of giving solace and diversion to the mind which might otherwise dwell too long on the gloomy side of things, and of beguiling the tedium of enforced solitude, or of confinement to a sick room. For an instance of this I may refer to the example of one of the most illustrious naturalists of the age. Mr. Charles Darwin has informed us that some of his most curious and interesting observations respecting the habits of climbing plants were made when he was a prisoner, night and day, to one room; and we cannot doubt that the occupation they afforded him not only served to lighten the weary hours, but occasioned him an amount of positive enjoyment which one less gifted might have failed to secure, though at liberty to participate in the ordinary pleasures of social life.

Such an example should encourage others to do likewise. Many particulars respecting the commonest of our wild plants, animals, and insects, are as yet imperfectly understood; and any woman who might select one of these creatures, and begin a series of patient observations on its habits, manner of feeding, of taking care of its young, of communicating with its kind, of guarding against danger, on its disposition and temper, and the difference in character between two individuals of the same species, would find such occupation not only exceedingly entertaining, but, if the observations were carefully noted, the result would be something of real, if not of great, scientific value. Gold is gold, whether our amount be an ingot or a spangle; and we need but to open our eyes, and carefully observe what is passing around us, to add perpetually to our store of the pure gold of knowledge.

No one should be deterred from either making or reporting original observations by the feeling that they are trifling or unimportant. Nothing that is real is considered insignificant by the naturalist, and observations apparently the most trifling have led to results which have turned the whole current of scientific thought. What could be a more trifling circumstance than the fall of an apple from a tree? Yet the appearances presented contained the clue that unravelled the mystery of the planetary movements. The law of gravitation maintains the stability of the universe, yet the fall of a pin to the ground is as truly a manifestation of this force as the movement of the earth in its orbit. With the sentiment of the poet in our hearts,—

> *"That very law that moulds a tear,*
> *And bids it trickle from its source,*
> *That law preserves the earth a sphere,*
> *And holds the planets in their course,"*—

we shall never regard any appearance as trifling which the tremendous forces of nature concur to produce.

How seemingly unimportant are the movements of insects, creeping in and out of flowers in search of the nectar on which they feed! If we saw a man spending his time in watching them, and in noting their flitting with curious eyes, we might be excused for imagining that he was amusing himself by idling an hour luxuriously in observing things which, though curious, were trifling. But how mistaken might we be in such an assumption! For these little winged messengers bear to the mind of the philosophical naturalist tidings of mysteries hitherto unrevealed; and as Newton saw the law of gravitation in the fall of the apple, Darwin found, in the connection between flies and flowers, some of the most important facts which support the theory he has promulgated respecting the modification of specific forms in animated beings.

It is true we are not Darwins nor Newtons, and cannot expect to make surprising discoveries; but we may be sure that these, and all other philosophers, have found an exquisite pleasure in tracing the workings of nature, and this enjoyment may be had by all who follow, however humbly, in their footsteps. And if we wish to understand their theories, it is refreshing to find our attention directed at the outset to pleasant and familiar natural objects—to varieties of pigeons, to humble-bees sucking clover flowers, to beetles swimming with their wings, to primroses and crimson flax, and grotesque orchids with their wild, weird beauty, setting traps for unwitting insects, and making them pay for their feast of honey by being the bearers of love-tokens from one flower to another—to be sent, in fact, to the Book of Nature, and bidden to read its wondrous stories with our own eyes.

Besides the addition to our store of positive knowledge, there is another important advantage to be derived from scientific study; namely, the cultivation of those habits of accuracy in speech and thought which are so absolutely necessary to its successful prosecution. One of the first lessons which a scientific student learns is, that he must not take a mere impression on his own mind as representing a positive fact, until he has carefully verified its accuracy by comparing it with the results of observation, and is prepared to state exactly on what grounds he entertains it. And when he hears an assertion made, he will pause before accepting it as true, for the mental inquiry whether the asserter is likely to be personally acquainted with the fact he alleges; and if not, what are his probable sources of information. On the answer to these expressed or unexpressed queries will depend the measure of credence to be given to the assertion in question. A reverence for accuracy of this kind would arrest many a baseless and painful rumour; and if it is the tendency of scientific investigation to conduce to such a tone of mind, the most inveterate sceptic as to the benefits of intellectual culture for women might be induced to confess that it is better that maids, old and young, should graduate in the School for Science, rather than in the School for Scandal.

If we turn from the consideration of the advantages women would gain from taking an active part in scientific pursuits, to the means accessible to them for prosecuting these studies, we perceive a very deplorable state of affairs.

The necessity for some common ground on which all interested in intellectual pursuits may meet, has been so strongly felt, that there exist all over the country institutions and societies, devoted either to literature and philosophy in general, or to the cultivation of special departments of knowledge. But most of these institutions, especially such as are devoted to the higher branches of scientific investigation, have one strange and injurious deficiency. They do not throw open such opportunities as they afford for acquiring knowledge freely to all who desire it; they draw an arbitrary line among scientific students, and say to one half of the human race, "You shall not enter into the advantages we have to offer; you shall not enjoy the facilities we possess of cultivating the tastes and faculties with which you may be endowed; and should any of you, in spite of this drawback, reach such a measure of attainments as would entitle one of us to the honour of membership or fellowship in any learned society, we will not, by conferring such distinctions on any of you, recognise your right to occupy your minds with such studies at all." It is no light mortification to a woman, who is desirous of prosecuting a study, to find that those best qualified to help her on her way are sedulous in affording her all the discouragement in their power, and that the doors of the high places of science are rigorously closed against her.

In order to have definite information on this head, I applied to the secretaries of one or two of the scientific societies of the metropolis, with the following result. Mr. White, Assistant-Secretary of the Royal Society, writes:—

"In answer to your inquiry as to what is the position of women with regard to the Royal Society, I beg leave to say that the Society is not open to women; that ladies are not admitted to the meetings, and have never been elected Fellows.

"Mrs. Somerville many years ago was elected honorary member of the Astronomical Society, but I am not aware that she has ever written F.R.A.S. after her name."

Mr. Henry Walter Bates, Secretary to the Royal Geographical Society, kindly furnished me with the following statement:—

1. Women are not entitled to become members or Fellows of the Royal Geographical Society. But they are allowed to attend the meetings as visitors introduced by Fellows, and, if they are teachers of geography, they can obtain a Council card of admission for the season.

2. There is no instance on record of the Society bestowing medals or other rewards on women. Lady Franklin received a medal on behalf of her deceased husband. Women have distinguished themselves as explorers, both singly—Madame Pfeiffer—and with their husbands—Lady Baker, Madame Helfer, Madame Semper—but I do not think it has been proposed in our Council to bestow a reward for geographical merit on a woman.*

3. Women are admitted as visitors to the Ethnological Society; but I am not aware that this is allowed or practised in any other scientific society.

"Ladies are Fellows of the Royal Horticultural Society; but this is not for scientific purposes, but to obtain admission to the gardens. Ladies are not generally invited even to the *soirées* of learned societies such as the Royal, the Linnæan, etc. Ours is an exception, a small number being invited as friends of the President. They are, however, invited freely to the *soirées* of the microscopical clubs and societies, and seem to avail themselves very largely of the privilege, and to look through microscopes quite as eagerly as the men. They are also invited freely to the *soirées* of the Society of Arts. I think if a lady was to offer a really good paper on a scientific subject to any of these societies, it would be accepted and published like a man's paper in their transactions. I have seen papers by ladies (I think) in the transactions of the Linnæan Society."

Mr. Bates speaks of a lady offering a really good paper on a scientific subject. But so long as ladies are shut out from the association of those who are engaged in such pursuits, it is hardly to be expected that they would have either the stimulus or the opportunity of producing much that was valuable; and he seems not quite certain that if they did, their papers would be accepted and published.

To the list of ladies enumerated by Mr. Bates, as having distinguished themselves in geographical exploration, I may add the name of Mademoiselle Alexandrine Tinné, who, a few years ago, fitted out a steamer at her own expense, to explore the Bahr el Ghazal, one of the tributaries of the White Nile, and accompanied the expedition, along with her mother and aunt. I remember that, at one of the meetings of the British Association, some one asked Sir Roderick Murchison, whether the Royal Geographical Society would mark its sense of her munificence and courage in geographical enterprise by electing her a Fellow of the Society. The learned president received the proposition with something very like disdain, making an observation to the effect that they never had conferred such a distinction upon a lady. The gentleman who asked the question read a letter from Mademoiselle Tinné, giving intelligence of the progress of the expedition, which, at that time, was tolerably prosperous, but subsequently became entangled among the dreary swamps of the equatorial Nile regions, and fever and disaster arrested its progress. But the enterprising lady is still bent on making further explorations, and when last heard of, in December 1868, she was on the point of setting off from Tripoli to Lake Tschad and the kingdom of Borran. The expeditions of these ladies in Central Africa have been often referred to in the proceedings of the Royal Geographical Society.

We cannot claim the honour of numbering any of these distinguished ladies among our countrywomen. Madame Helfer and Madame Semper are probably foreigners, and the ladies Tinné and Lady Baker certainly so. I call them distinguished because, though they have not attained the honours or distinctions bestowed on other explorers, they have done the deeds which merited such reward.

I have been informed that on one occasion the authorities of the Royal Astronomical Society had a discussion as to whether they should award their gold medal to Miss Caroline Herschel for her discovery of five comets. It was understood that it would undoubtedly have been given had the discoverer been a man. But they came to a determination akin to that of the Royal Geographical Society—not to recognise or reward services to science when rendered by a woman, and the medal was withheld.

When the Meteorological Society was formed it was decided to admit women, and four ladies were elected on the original foundation; among them the Countess of Lovelace—Byron's daughter "Ada." In a little while one of these ladies, the wife of an eminent meteorologist, wrote to say that she had been told it would be injurious to the Society to have women as members; she, therefore, thought it her duty to resign, and she hoped the other ladies would follow her example. One of them did so; but another, who could not be made to comprehend the necessity for maintaining the scientific disabilities of women, refused to withdraw, and no one even suggested the propriety of resignation to Lady Lovelace. But the two ladies who remained members are since dead, and no others have been elected; for it appears that the Royal Charter which was subsequently obtained would not have been granted to any Society which admitted women to participate in its advantages.

The story of the connection of women with the scientific societies of the metropolis being chiefly of a negative character, is thus soon told. That a different complexion would be given to the tale, were the advantages and the honours they possess open freely to all lovers of intellectual pursuits without invidious distinction, may be reasonably inferred from the results obtained where this principle has been acted on.

In illustration of this proposition I will read a report, which has been furnished me by a lady in Dublin, of the working of an institution which has been doing a good work for some years, and is now named the "Royal College of Science for Ireland." The lady to whom I am indebted for the information is herself a student, and has carried off some of the highest prizes.

Dublin, October 14th, 1868.

"MY DEAR MISS GOUGH,

"You have given me a very pleasant task in asking me to retrace the history of our College; it is endeared to us by the associations of many years. And we are proud of it because it is, I believe, the only one in the kingdom where, as our much-esteemed Dean Sir Robert Kane said to me a few days ago, 'woman is in her proper intellectual position, on a perfect equality with man.' I shall rejoice that the working of our College should be more widely known among the friends of our cause, and I think the facts I can bring forward will interest them, and will strengthen their hands.

"Sir Robert Kane has kindly given me every assistance by ordering all the reports and records of the College, and of the older institution to which it succeeded, to be placed in my hands for this purpose.

"Shortly after the Great Exhibition of 1851, a museum was established in Dublin by the Department of Science and Art, and called the Irish Industrial Museum. Sir Robert Kane, President of the Queen's College, Cork, was appointed Director. In 1854 a staff of professors was added, chosen from among the most distinguished professors of the University of Dublin, of the Royal College of Surgeons, and members of the Royal Society, and of the Royal Irish Academy, who gave courses of lectures on the following subjects:—Geology, botany, zoology, physical science, and theoretical chemistry. These lectures were partly free, the last twenty or thirty requiring the small fee of 3s. 6d. They were attended by large numbers of men and women of different classes of society. In the session of 1855–6, examinations at the end of each course were instituted, and prizes of £3, £2, or £1 in books, or in money, were given to the best three; certificates were given to those who, though not attaining to a prize, showed a fair degree of proficiency. The Department also granted its bronze medal to the winner of the first prize in each class.

"At the very first examination several gentlemen, and *three ladies*, presented themselves; one of the latter won the first prize in botany and zoology (which were united the first year), and the other two took good places in the same subjects and in geology. This step met with no opposition from any one connected with the place, but with every encouragement from the enlightened Director. At the public distribution of prizes at the close of the session, his address contained these words:—'We have been avoured by the presence here, as students in our classes, of several ladies (hear, hear), who have distinguished themselves not only by their attention and diligence, but have, also, at the competitive examinations held by the several professors, distinguished themselves in a very great degree. The result will, I trust, lead to very important consequences." And his excellency, the Lord-Lieutenant, who handed the prizes, when congratulating the students, spoke of them as 'my young friends, to whichever sex you may belong, for I am happy to find you are not exclusive in the award of your prizes.' So quietly was this important step taken; but we shall never cease to remember with gratitude the three ladies who took it—Miss Halgena Hare (now Mrs. Lewis), Miss Frances Armstrong (now Mrs. Whitsett), and Miss Kate Egan.**

"From that time till the October of last year, the School of Science in the Irish Industrial Museum continued to flourish. Large numbers of male students, and a considerable number of female students, attended the lectures. The latter so frequently won first or second prizes, that instead of enumerating them here I shall append to this the results of some of the examinations, writing opposite to each name the number of marks attained. The Dublin press always spoke most favourably of the union of the two sexes in this institution. Here is one of many extracts I might make; it is from a leading Dublin paper on the occasion of the distribution of prizes for the session 1857–8:—'A very interesting feature of the proceedings was that ladies entered the lists as competitors, and vindicated the genius of their sex by carrying off the highest prizes.' And successive Lord-Lieutenants spoke sensibly, courteously, and approvingly of the female students. Here is a short extract from one of Lord Carlisle's addresses on such an occasion:—'It is always a pleasing circumstance here, that whereas in almost every other country where we hear of classes, and lectures, and competitive examinations, the actors in these operations are almost exclusively of the rougher sex, while here without any departure from the rigid rule of impartiality, the lists are entered and the palm is, as we have frequently seen, carried off by lady aspirants.' And in 1858 Lord Eglinton said:—'I rejoice to find that among these students such a fair,—in the double acceptation of the word,—such a fair sprinkling of the gentler sex; not only have they attended all the classes, but have attained eminent distinction in them.'

"The people of Dublin showed their appreciation of the School of Science by crowding to the distribution of prizes, and applauding the successful student as each was presented to the Lord-Lieutenant; and when the successful student was a lady, she received a double meed of applause, so far were they from thinking that she had stepped out of her sphere.

"The examination papers of the School of Science can be had on applying for them, and they will bear comparison with those of any college or university in the kingdom.

"In the October of last year, '67, the School of Science was enlarged, more professorships were added, the instruction was made more systematic, the fees were raised from 3*s.* 6*d.* to £2 for each course of lectures; it received the name of the Royal College of Science for Ireland, and Sir Robert Kane was made Dean.

Some of the new professors, strangers to the institution, were startled at the idea of mixed classes, and wished to exclude the female students; but this opposition was really an advantage to our cause, for it elicited the gratifying fact that the Dean, and the original staff of professors, who had seen the working of the system for thirteen years, were unanimous for our admission as before, and overruled the opponents. The high fee is still a great barrier; however, two ladies entered, one competed at the examination in pure mathematics, and won the first prize.

"The success of the female students disturbed, of course, very much the preconceived notions of some people, who had always taken for granted that the female intellect was inferior to the male; and not being able to combat the stubborn facts that appeared from time to time in the newspapers when the results of the examinations were published, they tried to account for them. One manner of doing so was by stating that the female students had more leisure for study than the male students. This was not true as a general rule; three, at least, of the most successful female students I know to be engaged during the greater part of the day in supporting themselves by teaching, while their evenings were of necessity often occupied by those domestic duties from which men are free.

"Another way of accounting for their success was that those women were above the average, and that among the students were to be found first-class women, and only second or third class men. The women may have been above the average; the men, large numbers of them, certainly were. A few simple facts will show this. In 1859 a lady won the first prize in physical science; among her competitors was a gentleman who, about the same time, passed the Woolwich examinations most creditably. In 1862 a lady won the first prize in chemistry, and was 320 marks (the total being 1,000) a-head of the nearest male competitor, who had passed several examinations in Trinity College, Dublin, where he was a student, but had been attracted to the college by the fame of its chemical courses. In 1863 another lady took the first prize in chemistry, and the gentleman who took second prize was the best student that year in the chemical classes in the Royal College of Surgeons. In '65, '66, and '67 first prizes were taken by ladies, while among the male competitors were those who had won scholarships and exhibitions at the South Kensington May examination; and at the examination in June, when a lady won the first prize in pure mathematics, ALL her competitors were exhibitioners.

"When I say that some people have tried to *account* for the ladies' success, I wish it to be clearly understood that those cavillers were entirely unconnected with the Museum. The most perfect harmony, courtesy, and good-feeling has always existed there. We sit on the same benches in the lecture theatre, and read in the same library; and I have seen students of both sexes, after an examination, looking over the examination papers and asking each other which questions they had answered. I was the first lady who worked in the laboratory, and I found my fellow-students as ready to tender me any little civility I needed, as if I were in a drawing-room. They would lend me a piece of platinum wire, or a pair of crucible tongs, when my own were not as hand, as simply and as politely as they would have turned over the leaves of a piece of music. "M. C."

SESSION, 1855 AND 1856.*

GEOLOGY.

Prizes.

1. William Plunkett.
2. C. W. Bateman.
3. Edwin Birchall. *Certificates.*
    FRANCES ARMSTRONG (1st).
Then follow the names of five gentlemen.

BOTANY AND ZOOLOGY.

Prizes.

1. HALGENA HARE.
2. William Plunkett.
3. FRANCES ARMSTRONG.
    Seven certificates granted.
KATE EGAN (3rd).
FRANCES HARE (4th).

SESSION 1856 AND 1857.

Total number of marks, 1,000.

GEOLOGY.

Prizes.	*Marks.*
1. D. M'Cready	.871
2. FRANCES ARMSTRONG	.792
3. HALGENA HARE	.775

Then follow the names of twelve gentlemen.

CHEMISTRY.

Prizes.	*Marks.*
1. John Mulligan	.895
2. M. P. Dowling	.850
3. KATE EGAN	.730

Then follow eleven gentlemen.

PHYSICAL SCIENCE.

1. E. J. Wood	.930
2. FRANCES ARMSTRONG	.885
3. C. M'Cready	.780

Then follow six gentlemen.

ZOOLOGY.

1. FRANCES HARE	.895

Then follow five gentlemen.

BOTANY.

1. C. M'Cready	.895
2. E. J. Wood	.785
3. FRANCES HARE	.675

SESSION 1857 AND 1858.

CHEMISTRY.

Prizes.	Marks.
1. Not awarded	
2. HALGENA HARE	640
3. Alban Meredith	575

One certificate granted.

BOTANY.

1. William Corker	920
2. ZOE LEIGH CONEYS	770

. . . .

GENERAL EXAMINATION IN FIVE SUBJECTS.
Total, 3,000.

1. HALGENA HARE	1,830
2. Alban Meredith	1,331

PHYSICAL SCIENCE.

Prizes.	Marks.
1. J. P. Brophy	700
2. HALGENA HARE	585
3. Alban Meredith	555

ZOOLOGY.
Total, 2,000.

1. William Hartford	1,840
2. HARRIET HARMAN	1,824
3. A. F. Gordon	1,700

Certificates.

KATE EGAN	1,600
Alban Meredith	1,500
HESTER HARMAN	1,300
Mathew Higgin	1,250

SESSION 1858 AND 1859.

ZOOLOGY.

Prizes.	Marks.
1. HESTER HARMAN	930
2. William Corker	640
3. John Dowling	600

Two gentlemen follow.

GEOLOGY.

1. J. P. Brophy	890
2. KATE SEYMOUR	810
3. William Corker	765

Certificates.

J. F. Murray	640
MISS PALMER	590
MRS. MURRAY	545
J. O. Rearden	440
O. H. Brien	400

PRACTICAL ZOOLOGY.
Total, 2,500.

Prizes.	Marks.
1. A. Gordon	1,980
2. HARRIET HARMAN	1,890

Two gentlemen follow

PHYSICAL SCIENCE.

1. MATILDA CONEYS	796
2. S. Boileau	504
3. J. Donovan	480

Then follow three gentlemen.

BOTANY.

1. HARRIET HARMAN	850
2. J. F. Murray	680

Certificate.

MRS. MURRAY	370

SESSION 1859 AND 1860.

GEOLOGY.

Prizes.	Marks.
1. Not awarded	
2. A. Penny	.595
3. A. M'Alister	.523

Certificates.

MRS. MURRAY	.503
HARRIET HARMAN	.503
MISS UNDERWOOD	.420
J. F. O'Rearden	.402
MISS HARE	.385
C. H. Brien	.381

PRACTICAL ZOOLOGY.

Prizes.	Marks.
1. Not awarded	
2. HESTER A. HARMAN	.690
3. C. H. Brien	.600

Special prize for collection.

HESTER A. HARMAN

BOTANY.

1. HESTER A. HARMAN	.825
2. Philip Lyons	.685
3. J. Pierce	.595

Certificates.

Then follow four gentlemen.

SESSION 1861 AND 1862.

GEOLOGY.

Prizes.	Marks.
1. William Dudley	.688
2. SARAH G. KEOGH	.665
3. M. SIBTHORPE	.613

Then follow seven names, including one lady, who got fourth certificate.

CHEMISTRY.

Prizes.	Marks.
1. ZOE LEIGH CONEYS	.875
2. Henry Chute	.555
3. George Griffin	.515

Certificate.

R. Fitzgerald.

LABORATORY EXAMINATION.

MATILDA CONBYS	
James Cohill	
S. Johnson	
J. Laylor	

Equal, each attained 1,000 marks, which was the total.

SESSION 1862 AND 1863.

GEOLOGY.

For this class fifteen gentlemen and three ladies competed; fourth place was taken by

	Marks.
MISS LEEPER	.721

BOTANY.

Prizes.

1. MISS M'CLEAN	.900
2. MISS LEEPER	.670

Then follow seven gentlemen.

CHEMISTRY.

Prizes.	Marks.
1. MATILDA CONEYS	.936
2. John Benson	.911
3. S. Johnson	.903
A. S. Kerrison	.652
W. O. Really	.642
A. W. Scott	.538

SESSION 1863 AND 1864.

GEOLOGY.

Prizes.	Marks.
1. Bryan Clinche	.884
2. MISS LEEPER	.795
3. H. G. Penny	.783

Certificates.

Mr. Lloyd
Mr. Westropp

MISS QUINLAIN	.675
MISS M'KAY	.578
MISS A. SMITH	.555

Then follow three gentlemen.

SESSION 1864 AND 1865.

GEOLOGY.

Prizes.	Marks.
1. MISS SWAN	.957
2. William O'Donovan	.888
3. M. Smith	.865

Ten gentlemen follow, who get certificates.

BOTANY.

1. ADELINA RORKE	.956
2. R. D. Fennell	.819
3. T. O. Atkinson	.776

Five ladies and seven gentlemen follow.

SESSION 1865 AND 1866.

BOTANY.

Prizes.	Marks.
1. MARIAN SEARIGHT	.903
2. GRETA D. STRITCH	.859
3. ANNETTE SMITH	.847

Six gentlemen and two other ladies follow.

SESSION 1866 AND 1867.

BOTANY.

Prizes.	Marks.
1. MARIAN HAYES	.821
2. William Hunt	.800
3. GERTRUDE HAYES	.772

Four gentlemen and one lady follow.

SESSION 1867 AND 1868.

PURE MATHEMATICS. Total, 100.

Prizes.	Marks.
1. MATILDA CONEYS	.84
2. William Hunt	.76
3. James Kilroe	.74

* The names of the ladies are in small capitals.

The experience gained in this institution ought to afford encouragement to those who are seeking to raise the standard of education for women by opening to them the advantages of existing educational institutions. It proves, by practical experiment, that men and women can associate with as much mutual advantage in the classroom and examination-room as in the home and the drawing-room, and that there is no necessity to isolate them from the sympathy and encouragement each gives the other in their common pursuit. It points to the possibility of rendering all public institutions for education, national in the broadest sense of the word. The means provided for cultivating the mind of the nation should be freely accessible to all who have minds to cultivate; and the honours and rewards attaching to intellectual attainments, such as scholarships, fellowships, university degrees, and membership in learned societies, ought to be within the reach of either man or woman who has the taste to desire and the ability to earn them. Educational provisions ought to be strictly correlative to educational necessities, for needs and rights in this respect are convertible terms. These needs and rights have been admirably defined by Mr. J. G. Fitch in the very important Report presented by him to the Schools Inquiry Commission. If the doctrine which he enunciates with regard to women be extended so as to include both sexes, it will embody the principle which should guide every effort made for the promotion of national education, whether of an elementary, secondary, university, or scientific standard. "The true measure of a (man or) woman's right to knowledge is (his or) her capacity for receiving it, and not any theories of ours as to what (he or) she is fit for, or what use (he or) she is likely to make of it."

So long as intellectual pursuits were confined to a select few, and the masses of the people, men and women alike, cared nothing for these things, the disadvantages of the exclusion of women from participation in these pursuits were but slightly felt. There was nothing in this exclusion to cut them off from the sympathy of those with whom they lived, or to cause divergence or estrangement between their minds, and these of persons with whom they habitually associated. If they were on a low level intellectually, the men around them were on the same, and so the balance was preserved. But now we see symptoms of a change. Everywhere educational institutions of more or less pretensions and efficiency are springing into being. Scarcely a town but has its mechanics' institute, debating society, literary society, or some kindred mechanism for promoting intellectual activity. But almost invariably the efforts of those who promote these institutions are directed to producing a divorce between the thoughts and sympathies of those who should be mates and helps to each other in all the concerns of life. They see man and woman, ignorant and undeveloped, grinding at the mill of life's daily toil. They desire to lighten this toil, by affording a glimpse of something beyond the narrow horizon of each day's mechanical duties. They go to the man, and they open to him the vista of intellectual enjoyment, leaving his companion uncheered in her solitude, unthought of, uncared for. Should she cast a wistful glance at the prospect, and ask why she may not share in the good things they set forth, she is encountered by the assurance that the pathways that load to the higher regions of thought were never meant for her to tread, and with a contemptuous reproof for her presumption in wishing to stir a step beyond her appointed "sphere." So, bereft of the companionship of her partner in ignorance, the last condition of that woman is worse than the first. After profiting by the advantages denied to her, he returns elated with the consciousness of superior wisdom, and complacently propounds theories as to the "radical inequality of the sexes—the radical inferiority, physical, moral, and intellectual, of woman."

The danger of producing disunion in families by teaching all the men, and leaving all the women out in the cold, is no fanciful one. It is already beginning to be felt, and will increase with the success of every attempt to promote popular education that is not based on comprehensive principles. A very intelligent working man in one of the manufacturing towns of Lancashire, with whom I was conversing on the subject of a public movement in which he was greatly interested, informed me that it was a source of serious trouble to him that his wife had not kept up with the advance of his mind, and for want of knowledge and cultivation, was unable to understand the importance of the work on which he was engaged, and unwilling to see him devoting his attention to it, or to sympathise with his efforts for its advancement. He spoke in sorrow, as feeling it a real misfortune, and as if his were but a representative case as regards the effect on domestic happiness of the present one-sided system.

Some of the educational institutions so far recognise the existence of the other sex as to make a feeble effort to supplement their main provisions by the establishment of supernumerary "women's classes." I have not heard whether these well-meant but ill-advised efforts to combat the evil have done much good. The little I have heard leads me to the belief that the result has been what one might from the first have anticipated, and that the interest displayed in these classes has been languid. There are not a sufficient number of women as yet roused to the interest of such subjects to afford material for the promotion and continuance of such isolated classes, and the fact of their exclusion from the companionship of the other sex acts as a damper on their spirits. They would not care much for social pleasures if they were only admitted to women's balls, women's dinner parties, women's croquet parties, and women's concerts; and if they are only allowed to participate in intellectual pleasures on these exclusive terms, they will certainly not derive from them either the advantages or the healthful stimulus which these are capable of affording.

It seems to me a matter for sincere regret that any effort made to promote the intellectual activity of women should be based on this system of separation and exclusion. Whatever difficulties may be thought to stand in the way of studies conducted in concert, none can exist, even in imagination, when the proposal is simply that of simultaneous and identical examinations; the placing of all the papers together for judgment, and making out the class list in order of merit, with absolute impartiality and indifference as to whether the papers were the production of male or female students. The success of the local examinations in connection with the University of Cambridge, where no difference of any king is made in the examination of girls and boys, should point out the principle to be acted on in further efforts in the same direction. The only matter for regret in respect of these examinations is the treatment of the successful students, in the invidious distinction implied in the exclusion of girls from the class lists. The boys who pass honourably have their names published; the girls who pass honourably have their names suppressed. It is just as natural for a girl as for a boy to be pleased to see her name in a list of those who have done well. The University encourages the boys by marking the proficiency they have attained as something to be proud of; it discourages the girls by implying that the acquirements they have gained are something to conceal, or be ashamed of.

A still further departure from the principle of equality has been made by the University of London. They have instituted a special examination for women, to which no male student is admitted, and the recognition attached to success is a mere certificate of having passed, without the honours of a University degree.

Perhaps I ought to consider the step that has been taken by the London University not so much a departure from the principle of intellectual equality as an advance towards it. It is the pleasanter, and possibly the truer way. Certainly, before this concession was made, women were not allowed by the authorities to have any rights at all in the matter. Now that their eyes have become partly open to the needs of women in this respect, we may hope that the process will not stop till complete justice has been done.

From all that I can gather respecting the proposed examination, it is in no way inferior in what examinees call "stiffness" to that provided for the other sex. A woman who passes in any subject will do quite as much as a man who passes the men's examination corresponding in grade. But though she will have worked as hard and done as much as the men, she will not have equal honour. The men will say to her: "You are not on our level; you have only passed the women's examination;" and she will not be admitted as a graduate whatever the amount of intellectual power or attainments she displays. The whole arrangement proceeds on the principle that it is very womanly to work, but "unfeminine" to receive pay or reward for work. Women may be admitted to the course of study, but not to the honours or advantages to which that course of study leads men.

It will not be very wonderful if an experiment based on what seems a radically false principle should prove a failure, and if high-spirited and accomplished women who are conscious of no moral nor intellectual inferiority to the other sex, should refuse to enter an examination which does not place them on a level with others. It is only to be hoped that the possible failure of an experiment of this nature will not be used as an argument against better devised future attempts to extend the educational privileges of women.

The efforts made by the London University to help women up the ladder of learning remind me of the history told by Mr. Frank Buckland of his endeavours to facilitate the ascent of salmon up rivers, the natural course of which had been obstructed by weirs and dams placed there by man. After exhausting his ingenuity in providing a way for the salmon up these artificial barriers, by means of a contrivance "nicely adapted to their special tastes and capacities," he found, to his dismay, that his pains had been entirely thrown away, for "the ungrateful beasts wouldn't go in!" But Mr. Frank Buckland is a man of resources, and failure is no word in his vocabulary. He informed us that the only plan then available was to catch a salmon, and ask it what it wanted. Of course the creature very soon told him, and the moment arrangements were made in accordance with its real needs, off it went, like an arrow, up the stream, on its way to the mountains. Now, if those who are sincerely, but perhaps somewhat blindly, trying to open the way to a higher life for women, will be as wise as Mr. Frank Buckland in seeking to adapt their means to the real feelings and wishes of those whom they are striving to benefit, instead of to what they imagine women ought to feel and desire, they will be as successful as he was in setting the struggling creatures free, and in peopling the stream of life with fish worth catching, instead of leaving nothing for the angler but the minnows and sticklebacks "of the period."

Besides the special benefits to women themselves, results of a yet more important nature with respect to the happiness and welfare of mankind, would follow from making them acquainted with the results of scientific inquiry, and imbuing their minds with the principles on which such researches are based. The importance of scientific knowledge is not yet appreciated by the general public. A knowledge of science is frequently treated as if it were merely a branch of learning, like Latin or Greek, and the question of making it a part of general education is regarded as if it were simply a question of what course of study was best fitted to train the faculties or suit the taste of the student.

But surely there is a more important aspect of the study of science than that which regards it as merely a mass of curious and interesting information. Men and women constitute an integral portion of a universe governed by uniform and undeviating laws. It is the object of scientific explorers to discover these laws, a pursuit in which they may be said as yet to have hardly made a beginning. Every step gained in advance reveals something which can be turned to account in ameliorating the hardships and discomforts of life, and promoting the happiness of mankind. With complete knowledge of the conditions under which we live, and complete conformity to these conditions, we might hope to see most of the evils that afflict our race entirely disappear. This knowledge is presumably attainable by human faculties, if the search be conducted with sufficient perseverance, and based on right principles. The greater the number of minds that are impressed with this belief, the greater the encouragement that will be given to the inquiry, and the greater the probability and the proximity of success. When the conviction of the preventability of misery shall have become the prevailing one, men and women will cease to meet its existence chiefly with endeavours to palliate its effects, but will set resolutely to work to remove its causes. They will then no longer accuse either chance or Providence of sending the ills that afflict mankind, but perceive that they are traceable to the action of inexorable and undeviating law, and that most, if not all of them, may be averted or avoided by human foresight acting on human knowledge.

When the science, the practice, and the principles which lead to this habit of thought shall cease to be considered the exclusive privilege of the dominant sex, and become the heritage of humanity at large, the progress of the race will receive an impetus which shall carry it on at a pace hitherto undreamed of. The rate of advancement will be far more than doubled, because the untrained and stationary half of mankind necessarily acts as a drag on the other. We seek to unloose the locked wheels of the car, and to set free the imprisoned energies now pining for scope.

There is a strong tendency greatly to undervalue the extent and the intensity of the feeling that exists among women of dissatisfaction with their present condition, and with their exclusion from participation in the pursuits that interest and occupy men. It is assumed that the majority are contented, and that the desire for an amelioration of their lot is felt only by a few exceptional natures. But let not those in whose hands the power lies, pass over the cry so lightly. Many are the signs of the times which tell a different story. Among many voices, one has been raised that had no strength in itself, but in the truth of the note that it rang. That note has found an answering chord in thousands of women's hearts, and has come back from near and far, over the length and breadth of the land. Not in loud and turbulent cries, but in tones unmistakably clear to an ear attuned to catch the delicate harmonics that breathe from sorrowful and suffering souls. And not from our land alone, for women's hearts are everywhere the same. Voices have resounded from the Alps; signs have reached over the Pyrenees; echoes have bridged the Atlantic. No false note could awaken so deep and so wide a response; no harsh tone could evoke such loving sympathy.

The cry for equal rights for all human beings proceeds from the irrepressible consciousness of equal needs, and the possession of common feelings. The movement now hourly gaining strength for the social, educational, and political enfranchisement of women, arises from no spirit of opposition or rivalry with men, but from deep and intense sympathy in their noblest aims and aspirations.

—*Lydia Ernestine Becker.*

* Mr. Bates writes with regard to the ladies last-named in his statement:—"Madame Helfer accompanied her husband, Dr. Holfer, to Burmah and the Andaman Islands, and assisted him in his scientific investigations. Madame Semper travelled with her husband, Dr. Semper, in the Philippines. Their narrative is not yet published, but it will, I have no doubt, show how much Dr. Semper owed to the enterprise, endurance, courage, and scientific enthusiasm of his partner. They travelled in a small boat round the islands, dredging the sea bottom for marine animals, and had sometimes to run in ashore to escape from pirates. The result was a most magnificent collection of the animal productions of the Philippine archipelago."

** "Since writing the above I have found that a fourth lady entered for examination—Miss Frances Hare (now Mrs. Appleton), who has since been a successful student at the Female Medical College, London."

Source: Becker, Lydia. 1869. "On the Study of Science by Women." *Contemporary Review* 10: 386–404.

Boston doctor Samuel Gregory was responsible for opening the Boston Female Medical College in 1848, the first medical school for women in the world. In this article Gregory makes the case for educating women in medicine. While supporting women becoming "doctresses" he nevertheless believes that such women must necessarily be constrained by their feminine nature to the practice of medicine for women and children. Certainly many women wished to become doctors to alleviate the physical and mental suffering of women and children, however, his argument relegated women to a lower status of medical practice than that of male doctors. The article thus highlights how, even when men supported women's medical education, they were not keen to let women become equal partners in surgery, medical research, and university professorships beyond women's colleges.

Female Physicians

SAMUEL GREGORY

Women always have been and always will be physicians. Their sympathy with suffering, their quickness of perception, and their aptitude for the duties of the sick room, render them peculiarly adapted for the ministrations of the healing art. Let them have medical knowledge corresponding with their native abilities and they will excel, especially in the departments of practice which pertain to women and children.

The medical profession is incomplete and ineffective without female co-workers in promoting health and relieving sickness and suffering. While the doctor cannot be dispensed with, the doctress is no less essential to the physical well-being of society; and as three-fourths, probably, of the duties of the medical profession relate to women and children, there should be at least as many female as male physicians.

The preservation of health is a matter of more importance than its restoration; sanitary knowledge of more value than curative. In all domestic sanitary arrangements and household hygiene women must necessarily be the chief agents, and they ought to be intelligent and efficient ones—a *cordon sanitaire*, ever on guard to preserve their own health, and secure the constitutional well-being of the rising race. Now, who can so advantageously and successfully instruct girls, young women, and mothers, in all sanitary, physiological, and hygienic knowledge as thoroughly educated lady physicians? Though there are Ladies' Sanitary Associations, they have to depend chiefly upon men to write their tracts and lecture to them. It is very reasonable that professional men should perform a good portion of the writing and lecturing upon these subjects, but female physicians can impart to women indispensable information which a natural reserve would prevent medical men from communicating.

As the public become more enlightened in reference to the principles upon which health is to be preserved, and the rational methods by which it is to be restored when lost, the relation of the medical profession to society must necessarily be modified. Ignorance on the part of the patient and mystery on the part of the physician will recede together; and already some of the most intelligent medical men are giving proof of a higher regard for the welfare of society than for the interests of the profession, as it is obvious that the more there is accomplished in the preservation of individual and public health the less will be the demand for the services of the physician—the more of nature, the less of art.

Among the eminent pioneers in this reform is Dr. Jacob Bigelow, of Boston, who has written ably in favour of rational medicine and a reliance upon nature in the cure of disease. "It is," says he, "the part of rational medicine to enlighten the public and the profession in regard to the true powers of the healing art. The community require to be undeceived and re-educated, so far as to know what is true and trustworthy from what is gratuitous, unfounded, and fallacious. And the profession themselves will proceed with confidence, self-approval, and success, in proportion as they shall have informed mankind on these important subjects. The exaggerated impressions now prevalent in the world in regard to the powers of medicine serve only to keep the profession and the public in a false position, to encourage imposture, to augment the number of candidates struggling for employment, to burden and disappoint the community already overtaxed, to lower the standard of professional character, and raise empirics to the level of honest and enlightened physicians."

In England, Sir John Forbes has given the weight of his great medical learning and influence in this direction. In an article published as long ago as 1846, he enjoined it upon the profession "to direct redoubled attention to hygiene, public and private, with the view of preventing diseases on the large scale, and individually in our own sphere of practice. Here the surest and most glorious triumphs of medical science are achieving and to be achieved. To inculcate generally a milder and less energetic mode of practice, both in acute and chronic diseases. To make every effort, not merely to destroy the prevalent system of giving a vast quantity and variety of unnecessary and useless drugs—to say the least of them—but to encourage extreme simplicity in the prescription of medicines that seem to be requisite. To place in a more prominent point of view the great value and importance of what may be termed the physiological, hygienic, or natural system of curing diseases, especially chronic diseases, in contradistinction to the pharmaceutical or empirical drug plan generally prevalent. To endeavor to enlighten the public as to the actual powers of medicines, with a view to reconciling them to simpler and milder plans of treatment. To teach them the great importance of having their diseases treated in their earliest stages, in order to obtain a speedy and efficient cure; and, by some modification in the relations between the patient and practitioner, to encourage and facilitate this early application for relief."

This tendency of things has an important bearing upon the introduction of women into the medical profession; for while they, as the handmaids of nature, possessing all the qualities for good nursing, are predisposed to the natural and rational modes of dealing with disease, many might be deterred from becoming healers of the sick, by the formidable task of comprehending and working the complicated and unwieldy machinery of the system, and by their repugnance to so much of the experimental, the artistic, and heroic, as now prevails, to the reproach of the profession, and the detriment of the public. Had the family of Æsculapius consisted of daughters as well as sons, these milder methods of treatment, this co-operation with nature, recommended by those eminent medical gentlemen, would doubtless have ever prevailed.

Women physicians are especially needed in the female wards of hospitals, insane asylums, almshouses, prisons, and reformatory institutions for females, where the professional skill of women could be so properly and advantageously employed in the investigation and treatment of disease, and their kindly ministrations and healing influence would do so much to restore mental and moral health to the afflicted and the erring. And to provide none but male physicians for the female patients of these various institutions is a grave error, and one that should be corrected as soon as practicable. Female seminaries should also be provided with female physicians to act as teachers of physiology and hygiene, and supervisors of health, as well as medical attendants.

One of the evils of the present system of having men only in the medical profession is, that the benefits of medical science and skill are to a great extent lost to the female portion of the public. This point is well presented by Professor

Meigs, of the Jefferson Medical College, Philadelphia, one of the most numer-
ously attended medical institutions in the United States. Dr. Meigs is a physician
of extensive practice and great experience, and author of large medical works.
In his volume on the diseases of women, he speaks as follows:—

"The relations between the sexes are of so delicate a character that the
duties of the medical practitioner are necessarily more difficult when he comes
to take charge of a patient laboring under any one of the great host of female
complaints than when he is called upon to treat the more general disorders, such
as fevers, inflammations, the exanthemata, &c. . . . It is to be confessed that a
very general opinion exists as to the difficulty of effectually curing many of the
diseases of women; and it is mortifying, as it is true, that we see cases of these
disorders going the whole round of the profession, in any village, town, or city,
and falling at last into the hands of the quack; either ending in some surprising
cure, or leading the victim, by gradual lapses of health and strength, down to the
grave, the last refuge of the incurable, or rather uncured. I say uncured, for it is
a very clear and well-known truth, that many of these cases are, in their beginning, of light and trifling importance. All these evils of medical practice spring
not, in the main, from any want of competence in medicines or in medical men,
but from the delicacy of the relations existing between the sexes, and in a good
degree from a want of information among the population in general as to the
import, and meaning, and tendency of disorders manifested by a certain train of
symptoms.

"It is an interesting question as to what can be done to obviate the perpe-
tuity of such evils—evils that have existed for ages. Is there any recourse by
means of which the amount of suffering endured by women may be greatly lessened? I am of opinion that the answer ought to be in the affirmative; for I believe
that, if a medical practitioner know how to obtain the entire confidence of the
class of persons who habitually consult him; if he be endowed with a clear per-
ceptive power, a sound judgment, a real probity, and a proper degree of intelligence, and a familiarity with the doctrines of a good medical school, he will, so
far as to the extent of his particular sphere of action, be found capable of greatly
lessening the evils of which complaint is here made; and if these qualities are
generally attached to physicians, then it is in their power to abate the evil
throughout the population in general."

Here we have a statement of the evils and the remedy. If such and such
qualities and qualifications are combined in medical men, and they know how to
obtain the entire confidence of their female patients, the Professor believes it is
in their power to abate the evil. There is, however, a simple, natural, and effectual remedy to which Dr. Meigs does not allude. He says these evils arise mainly
"from the delicacy of the relations existing between the sexes." Let, then, those

relations be dispensed with, in these matters, and let females have physicians of their *own* sex. This remedy will moreover, so far as females are concerned; meet a point suggested by Sir John Forbes, in speaking of the great importance of having diseases treated in their earliest stages, in order to obtain a speedy and efficient cure—namely, will encourage and facilítate an early application for relief—by removing embarrassments and obstacles which now frequently prevent application at all, or till too late for effectual relief. Humanity, morality, and the physical well-being of society demand the introduction of women into the medical profession.

There is one department of professional duty so peculiarly feminine, that in past times in all nations it has, with hardly any exceptions, been performed by women; and at the present time in no country has it been wholly wrested from them, the duty of assisting women in childbirth. It would seem that if there is any "appropriate sphere" for women, beyond that which is inseparable from her sex, it is this. The "midwives" are spoken of with commendation in Scripture; in Egypt, Greece, and Rome, they were a recognised class; in China, Japan, India, and Turkey, at the present day, this service is performed by women. In most, if not all, of the Continental countries of Europe they are regularly educated in schools provided by the Governments, trained in the public hospitals, and duly licensed to practise.

In a paragraph in the *Boston Medical and Surgical Journal*, in 1856, it was stated that the medical profession in Austria consisted of 6,398 physicians, 6,148 surgeons, 18,798 midwives, and 2,951 apothecaries—the women numbering 3,307 more than the men in their three departments.

In Great Britain and the United States, where kindred customs prevail, the encroachments of men upon this department of female service have proceeded to a greater extent than anywhere else. The displacement of women has been very gradual and has resulted from the fact that the medical schools and the hospital practice have been appropriated by men, while women have been left in ignorance, and have consequently been set aside as incompetent. The intrusion of men into this office began in France about two centuries ago, in England thirty or forty years later, and in this country about a century ago. In France the *sages femmes* are still systematically educated and extensively employed. In Great Britain this class of women has not died out—the census of 1851 returning 2,882 midwives; and in the United States many times that number must be practising without special training for the office.

The following inscription, from a gravestone in our neighboring city of Charlestown, gives an idea of the position of these professional women, and of the estimation in which they were held at the period indicated. The quaint simplicity of the record and its conspicuous publishment give proof that along with delicate customs there existed a freedom from exquisite and affected refinements—things sadly reversed in our day.

"Here lyes Interred the Body of Mrs. Elizabeth Phillips, wife to Mr. John Phillips, who was Born in *Westminster*, in Great Britain, and Commissioned by John, Lord Bishop of *London*, in the year 1718; to the office of a Midwife, and came to this country in the Year 1719; by the blessing of God, has Brought into this world above 3000 children. Died May 6th, 1761, aged 76 Years."

The writer has before him a volume of 471 pages, "A Treatise on the Art of Midwifery," &c. "By Mrs. Elizabeth Nicholl, professed Midwife," published in London in 1760. Speaking of the invasion of men into her profession, she says, "Besides, it is even ridiculous to confine the practice of midwifery by females only to early ages. Who does not know that it was so in all ages, and in all countries, till just the present one, in which the innovation has crept into something of a fashion in two or three countries? The exceptions before, or anywhere else, to the general rule are sofew, that they are scarce worth mentioning."

In 1759 Sterne employed his satirical pen against "the scientific operators" and their "improvements," in "The Life and Opinions of Tristram Shandy, Gent.," in which the worthy Dr. Slop is consecrated to immortality. In fact, the transfer of this vocation from women to men has, from its inception to the present moment, encountered earnest remonstrance, and steady opposition, arising from the general sentiment that it was unnatural and wrong. The argument of superior qualifications of male physicians, and the consequent greater safety in employing them, has, however, overborne the weighty considerations on the other side, and temporarily installed men in an office which obviously belongs to the other sex.

The question now to be solved is, whether women can be so qualified by education and training as to render the practice in their hands as *safe* and *successful* as in the hands of men—all other considerations, of course, being in favor of female practitioners. It is believed that women can be so qualified as not only to equal men, but that, with the advantages of sex and natural aptitude, they will greatly excel them in the exercise of this vocation. But to secure this end, women must have a complete and thorough medical education. The plan of giving them a narrow and partial training, as being sufficient for the ordinary routine of the art, keeps them in an inferior professional position, and diminishes the confidence of the public in their abilities. These specially trained midwives should, however, be encouraged till female physicians can be provided. In fact, even with their limited professional education they can, with rare exceptions, manage these matters with greater safety and success than medical men, however extensive their scientific attainments. Abundant statistics of hospital and private practice might be presented in proof of this statement. It is a well-known fact that the attendance of male practitioners has often a very embarrassing, disturbing effect, causing disastrous and not infrequent fatalities to mothers or infants, when there was not the least necessary occasion for such a result.

But it sometimes happens that complications and difficulties arise, and the doctor must be called; or medical advice and treatment are needed, before, at the time, or subsequently; and this will be an ever-ready and, to the minds of many, an unanswerable argument in favor of dispensing entirely with the female subordinate, and employing the doctor throughout. And hence the need of fully educated female physicians for this, as for other departments of female practice.

It is objected that, as woman's sphere is home and its duties, she cannot, like man, devote herself uninterruptedly to the profession, and therefore must be unsuccessful. To make the objection as strong as possible, let us suppose that every woman is to be married and become the mistress of a home. According to the census of Great Britain for 1851, the average number of children to a family was two, minus a fraction of five one-hundredths. As a medical education would be a most valuable qualification for the maternal head of a family, suppose large numbers of young women should study medicine, commence practice, and then be diverted wholly or in part for a few years; they could then resume their vocation, with additional qualifications, and pursue it for ten, twenty, or thirty years. The wife is often obliged to aid in supporting the family, and sometimes does it wholly, by manual or intellectual labor; and why not by the practice of the healing art?

But, from the census alluded to, it appeared that in about one-fifth of the families, in one thousand in five thousand, there were no children to absorb the attention of the mistress of the house; and further that there were in Great Britain, not including Ireland, 795,590 widows, many thousands of whom of course need some employment for self-support. Again, it appeared that there were above half a million more females than males, and that one hundred women in every eight hundred remained single. In an article on "Female Industry," in the *Edinburgh Review* for 1859, it is stated, that "out of six millions of women above twenty years of age, in Great Britain, exclusive of Ireland, and of course the colonies, no less than half are industrial in their mode of life. More than a third—more than two million—are independent in their industry, are self-supporting, like men." The number of men returned by the census, under the head of "Medical Profession," was 22,383. To supply half of the profession with women would therefore make but a slight draught upon the vast available number.

Similar calculations would apply to other countries, though, from the extensive colonization and other disturbing causes, the surplus of females in Great Britain is unusually large. In most countries, however, there appears to be an excess in the number of females over that of males at certain periods of life.

In a paper prepared by John Robertson, and published by the Manchester (Eng.) Statistical Society, in 1854, the author says, "A number of years ago, in a paper read before this Society, entitled 'Thoughts on the Excess of Adult Females in the Population of Great Britain, with reference to its Causes and Consequences,' I endeavored to show that the female sex, in Christian countries, are probably designed for duties more in number and importance than have yet been assigned them. The reasons were, that above the twentieth year, in all fully-peopled States, whether in Europe or in North America, women considerably outnumbered the other sex; and that, as this excess is produced by causes which remain in steady operation, we detect therein a natural law, and may allowably infer that it exists for beneficent social ends."

The number of physicians in the United States, according to the census of 1850, was 40,564, and is now probably 50,000. But there is an immense multitude of unemployed women to supply co-laborers in the profession.

There is one disadvantage under which this enterprise must labor for a time; that is, the lower standard of female education and mental discipline, as compared with that of males.

Women have, however, a quickness of comprehension, a ready intuition and tact for the study and practice of the healing art, which compensate for the defect; and the defect is in the course of being removed. Indeed, there are now enough of well-bred and well-educated women to supply the profession many times over, who might and who ought to volunteer for the good of their sex and their kind.

It is sometimes objected, that this is a masculine occupation, and that to go through, the disagreeable process of obtaining a medical education is improper and indelicate for a woman. The writer has as little disposition to see women in men's places as men in women's. He is not one of those who take extreme views on the question of "women's rights," so called. In the medical profession itself there are departments as unwomanly as others are unmanly. Even the matter of the *title* should not be disregarded: the masculine appellation, of Doctor belongs exclusively to men, and the feminine correlative, Doctoress, both convenience and propriety assign to the lady physician. But to take the ground that it is indelicate and unfeminine to study the structure of the human system, with a view to understand its conditions of health and disease, and thereby to alleviate suffering and save life, is more fastidious than sensible. It is surely more modest for one woman in a thousand to study medicine and take charge of the health of the nine hundred and ninety-nine, than for the whole to remain ignorant and helpless, and depend on men for information and treatment in all cases and circumstances. No one who approves of female nurses for men, especially in military hospitals, can with a shadow of consistency object to the education of female physicians and their practice among women and children.

In the United States the plan of introducing women into the medical profession has fairly commenced and is making good progress.

The New England Female Medical College, located in Boston, commenced in 1848, the germ being a school with two lecturers and twelve pupils, and the course of instruction, not extending beyond midwifery and the diseases of women and children. In the same year an association was organized to carry forward the object, in the language of its constitution, "to educate midwives, nurses, and (so far as the wants of the public require) female physicians." In 1850 the association was incorporated by the Massachusetts Legislature, under the name of the "Female Medical Education Society." In 1852 the number of professors was increased and a full course of medical education was given. In 1854 the Legislature made a grant of 5000 dollars for scholarships; in 1855, another grant of 10,000 dollars for other purposes; and in 1856, a full college charter was conferred. The course of education is similar to that in other medical colleges in the country. The number of graduates to the present time is thirty-four.

The College has been sustained mainly by donations and State aid, but in 1858 Hon. John Wade, of Woburn, left a bequest of 20,000 dollars as a scholarship fund, "for the support and medical education of worthy and moral indigent females." The Wade Scholarship Fund is now available for students. He also left about 5000 dollars which is to accumulate to 10,000 dollars, and then be paid over to the college to found a professorship. A bequest of 7000 dollars has also been left to the college, but it is not likely to be realized for many years, though it will be largely increased by the accumulation of interest.

In 1849 Miss Elizabeth Blackwell graduated from the medical school in Geneva, New York, being the first lady in the country to receive a medical degree. This incident attracted public attention and helped to increase the interest in the movement already in progress. In 1850 the Female Medical College was opened in Philadelphia, with a State charter, and a fully organized faculty of instruction. In 1853 the Penn Medical University was started in Philadelphia, with separate departments of instruction for males and for females. The Eclectic Medical Institute in Cincinnati, Ohio, has graduated numerous ladies; a few have taken degrees at the two colleges—the regular and the homœopathic—in Cleveland, Ohio, and perhaps from other medical colleges in the country. There are, as the writer has ascertained, above two hundred graduated female physicians in the United States.

As all of these are comparatively beginners, and most of them have been but from one to five years in service, and it usually requires a long time for any young physician to build up an extensive practice, it cannot be expected that marvellous things should have yet been achieved in their professional career. Many of them are, however, making themselves very useful to the public, and receiving a good remuneration, while others are laying the foundation for future success. Some have become public lecturers to female audiences, and are thus disseminating valuable knowledge where it is most needed. A graduate of this college has given lectures on anatomy, physiology, and health, in the four State Normal Schools of Massachusetts, to the young ladies preparing to be teachers, thus aiding them in preserving their own health and that of the children and youth of the public schools. Another of the graduates is physician in the Mount Holyoke Female Seminary, at South Hadly, in this State, where there are near three hundred young women to receive the benefit of her teachings, and of her medical advice and treatment when needed. Thus she combines the office of physician with the more important one of supervisor of health to this female household, an admirable position for a doctress, but one that a doctor would awkwardly fill. All such seminaries ought to be thus supplied. And what an interesting field of usefulness these female schools and seminaries open for women of literary and medical education!

For the purpose of promoting their success in the profession, the graduates of this College four years ago formed an association, called the New England Female Medical Society, now numbering twenty-five members, graduates from this and other colleges. Communications, verbal and written, are made at their meetings, and as their experience and observation extend they will be able to contribute more and more to the common stock for mutual improvement.

There are some persons who think there should be no separate medical schools for females, but that the sexes should be educated together. If the argument of propriety, urged in favor of female physicians for their own sex, has any force, it holds good in favor of separate schools for their education. That the experiment of admitting female students to male medical colleges has proved unsatisfactory may be inferred from the circumstance that in most or all of the instances of the kind the practice has been discontinued, and applications from ladies are rejected on the very reasonable ground, that there are now medical colleges expressly for females which it is more proper that they should attend. For a time it was of course necessary to employ male professors only, there being no others; but of the six instructors in the college in Boston, three are now ladies; there are now also three in the Female College in Philadelphia.*

In regard to hospital practice, there seems to be no good reason why female students should not obtain it in existing hospitals. In lying-in hospitals female physicians are certainly the proper attendants; and female students are the proper persons to assist and receive from them clinical instruction in the obstetric art. Madame Boivin and Madame Lachapelle, learned and skilful physicians, superintended above twenty thousand births each in the Hospital of Maternity in Paris, and with unequalled success. The women and children's wards in general hospitals, if not at present under the exclusive management of women physicians, could at specified times be attended by female students, by themselves, with lady professors to give the clinical instruction.

The important movement now in progress for educating nurses would be greatly facilitated and advanced by the co-operation of female physicians, who could more appropriately and more conveniently, and therefore more success- fully, than male physicians, instruct and train nurses in the care of lying-in and other female patients.

That this is an enterprise of great magnitude, requiring labor and patience to carry it forward, all will concede. But what ought to be done can be done. "Time and I against any two," said Philip of Macedon. So time and the spirit of progress will overcome all obstacles; and the current once turned will move on of itself, broader and deeper. The profession will find their female colaborers gradually multiplying, and in the process of time the proportions will be duly adjusted.

The progress of the cause must of course depend mainly upon women themselves. They alone, by earnest and patient endeavor and actual success, can practically solve the doubts and misgivings of well-wishers, remove the want of confidence of women in the abilities of their own sex, and overcome prejudice, interested opposition, and the tenacity of custom. Hitherto the men have taken the lead and shown the greater interest in this movement, women having natu- rally waited a little for the clearer sanction of the public voice. But they will not long hesitate where duty and humanity call.

Any demonstration of the principle and of the success of the enterprise in one country of course gives it an impulse in every other enlightened nation. The cause has made some progress in America, but it needs the reacting influence of successful European experiment—especially from our fatherland. It is certainly time that England, in her great metropolis, had at least one medical college for women.

* It is, however, obvious that in a country where no female medical schools exist the experiment cannot be made unless the first students be allowed entrance to a male med- ical college or hospital, as was done in the case of Miss Blackwell, and with no undesir- able result.

Source: Gregory, Samuel. 1862. "Female Physicians." *The Englishwomen's Journal* (March): 1–11.

These two pieces highlight the significant number of possible job opportunities for women in science in the first half of the twentieth century. Yet both pieces also suggest the ways in which women could be at a disadvantage if they entered the science profession. The very existence of the vocational guide suggests that there was an impetus to encourage women to take up positions in science and the large range of scientific opportunities for women suggests that plenty of jobs were available. Similarly, Ruby Worner's report points out that there are many and varied work opportunities in chemistry. The guide and Worner's report, however, caution that women may confront discrimination against them and suggest that many of the positions available to women are of low status and low pay compared with those available to men. In addition, Worner points out that fewer women are hired in almost every area of chemistry work than men. Although women were clearly moving into the science profession and being hired to do paid scientific work they were largely hired into feminized scientific positions and had less of a chance of promotion than their male counterparts.

Opportunities for Women in Scientific Work

UNITED STATES VOCATIONAL GUIDE

Fields of Work

> Anthropology,
> Bacteriology,
> Biology,
> Botany,
> Chemistry,
> Agricultural Chemistry,
> Analytical Chemistry,
> Industrial Chemistry,
> Entomology,
> Eugenics,
> Micro Analysis,
> Plant Pathology.

Where They Are

> United States Bureau of Chemistry.
> United States Bureau of Plant Industry.
> United States Bureau of Department of Agriculture.
> > Dairy Division.
> > Food Division.

United States Bureau of Agricultural Experiment Stations.

State Agricultural Experiment Stations.

> The departments of agriculture employ chemists, bacteriologists and entomologists.

State Agricultural Boards of Health.

> Bacteriological Laboratories and Research Laboratories.

City Boards of Health:

> Research Laboratories.

Commercial Laboratories:

> Business and Manufacturing firms;
>
> Industrial plants having Research Laboratories.
>
> Large milk contractors maintain bacteriological laboratories for bacterial tests of milk. This is necessary to conform to city and state regulations for milk standards.

Hospitals: Research Laboratories.

Research Laboratories (Private).

Colleges and Universities: Assistantships to professors in biology, zoology, botany, entomology, etc.

State Museums.

Museums of Natural History.

Institutions (state or private) conducting research work in eugenics.

Physicians (private) :

> Bacteriological Laboratories.

How to Get Them

By:

1. Civil service examination for federal, state and city departmental positions.
2. Volunteer work in laboratories in order to qualify for civil service examination.
3. Application to chief of laboratory or department in institution or museum.
4. Preliminary service as assistant without pay.
5. Application with recommendation to commercial and industrial establishments.
6. Application to college professors for assistantships in zoology, botany, etc.
7. Application to hospitals and institutions where work in bacteriological diagnosis is required.

8. Application to chief of departments in institutions which require a secretary or librarian having some scientific training.

9. Advertisements in scientific magazines.

What They Pay

Laboratory assistants and research assistants in federal and state departments $600–$900

Research positions in federal and state departments $1,000–$2,500

City Boards of Health, Bacteriological Laboratories $700–$1,800

Administrative positions in Boards of Health requiring a medical degree $3,000

Hospital chemist $1,200

State museums: assistant entomologist $780

Agricultural experiment stations: Bacteriologist $900–$1,260

Manufacturing firms $600–$1,000–$1,500

Industrial plants $600–$1,000–$1,500

Business firms $600–$1,000–$1,500

Assistants to college professors in biology, zoology, etc. $500–$900

Field workers in eugenics $900–$1,000

These are the average salaries. They vary, of course, in different states and with different commercial establishments. Only the highly trained worker can command a large salary.

The ordinary college course in Science will fit a girl for an assistantship, but she must have special training to qualify for a better position.

In some state or city Boards of Health volunteer service makes the worker eligible for the civil service examination for higher positions.

In some commercial laboratories there is a chance for advancement as the worker becomes expert in the special branch of work. Executive ability also leads to better positions in these establishments.

Nature of the Work

Bacteriological laboratories in United States government departments offer two lines of work—routine work as in a city laboratory and research work which requires more training and pays larger salaries.

Colleges and Museums:

An assistant in zoology or biology must prepare material and make slides for microscopical study. Drawing also is very necessary.

Museums of Natural History:

A museum assistant often has to learn to make artificial flowers and back-grounds for exhibition groups.

A research worker in any branch of science should be able to read technical French, German, Italian and Spanish.

There is a wide field in industrial chemistry.

In analytical work there is the discovery of new combinations and means of reducing the cost of production.

Hospitals and physicians maintain laboratories for special pathological research work.

State Agricultural Experiment Stations and State Boards of Health:

Every state agricultural experiment station has an income for scientific research in

Agriculture,

Chemistry,

Plant physiology,

Animal nutrition.

The assistant has routine work at first in the inspection analysis of

Foods,

Feeding stuffs,

Fertilizers.

For study in Pathology, the worker must have a good knowledge of chemistry and bacteriology.

Eugenics:

Field workers in eugenics are employed by hospitals for the insane and feeble-minded, to study the family history of the patients.

City Boards of Health:

Bacteriological laboratories employ

Helpers,

Chemists,

Bacteriologists,

Clinical workers.

There is much routine work as well as research work.

The work consists of

Food examination,

Milk standard tests,

Sanitary water analysis,

Bacteriological diagnosis of diseases,

Examination of pathological specimens.

Assistants prepare the media in which different bacteria are developed.

Makers of media require slight skill.

Makers of cultures must be bacteriologists.

Bacteriology can be learned in the laboratory if the worker has had college training in the allied sciences. The bacteriologist must be able to read technical French and German.

Administrative positions are open only to women with medical degrees.

Are Women Wanted in This Field?

There is a growing demand for women in research laboratories in federal, state and municipal laboratories, as well as in commercial houses.

Some positions discriminate against women but, if well qualified, they stand a fair chance when they have once demonstrated their ability.

In many positions in plant pathology and entomology men are preferred on account of the difficult nature of the field work.

There is a good opportunity for women in state agricultural experiment stations.

There are many board of health positions open to women.

Commercial and industrial laboratories are sometimes conservative about employing women, but they are constantly employing more chemists, and there are often good openings for well-trained women.

Women with some scientific training are in demand as secretaries, assistants and librarians in colleges and various institutions doing scientific work.

Secretaries must have a knowledge of scientific French and German as well as stenography and typewriting. Librarians must have some knowledge of library methods.

Openings for women as chemists in hospitals are increasing and women are wanted in this work.

Many manufacturing establishments are willing to employ women as chemists, except where the chemist must actually work in factories or use furnaces. A woman is desirable in a business house if she can combine executive ability with scientific training.

Women are wanted as assistants in scientific departments of colleges and museums.

Training Required. Mathematics are essential in most scientific work.

A college course in science is necessary for assistants.

Special training is necessary for research work.

Research workers in any branch of science must have a knowledge of modern languages, French, German, Italian and Spanish at least, to read scientific books and periodicals.

The bacteriologist must have a solid foundation in chemistry, general, organic and physiological, as well as in biology, zoology and physiology.

The micro-analyst must have training in histology (animal and plant), and chemistry, organic and inorganic.

A chemist in a hospital must have training in sanitary chemistry, bacteriology, zoology, chemistry, quantitative and qualitative analysis.

An anthropologist must cover the field of anthropology broadly at first in order to specialize in some one branch such as basketry or pottery. She must be a constant student of her subject in English and foreign publications.

A zoologist must have training in zoology, botany, biology, microscopy and art, especially drawing. To enter this work some biological center like Wood's Hole is a good place to make enquiries for assistantships.

An entomologist must have biology, zoology, economic and systematic entomology, French, German and drawing.

In commercial houses: If a woman has executive ability and an introductory scientific training she will find commercial houses ready to pay her salary while she acquires her special training with them. There is always room for a careful, conscientious worker, however, who has special training without executive ability.

If she can do the work herself and also teach untrained labor to produce perfect results in large quantities, she will be invaluable to her firm.

Qualifications Other Than Scientific Training

Good Health: Much of the work in laboratories is standing and the hours are usually long.

The worker in any branch of science must have a strong and genuine love of the science chosen. This work will not respond to a desire for salary primarily. The salaries are always small for the inexperienced worker, and the better paid positions demand a high degree of specialized training as well as native ability.

In each branch there are certain qualifications necessary for success. The anthropologist, for instance, must have a natural sympathy with primitive peoples and a desire to understand them and enter into their point of view, as far as possible. The worker in a Board of Health laboratory must have enthusiasm and interest in the painstaking work which is being done to protect the public health of to-day and in the advancement of science to that end.

There are certain general qualities which workers in all the different branches of science agree are necessary to success:

Enthusiasm.

Patience and love of detail and routine.

Accuracy and concentration on the work.

Ambition to read and study constantly in order to keep up with new methods and discoveries.

Systematic habits of work and management.

One must always be a faithful and accurate student, careful in details of workmanship and the interpretation of results. The responsibility is tremendous. Human lives may depend on decision. This is especially true of positions in bacteriological laboratories.

A broad vision as well as a penetrating one.

Unlimited adaptability and the research spirit.

Originality, independence and persistence.

Executive ability is important for some positions.

Disadvantages

Small salaries as assistants and little opportunity to rise without giving much time to special training.

Salaries for research workers are small considering the amount of training required to obtain such positions.

In laboratories especially the disadvantages are:

Long hours,

Confining work,

Too much routine work,

Very short vacations,

Promotion slow.

Advantages

If a woman does not mind confining work there is no limit to its interest and growth.

There is wonderful opportunity for service.

Agricultural and biological chemistry are of extreme interest.

The work is intellectually stimulating and constantly progressing.

Always a possibility of rise with increase of salary and more responsible work and research work.

Getting in touch with specialists is a constant education.

The salaries are fairly good as compared with other occupations.

Source: Anon. 1920?. *Opportunities for Women in Scientific Work* [United States Vocational Guide. Pp. 1–44. [From: History of Women Series, Reel 944, No. 8660. Microfilm. Woodbridge, CT: Research Publications, 1977. Filmed from the holdings of the Schlesinger Library, Radcliffe College, Cambridge, MA.]

Opportunities for Women Chemists in Washington

Ruby K. Worner

When this title was selected, I had it in mind to assume the role of the inquiring reporter and to visit women at work and also those who have had experience in placing and observing women chemists in Washington. I happened to meet Mr. Boutell, who is in charge of our Information Section at the National Bureau of Standards and who generally knows the answers, so I asked him what he thought of opportunities for women chemists in Washington. "Well," he smiled, "I have heard that the women so far outnumber the men here that they don't have many opportunities."

However, an examination of the situation indicates that this bit of logic may not be applicable to women chemists.

Let us consider first the various positions in Washington that require chemical training. The different Government agencies employ the largest number, and these will be discussed more particularly. Since Washington has certain unique advantages, there is considerable variety in the other agencies that employ chemists. The Geophysical Laboratory of the Carnegie Institution of Washington is here. A number of trade associations have offices and some laboratories, for example, the Institute of Paint and Varnish Manufacturers and the National Association Institute of Dyeing and Cleaning. Editorial offices of some of the scientific journals, such as *Industrial and Engineering Chemistry, Journal of the American Pharmaceutical Association*, and *Science Service* are located here. Some business concerns and patent lawyers station chemists here to do library research. There is more than the usual number of opportunities for teaching in colleges as well as in secondary schools, for in and near Washington are American University, Catholic University, George Washington University, Georgetown University, the University of Maryland, and others. Coöperative researches with industry are invited by some of the Government departments; for example, at the National Bureau of Standards there is a research associate plan which operates somewhat similar to that at the Mellon Institute of Industrial Research. A large number of organizations have taken advantage of this plan, including the American Association of Textile Chemists and Colorists, the American Petroleum Institute, and more recently, the Textile Foundation.

In the Federal service, practically every type of chemistry is utilized. According to the latest classified list of members of the Chemical Society of Washington, the majority of chemists in the classified service are employed in six of the major departments of which Agriculture and Commerce lead in numbers. It would only be confusing to name the twenty-seven or more bureaus represented, but you will be interested in knowing that women chemists are working in the Bureaus of Animal Industry, Chemistry and Soils, Dairy Industry, Home Economics, and Food and Drug Administration in the Department of Agriculture; in the National Bureau of Standards, the Geological Survey, the National Institute of Health, and the United States National Museum. There are probably others, but it is difficult to obtain complete figures.

In all the Bureaus mentioned, the men far outnumber the women, with the exception of the Bureau of Home Economics. The Chief of this Bureau is Dr. Louise Stanley, who is a chemist, as are also the heads of the nutrition and textile divisions. In the Bureau of Home Economics, women are definitely given a break, for the Bureau is practically "manned" throughout by women. Moreover, the history of this Bureau shows the ability of women to work together, which is contrary to Dr. Landis' contention.

Some of you are probably wondering how one gets a position in the Government. With the exception of establishments outside the competitive classified service, all positions in the Federal Government are obtained through Civil Service examinations. Technically speaking, there are no provisions discriminating against women in the laws and rules governing the Federal Service. Moreover, the regulations of the Commission provide for "equal compensation irrespective of sex." Examinations for positions paying an annual salary of $2600 or less are designated "assembled" and require the applicant to take a written examination on subject matter appropriate for the position to be filled. Although similar examinations may be held for positions paying higher salaries, these examinations are usually "unassembled" and require the applicant to furnish information on his training and experience and to submit lists of publications. The rating of the applicants by the Commission is based on the grade received in the examination, although credit is also given for military preference. When an opening in any agency occurs, the three highest applicants on the appropriate register are certified to the agency which is then free to choose among them. In asking for certifications, the agency may specify if men or women are preferred.

Advance notices of these examinations are posted in first and second class post offices and are published in appropriate journals. Any person interested in particular examination may file a card with the Commission and then receive direct notification when the examination is given.

And now, a little about the training required for Government service. As stated above, practically every type of chemistry is required in some part of the service, so the best training will depend upon the particular job to be filled. In general, however, it may be stated that for the sub-professional grades which pay salaries from $1020 to $2000, a college degree is not required, whereas one is necessary for the professional grades which pay $2000 and up. However, large numbers apply for these examinations: for example, 4495 applied for the last Junior Chemist's examination, 1945 the Assistant Chemist's; and 1246 the Associate Chemist's examinations. Thus, competition is keen, so usually it is the person with a college degree who gets the job in the sub-professional class, whereas a Ph.D. is a definite asset in even the lower professional grades. In addition to sound chemical training, a knowledge of physics, mathematics, statistics, English, and foreign languages may come in handy. Other branches of science, art, and literature may also be helpful. In addition, the personality and appearance of the applicant, which cannot always be evaluated, are important, particularly in regard to advancement.

Assuming a woman obtains a position in the Government, what are her chances for advancement? From personal observations and discussion with others, it appears that women are rather less likely to advance in the service than are men, assuming equal ability and application. There seem to be a number of reasons for this. There is always the economic reason, that man is the head of the household and, therefore, the support of a family. But it is not generally recognized that women often have similar responsibilities and that not all men are supporting families. Women are more readily accepted in positions of a routine or assistant nature and may not have an opportunity to develop or exhibit their executive abilities. There is also the difficulty of reallocating positions to higher grades after they are once fixed. There are still men who object to working on an equal basis or under women. It is possible that some of these men could have their minds changed by appropriate circumstances. In general, it is true that women are not as adept at creating and repairing equipment as men, largely because they have not been trained in that direction. Most of them were playing with dolls and toy baking sets, while their brothers hammered and sawed, or took the family "Lizzie" apart. But this creative ability can be developed or compensated for in other qualifications.

One gentleman who is in an excellent position to observe women in the Federal service said that he felt the greatest need among the women was organization; that, at present, they do not properly appreciate or evaluate their capabilities and consequently do not receive the recognition they deserve.

In this connection, you will be interested in Miss Ruth O'Brien's opinion expressed last spring before the Institute of Women's Professional Relations. As you probably all know, Miss O'Brien is in charge of the Division of Textiles and Clothing in the Bureau of Home Economics.

"The general feeling seems to prevail that women scientists have little chance of appointment in the Government service. True, the older bureaus have apparently not looked with favor upon the ladies as desirable workers in other than stenographic and clerical positions. However, as a representative of one of the newer bureaus and one which follows the general policy of employing women where qualified ones are available, I am more impressed by the indifference of women scientists to the Civil Service announcements than by the indifference of appointing officers to women's professional qualifications. Time and time again we call for certification of eligible women for positions and either receive the reply that none are on the present rolls, or the names certified are women who have rated so low that they are obviously undesirable. Granted that the number of openings is not over-whelmingly large, there are positions available for women who have the energy to watch for the announcements and take the examination."

In general, it appears to be much easier for women to enter the service in a higher classification than to advance from a lower grade to a higher one.

As regards the relative number of women chemists in Washington, it is difficult to give exact figures without making an exhaustive survey and study of individual cases. Some positions in applied chemistry might be overlooked for they use other titles than "chemist." So far as I know, unless it has been done by the Institute of Women's Professional Relations, complete data have not been compiled. Among the available data, it is interesting to note that according to the records of the Secretary of the Chemical Society of Washington, the total membership of six hundred eighty includes forty-five women. At the National Bureau of Standards, there are seven women chemists and one hundred twenty-three men. Data obtained last spring by Miss O'Brien showed thirty women chemists among two hundred sixty-five professional women in the Department of Agriculture. Undoubtedly some of the rest of the two hundred sixty-five might have been included from the standpoint of "applied chemistry." In the Bureau of Home Economics, there were sixteen chemists, all women; in the Bureau of Chemistry and Soils, five of the one hundred fifty chemists were women; in the National Institute of Health, there were five women chemists, but there were also twenty-five other women scientists, most of whom probably use some chemistry in their work.

It is evident, then, that with the exception of the Bureau of Home Economics, there are relatively few women chemists in the Federal service. One might conclude from this that the opportunities for women are also few, or at least very limited. However, the scattering of women in the various agencies might lead to the more optimistic view that women are just beginning to show their abilities and to be accepted in this field; hence the future potential opportunities are limitless. For the present, unless there is some very real change, I should not like to encourage anyone entering this field if she expects rapid advancement in it; but the experience to be gained in Federal service may make her more valuable elsewhere. It is not easy to get in; but, once in, Washington offers unusually fine opportunity for study and development, for recreation, for general culture, and for seeing first-hand some of those who are making history. The Government makes generous provision for annual leave, sickness, and retirement; and there may be more security in a Government position than in industry. For many, these other opportunities are ample compensation.

[1]Contribution to the Symposium on Training and Opportunities for Women in Chemistry, conducted by the Division of Chemical Education at the ninety-eighth meeting of the A. C. S., Boston, Mass., September 14, 1939.

Source: Worner, Ruby K. 1939. "Opportunities for Women Chemists in Washington." *Journal of Chemical Education* 16, 12: 583–585. (Reprinted with permission of *Journal of Chemical Education.*)

Scientist Evelyn Fox Keller has been at the forefront of feminist studies in science. In this biography of Barbara McClintock she presents the life of a woman scientist who achieved the Noble Prize by doing science, in part, somewhat differently from other (mostly male) scientists. Although Fox Keller, and McClintock herself, emphasize that McClintock follows the scientific method, feminists have held up McClintock's "feeling for the organism" approach to science as a specifically woman's approach to the study the natural world. In this excerpt from the biography, Fox Keller attempts to outline McClintock's scientific approach, balancing her vigorous scientific method with her desire to study and thus know each plant as an individual organism, as an approach that would be useful for all scientists to follow, male and female alike.

A Feeling for the Organism: The Life and Work of Barbara McClintock

EVELYN FOX KELLER

> *There are two equally dangerous extremes—*
> *to shut reason out, and to let nothing else in.*
> PASCAL

If Barbara McClintock's story illustrates the fallibility of science, it also bears witness to the underlying health of the scientific enterprise. Her eventual vindication demonstrates the capacity of science to overcome its own characteristic kinds of myopia, reminding us that its limitations do not reinforce themselves indefinitely. Their own methodology allows, even obliges, scientists to continually reencounter phenomena even their best theories cannot accommodate. Or—to look at it from the other side—however severely communication between science and nature may be impeded by the preconceptions of a particular time, some channels always remain open; and, through them, nature finds ways of reasserting itself.

But the story of McClintock's contributions to biology has another, less accessible, aspect. What is it in an individual scientist's relation to nature that facilitates the kind of seeing that eventually leads to productive discourse? What enabled McClintock to see further and deeper into the mysteries of genetics than her colleagues?

Her answer is simple. Over and over again, she tells us one must have the time to look, the patience to "hear what the material has to say to you," the openness to "let it come to you." Above all, one must have "a feeling for the organism."

One must understand "how it grows, understand its parts, understand when something is going wrong with it. [An organism] isn't just a piece of plastic, it's something that is constantly being affected by the environment, constantly showing attributes of that. . . . You need to know those plants well enough so that if anything changes, . . . you [can] look at the plant and right away you know what this damage you see is from—something that scraped across it or something that bit it or something that the wind did." You need to have a feeling for every individual plant.

"No two plants are exactly alike. They're all different, and as a consequence, you have to know that difference," she explains. "I start with the seedling, and I don't want to leave it. I don't feel I really know the story if I don't watch the plant all the way along. So I know every plant in the field. I know them intimately, and I find it a great pleasure to know them."

This intimate knowledge, made possible by years of close association with the organism she studies, is a prerequisite for her extraordinary perspicacity. "I have learned so much about the corn plant that when I see things, I can interpret [them] right away." Both literally and figuratively, her "feeling for the organism" has extended her vision. At the same time, it has sustained her through a lifetime of lonely endeavor, unrelieved by the solace of human intimacy or even by the embrace of her profession.

Good science cannot proceed without a deep emotional investment on the part of the scientist. It is that emotional investment that provides the motivating force for the endless hours of intense, often grueling, labor. Einstein wrote: " . . .what deep longing to understand even a faint reflexion of the reason revealed in this world had to be alive in Kepler and Newton so that they could in lonely work for many years disentangle the mechanism of celestial mechanics?" But McClintock's feeling for the organism is not simply a longing to behold the "reason revealed in this world." It is a longing to embrace the world in its very being, through reason and beyond.

For McClintock, reason—at least in the conventional sense of the word—is not by itself adequate to describe the vast complexity—even mystery—of living forms. Organisms have a life and order of their own that scientists can only partially fathom. No models we invent can begin to do full justice to the prodigious capacity of organisms to devise means for guaranteeing their own survival. On the contrary, "anything you can think of you will find." In comparison with the ingenuity of nature, our scientific intelligence seems pallid.

For her, the discovery of transposition was above all a key to the complexity of genetic organization—an indicator of the subtlety with which cytoplasm, membranes, and DNA are integrated into a single structure. It is the overall organization, or orchestration, that enables the organism to meet its needs, whatever they might be, in ways that never cease to surprise us. That capacity for surprise gives McClintock immense pleasure. She recalls, for example, the early post–World War II studies of the effect of radiation on *Drosophila:* "It turned out that the flies that had been under constant radiation were more vigorous than those that were standard. Well, it was hilarious; it was absolutely against everything that had been thought about earlier. I thought it was terribly funny; I was utterly delighted. Our experience with DDT has been similar. It was thought that insects could be readily killed off with the spraying of DDT. But the insects began to thumb their noses at anything you tried to do to them."

Our surprise is a measure of our tendency to underestimate the flexibility of living organisms. The adaptability of plants tends to be especially unappreciated. "Animals can walk around, but plants have to stay still to do the same things, with ingenious mechanisms. . . . Plants are extraordinary. For instance, . . . if you pinch a leaf of a plant you set off electric pulses. You can't touch a plant without setting off an electric pulse . . . There is no question that plants have [all] kinds of sensitivities. They do a lot of responding to their environment. They can do almost anything you can think of. But just because they sit there, anybody walking down the road considers them just a plastic area to look at, [as if] they're not really alive."

An attentive observer knows better. At any time, for any plant, one who has sufficient patience and interest can see the myriad signs of life that a casual eye misses: "In the summer-time, when you walk down the road, you'll see that the tulip leaves, if it's a little warm, turn themselves around so their backs are toward the sun. You can just see where the sun hits them and where the sun doesn't hit. . . . [Actually], within the restricted areas in which they live, they move around a great deal." These organisms "are fantastically beyond our wildest expectations."

For all of us, it is need and interest above all that induce the growth of our abilities; a motivated observer develops faculties that a casual spectator may never be aware of. Over the years, a special kind of sympathetic understanding grew in McClintock, heightening her powers of discernment, until finally, the objects of her study have become subjects in their own right; they claim from her a kind of attention that most of us experience only in relation to other persons. "Organism" is for her a code word—not simply a plant or animal ("Every component of the organism is as much of an organism as every other part")—but the name of a living form, of object-as-subject. With an uncharacteristic lapse into hyperbole, she adds: "Every time I walk on grass I feel sorry because I know the grass is screaming at me."

A bit of poetic license, perhaps, but McClintock is not a poet; she is a scientist. What marks her as such is her unwavering confidence in the underlying order of living forms, her use of the apparatus of science to gain access to that order, and her commitment to bringing back her insights into the shared language of science—even if doing so might require that language to change. The irregularities or surprises molecular biologists are now uncovering in the organization and behavior of DNA are not indications of a breakdown of order, but only of the inadequacies of our models in the face of the complexity of nature's actual order. Cells, and organisms, have an organization of their own in which nothing is random.

In short, McClintock shares with all other natural scientists the credo that nature is lawful, and the dedication to the task of articulating those laws. And she shares, with at least some, the additional awareness that reason and experiment, generally claimed to be the principal means of this pursuit, do not suffice. To quote Einstein again, " . . . only intuition, resting on sympathetic understanding, can lead to [these laws]; . . . the daily effort comes from no deliberate intention or program, but straight from the heart."

A deep reverence for nature, a capacity for union with that which is to be known—these reflect a different image of science from that of a purely rational enterprise. Yet the two images have coexisted throughout history. We are familiar with the idea that a form of mysticism—a commitment to the unity of experience, the oneness of nature, the fundamental mystery underlying the laws of nature—plays an essential role in the process of scientific discovery. Einstein called it "cosmic religiosity." In turn, the experience of creative insight reinforces these commitments, fostering a sense of the limitations of the scientific method, and an appreciation of other ways of knowing. In all of this, McClintock is no exception. What is exceptional is her forthrightness of expression—the pride she takes in holding, and voicing, attitudes that run counter to our more customary ideas about science. In her mind, what we call the scientific method cannot by itself give us "real understanding." "It gives us relationships which are useful, valid, and technically marvelous; however, they are not the truth." And it is by no means the only way of acquiring knowledge.

Source: Fox Keller, Evelyn. 1983. "A Feeling for the Organism." Pp. 197–207 in *A Feeling for the Organism: The Life and Work of Barbara McClintock.* New York: W. H. Freeman and Company. (Reprinted with permission of Henry Holt and Company, LLC.)

Have Only Men Evolved?

RUTH HUBBARD

The theory of sexual selection went into a decline during the first half of this century, as efforts to verify some of Darwin's examples showed that many of the features he had thought were related to success in mating could not be legitimately regarded in that way. But it has lately regained its respectability, and contemporary discussions of reproductive fitness often cite examples of sexual selection. Therefore, before we go on to discuss human evolution, it is helpful to look at contemporary views of sexual selection and sex roles among animals (and even plants).

Let us start with a lowly alga that one might think impossible to stereotype by sex. Wolfgang Wickler, an ethologist at the University of Munich, writes in his book on sexual behavior patterns (a topic which Konrad Lorenz tells us in the Introduction is crucial in deciding which sexual behaviors to consider healthy and which diseased):

> Even among very simple organisms such as algae, which have threadlike rows of cells one behind the other, one can observe that during copulation the cells of one thread act as males with regard to the cells of a second thread, but as females with regard to the cells of a third thread. The mark of male behavior is that the cell actively crawls or swims over to the other; the female cell remains passive.

The circle is simple to construct: one starts with the Victorian stereotype of the active male and the passive female, then looks at animals, algae, bacteria, people, and calls all passive behavior feminine, active or goal-oriented behavior masculine. And it works! The Victorian stereotype is biologically determined: even algae behave that way.

But let us see what Wickler has to say about Rocky Mountain Bighorn sheep, in which the sexes cannot be distinguished on sight. He finds it "curious":

> that between the extremes of rams over eight years old and lambs less than a year old one finds every possible transition in age, but no other differences whatever; the bodily form, the structure of the horns, and the color of the coat are the same for both sexes.

Now note: " . . . the typical female behavior is absent from this pattern." Typical of what? Obviously not of Bighorn sheep. In fact we are told that "even the males often cannot recognize a female," indeed, "the females are only of interest to the males during rutting season." How does he know that the males do *not* recognize the females? Maybe these sheep are so weird that most of the time they relate to a female as though she were just another sheep, and whistle at her (my free translation of "taking an interest") only when it is a question of mating. But let us get at last to how the *females* behave. That is astonishing, for it turns out:

> that *both* sexes play two roles, either that of the male or that of the young male. Outside the rutting season the females behave like young males, during the rutting season like aggressive older males. (Wickler's italics)

In fact:

> There is a line of development leading from the lamb to the high ranking ram, and the female animals . . . behave exactly as though they were in fact males . . . whose development was retarded. . . . We can say that the only fully developed mountain sheep are the powerful rams. . . .

At last the androcentric paradigm is out in the open: females are always measured against the standard of the male. Sometimes they are like young males, sometimes like older ones; but never do they reach what Wickler calls "the final stage of fully mature physical structure and behavior possible to this species." That, in his view, is reserved for the rams.

Wickler bases this discussion on observations by Valerius Geist, whose book, *Mountain Sheep*, contains many examples of how androcentric biases can color observations as well as interpretations and restrict the imagination to stereotypes. One of the most interesting is the following:

> Matched rams, usually strangers, begin to treat each other like females and clash until one acts like a female. This is the loser in the fight. The rams confront each other with displays, kick each other, threat jump, and clash till

one turns and accepts the kicks, displays, and occasional mounts of the larger without aggressive displays. The loser is not chased away. The point of the fight is not to kill, maim, or even drive the rival off, but to treat him like a female.

This description would be quite different if the interaction were interpreted as something other than a fight, say as a homosexual encounter, a game, or a ritual dance. The fact is that it contains none of the elements that we commonly associate with fighting. Yet because Geist casts it into the imagery of heterosexuality and aggression, it becomes perplexing.

There would be no reason to discuss these examples if their treatments of sex differences or of male/female behavior were exceptional. But they are in the mainstream of contemporary sociobiology, ethology, and evolutionary biology.

A book that has become a standard reference is George Williams's *Sex and Evolution*. It abounds in blatantly biased statements that describe as "careful" and "enlightened" research reports that support the androcentric paradigm, and as questionable or erroneous those that contradict it. Masculinity and femininity are discussed with reference to the behavior of pipefish and seahorses; and cichlids and catfish are judged down-right abnormal because both sexes guard the young. For present purposes it is sufficient to discuss a few points that are raised in the chapter entitled "Why Are Males Masculine and Females Feminine and, Occasionally, Vice-Versa?"

The very title gives one pause, for if the words masculine and feminine do not mean of, or pertaining, respectively, to males and females, what *do* they mean—particularly in a scientific context? So let us read.

On the first page we find:

Males of the more familiar higher animals take less of an interest in the young. In courtship they take a more active role, are less discriminating in choice of mates, more inclined toward promiscuity and polygamy, and more contentious among themselves.

We are back with Darwin. The data are flimsy as ever, but doesn't it sound like a description of the families on your block?

The important question is who are these "more familiar higher animals?" Is their behavior typical, or are we familiar with them because, for over a century, androcentric biologists have paid disproportionate attention to animals whose behavior resembles those human social traits that they would like to interpret as biologically determined and hence out of our control?

Williams's generalization quoted above gives rise to the paradox that becomes his chief theoretical problem:

> Why, if each individual is maximizing its own genetic survival should the female
> be less anxious to have her eggs fertilized than a male is to fertilize them, and
> why should the young be of greater interest to one than to the other?

Let me translate this sentence for the benefit of those unfamiliar with current evolutionary theory. The first point is that an individual's *fitness* is measured by the number of her or his offspring that survive to reproductive age. The phrase, "the survival of the fittest," therefore signifies the fact that evolutionary history is the sum of the stories of those who leave the greatest numbers of descendants. What is meant by each individual "maximizing its own genetic survival" is that every one tries to leave as many viable offspring as possible. (Note the implication of conscious intent. Such intent is not exhibited by the increasing number of humans who intentionally *limit* the numbers of their offspring. Nor is one, of course, justified in ascribing it to other animals.)

One might therefore think that in animals in which each parent contributes half of each offspring's genes, females and males would exert themselves equally to maximize the number of offspring. However, we know that according to the patriarchal paradigm, males are active in courtship, whereas females wait passively. This is what Williams means by females being "less anxious" to procreate than males. And of course we also know that "normally" females have a disproportionate share in the care of their young.

So why these asymmetries? The explanation: "The *essential* difference between the sexes is that females produce large immobile gametes and males produce small mobile ones" (my italics). This is what determines their "different optimal strategies." So if you have wondered why men are promiscuous and women faithfully stay home and care for the babies, the reason is that males "can quickly replace wasted gametes and be ready for another mate," whereas females "can not so readily replace a mass of yolky eggs or find a substitute father for an expected litter." Therefore females must "show a much greater degree of caution" in the choice of a mate than males.

E. O. Wilson says that same thing somewhat differently:

> One gamete, the egg, is relatively very large and sessile; the other, the sperm,
> is small and motile. . . . The egg possesses the yolk required to launch the
> embryo into an advanced state of development. Because it represents a con-
> siderable energetic investment on the part of the mother the embryo is often
> sequestered and protected, and sometimes its care is extended into the post-

natal period. *This is the reason why* parental care is *normally* provided by the female. . . . [my italics]

Though these descriptions fit only some of the animal species that reproduce sexually, and are rapidly ceasing to fit human domestic arrangements in many portions of the globe, they do fit the patriarchal model of the household. Clearly, androcentric biology is busy as ever trying to provide biological "reasons" for a particular set of human social arrangements.

The ethnocentrism of this individualistic, capitalistic model of evolutionary biology and sociobiology with its emphasis on competition and "investments," is discussed by Sahlins in his monograph, *The Use and Abuse of Biology.* He gives many examples from other cultures to show how these theories reflect a narrow bias that disqualifies them from masquerading as descriptions of universals in biology. But, like other male critics, Sahlins fails to notice the obvious androcentrism.

About thirty years ago, Ruth Herschberger wrote a delightfully funny book called *Adam's Rib*, in which she spoofed the then current androcentric myths regarding sex differences. When it was reissued in 1970, the book was not out of date. In the chapter entitled "Society Writes Biology," she juxtaposes the then (and now) current patriarchal scenario of the dauntless voyage of the active, agile sperm toward the passively receptive, sessile egg to an improvised "matriarchal" account. In it the large, competent egg plays the central role and we can feel only pity for the many millions of minuscule, fragile sperm most of which are too feeble to make it to fertilization.

This brings me to a question that always puzzles me when I read about the female's larger energetic investment in her egg than the male's in his sperm: there is an enormous disproportion in the *numbers* of eggs and sperms that participate in the act of fertilization. Does it really take more "energy" to generate the one or relatively few eggs than the large excess of sperms required to achieve fertilization? In humans the disproportion is enormous. In her life time, an average woman produces about four hundred eggs, of which in present-day Western countries, she will "invest" only in about 2.2 Mean-while the average man generates several billions of sperms to secure those same 2.2 investments!

Needless to say, I have no idea how much "energy" is involved in producing, equipping and ejaculating a sperm cell along with the other necessary components of the ejaculum that enable it to fertilize an egg, nor how much is involved in releasing an egg from the ovary, reabsorbing it in the oviduct if unfertilized (a partial dividend on the investment), or incubating 2.2 of them to birth. But neither do those who propound the existence and importance of women's disproportionate energetic investments. Furthermore, I attach no significance to these questions, since I do not believe that the details of our economic and social arrangements reflect our evolutionary history. I am only trying to show how feeble is the "evidence" that is being put forward to argue the evolutionary basis (hence *naturalness*) of woman's role as homemaker.

The recent resurrection of the theory of sexual selection and the ascription of asymmetry to the "parental investments" of males and females are probably not unrelated to the rebirth of the women's movement. We should remember that Darwin's theory of sexual selection was put forward in the midst of the first wave of feminism. It seems that when women threaten to enter as equals into the world of affairs, androcentric scientists rally to point out that our *natural* place is in the home.

Source: Hubbard, Ruth. 2002. "Have Only Men Evolved?" Pp. 161–164 in Janet A. Kourany (ed.) *The Gender of Science.* New Jersey: Prentice Hall.

These two excerpts from Angela M. Pattatucci's Women in Science: Meeting Career Challenges *represent the positive and negative experiences of women in science today. Both women feel they have to struggle against prejudice to obtain their scientific goals. These women's firsthand accounts raise important issues about women's involvement in science. They address such concerns as the need for supportive family and friends, an educational environment that accommodates women's ways of learning and life cycles, interaction with teachers who are willing to mentor women and assist them in networking with colleagues, and a work environment that enables women to feel included. These two women's experiences stand as examples of how women scientists can feel as though they are being pushed out of the profession but also of how women can succeed despite the odds against them and contribute substantially to science.*

Science and Women: From the Vantage Point of a "Leak in the Pipeline"

Minna Mahlab

My high school physics experience was probably just like everyone else's: very dry, very boring. Chemistry was the fun class, so I decided I wanted to be a chemist. You got to mix together strange substances, and scribble down odd combinations of numbers and letters that seem more like a secret code than anything else. I expressed this preference on my SAT forms, and received many letters and brochures from engineering and technical schools, but I chose to attend

Bryn Mawr College, a small, liberal arts, women's college located in a suburb of Philadelphia. Each graduating class is approximately 300 students; this small size contributes to a strong sense of community.

Despite my lackluster experience in high school, I enrolled in first-year physics and was immediately hooked. What I liked best about college physics was that I was fully engaged in the process. It required bringing together everything I knew—not just facts of science but the convoluted ways of thinking that I used in my literature classes as well. Physics was always a challenge; it was never boring. An adrenaline rush accompanied completing every problem and figuring out every new concept. Equally exciting was explaining it to someone else. My classmates and I worked closely together and I was a tutor or teaching assistant throughout my college career. Explaining difficult ideas to others and guiding them to a level of understanding was something at which I excelled and in which I took great pride. After four very successful years as a physics undergraduate, I decided I had found my goal. I was going to do physics for the rest of my life and teach other people to enjoy it as much as I did.

Slowly, I began to realize that all was not as it seemed. Science does not operate in a social or cultural vacuum. Even at a small institution such as Bryn Mawr, there were inter- and intradepartmental politics with which to deal. At larger schools and industries, it was even worse. I became painfully aware of how few women there are in the field, and how much I stood out for the sole reason that I am a woman. On my first day at one summer position, the student group was given a tour that halted awkwardly at the changing station. The director, who was giving the tour, did not know where the women's changing room was located, and finally just left me outside to wait while the rest of the group went in to be instructed in the proper procedure for putting on and taking off the special garb. When they emerged, the tour continued, and I was told that at some point they would find one of the other women to show me.

On graduating from college, I had the opportunity to travel abroad and participate in a political science program. After a great deal of thought, I deferred my entry into graduate school and joined the program. However, even in this seemingly unrelated academic setting, I quickly availed myself of the opportunity to conduct a research project, and spent much of my free time as a research assistant in the physics department at the institution at which the program was administered. Following completion of the program, I returned to the United States and began doctoral studies in physics at the University of California at San Diego. UCSD is a large research-oriented institution located in La Jolla, California, near San Diego. The physics faculty number about 55, with 4 tenured women, and there are active research groups in many areas of specialty. It was strange to walk into the orientation session the first day and find myself to be 1 of 6 women among 50 classmates. The department chair announced that he was pleased that women were so "well represented" in our class. I thought he was being sarcastic, but I was the only person in the room who laughed.

My graduate school experience started off on the wrong foot when I went to my first faculty advisor session. I was nervous, having performed poorly on the entrance exam, and expressed concern that my preparation was weak in certain areas. I suggested taking a couple of advanced undergraduate courses before jumping into the graduate curricula. The advisor replied, "You know, I've looked over your record, and I think that you can handle this. You can always come in for extra help, if necessary."

Three weeks later, I found myself floundering, struggling to keep my head above water, and failing miserably. I went to see my advisor. "Well," he said, "I didn't really expect anything else. You went to that girls' school, didn't you? I'm sure they could not have adequately trained you there." I was both shocked and offended by his blunt arrogance. Going to a women's college had insured my eventual failure? He continued by informing me that had I tried to major in physics at a "real" school, I would not have succeeded.

I found myself completely at a loss. Having no support network to fall back on, and finding myself unable to catch up and succeed, I fell into the trap of believing him. And once *I* believed I could not do it, I was doomed to fail.

My undergraduate experience left me accustomed to asking questions during lectures, but questions were not encouraged by professors or tolerated by most classmates at UCSD. When I asked for further explanations during a lecture, some of my classmates would grumble and wonder why I was wasting their time. Eventually, I stopped asking questions. I had been accustomed at Bryn Mawr to accessible and encouraging professors and teaching assistants. Although some of the professors were accessible at UCSD, they certainly were not supportive. Eventually I stopped asking for help. And throughout all of this, the idea that if I were not a woman, I would not be failing echoed loudly.

Most of the men in my class did not show that they were confused about the material. Instead, they behaved as though everything was crystal clear. They swaggered around in groups, loudly deriding the difficulty levels of problems that took me hours to solve. They were all succeeding, and I was not. Or so it seemed to me. It was only much later that I learned that many of them were performing at the same level as I was. However, they exuded confidence, and I did not perceive that it was usually false. And when confronted with that confidence, I think that all the years of messages that women could not do physics that I had ignored but absorbed internally finally surfaced and took hold. I failed.

When I considered whether or not trying again would be worthwhile, I struggled with several questions. What would I do differently next time? What mistakes would I try to avoid repeating? As I reflected, I began to wonder if the mistakes were totally mine in the first place.

At first, I enjoyed the notoriety that comes along with studying math and physics in general, and as a woman in particular. Most people report that they "hated physics in high school" and claim that they "don't know how to balance a checkbook." Many questioned my interest in physics, as well as my decision to be a physicist and accepted my career choice with some distaste accompanied by a great deal of awe. However, it did not take long before I tired of feeling different. I soon learned that a graduate student in English is received more graciously at a gathering, or in casual conversation on an airplane, than a graduate student in physics—that is, if she is a woman. Therefore, I sometimes lied about what I did. This was easier than trying to explain my choices.

The greatest outright disapproval I have received is from women in their 50s or older. Physics is an enigma to many of them and it is definitely not the sort of thing that might lure prospective husbands. From the perspective of some of these women, securing a husband is the bottom-line criterion a woman should use in making life choices. To illustrate, as an undergraduate I attended religious services in a community a few towns away from my school. More than once, on learning that I was attending Bryn Mawr, women in the congregation would say, "Oh, very nice. And what are you majoring in, dear? English?" "No, physics and math," I would reply. It was a great way to stop conversation—complete, dead silence usually followed that remark. The responses to my proclamation ranged from a weak, "That's nice, dear," to a blunt "Well, you're certainly not going to find a husband that way."

As I became older, the slant of these comments changed. It was no longer, "Why not switch to something easier," but, "Why aren't you married in addition to whatever nonsense you are involved in." I attended the wedding of a grade school friend after two years of graduate school. She and I had gone to a small private school together; many of the guests were the parents of our other classmates who had known us since the age of five.

"So," they said to me, "We hear you're studying physics." "Very nice, I'm sure." "Isn't this a lovely wedding?" "So tell me, when are you getting married?" "Oh, never mind about school, you can always get your degree on the side."

On the side? A PhD program in physics is not a casual, offhand experience even when you do not have the pressures of a family or relationship. But for me, because I am a woman, it was something I could or should do on the side. I am willing to bet that none of my male classmates had ever heard anything like that. But *my* priorities in life should not include an advanced degree in physics. That should be an afterthought, something in addition to the "real" part of my life.

Perhaps the worst part of my graduate school experience was that I spent so much time and energy questioning myself: my abilities, my knowledge, my self-esteem, my self-confidence. When I try to describe how the culture of physics as a whole made me feel unworthy, I can sense the lack of comprehension, and sometimes the disbelief on the part of listeners. I have no statistics that can convey the isolation women experience in the sciences and present a concrete picture. One of the most disturbing things is that condescending statements made about women typically go unchallenged. They are accepted as a matter of course, part of the daily dialog, and therefore in a bizarre twist they "make sense" to the listener.

I was asked recently if I have given up on physics. I have not. I still want to be a physicist. The problem is, I do not think I can be one in this system. For example, even after I had passed the first hurdle in graduate school and became part of a research group, there were still problems. My desk was in a large open room right next door to that of a professor. His undergraduate students would regularly poke their heads in and ask where he was. If I told them I did not know and to check with his secretary, they would invariably say, "Well, aren't you the secretary?" There were eight graduate students housed in that room. I never observed any other student being asked if he were the secretary. I am not surprised. Other than me, they were all men. I suppose it is "natural" to assume that the only woman in the room is a secretary—even if she is wearing scruffy jeans and a t-shirt, hunched over a physics text or scanning equations on a computer screen.

Even going to the bathroom was an isolating experience. With the exception of those near the secretary's offices on the first floor, there are no bathrooms for women in the science building at UCSD. Thus, for bathroom breaks, when the guys went down the hall, I had to run down three flights of stairs. At night, in a deserted building, that can be a frightening experience. Men do not often worry about their safety when walking alone from the library or laboratory to the parking lot late at night. Conversely, women must always be aware and on the defensive.

Is it harder for women to succeed in the sciences? Perhaps I cannot give a definitive answer, but I can offer the following story. I will never forget the words of my junior-level quantum mechanics professor at Bryn Mawr—a woman—looking at us wearily after a marathon problem-solving session that ran well into the evening and contemplating the rest of her night. Four hours of dealing with a stressed group of college students did not get her off the hook. She still had to go home and cook dinner for her family. As she was piling her books and notes together, she sighed wearily. "You know, sometimes I wish I had a wife."

Even when recognized, women's achievement is characteristically acknowledged within the context of serendipity rather than ability (Frieze, Whitley, & McHugh, 1982). What is skill for the male is considered luck for the female. This characterization severely undermines women's confidence and fosters an internal belief that we cannot trust our successes. After all, if our achievements are a matter of luck, then the odds are that our luck will change. However, if our accomplishments are a direct result of talent and ability, then each success lays a foundation on which subsequent achievements can be built.

Jacquelynne Eccles, professor of psychology at the University of Michigan, directed a series of studies designed to compare objective measures of students' abilities, as determined by standardized tests, grades in school, and so forth, to the subjective assessments and expectations of their parents. A goal of this research was to determine if a general belief that males have superior mathematics and athletic ability correlated with parents' distorted opinion of their child's competency. Eccles and her team found that parents amplified the ability of sons above, and downplayed the ability of daughters below, what the objective measures indicated. Eccles explained these results:

> If I believe that a girl is not going to do as well, then I am likely to believe she is not as talented as her grades would suggest. And it is that latter belief that is the important one. They know that their daughter is in the 98th percentile on performance. But they explain that by focusing on her hard work rather than her talent. It's very subtle, but it has the impact of undermining her own confidence in her talent. She starts to believe that she is doing well because she is a hard worker rather than because "I am good at this." . . . If parents believe the stereotype, they say to the girl, "It's because you're working hard." And to the boy they say, "You are working hard and you are talented." Over time he will come away with more confidence in his ability. (Mann, 1994, pp. 99–100)

A step in the right direction would be for parents, as well as teachers, to acknowledge a child's talent when it manifests. At the same time, resist the temptation to overdramatize achievements. It sends the wrong message to girls when they are overly praised for achievements that are treated as a matter of course for boys. As Eccles advised, stop recognizing girls' achievements solely in the context of hard work and determination. This characterization sends an inappropriate message to girls that they lack the requisite talent and ability to excel. Instead, hard work and determination should be acknowledged within the context of their *complementing* talent and ability. Finally attach a future to childhood interests and talent. Aptitude in a specific area can usually be applied generally to several career endeavors. For example, an outfielder on a high school girls' softball team had just made a spectacular running grab of a fly ball, making the third out and retiring the sides. As she approached the bench from the field, she was welcomed with pats on her back and compliments such as, "Great catch!" I was substituting for the regular coach, who had been unexpectedly called out of town, and as I congratulated the player I remarked, "You'd make a great physicist!" The entire team was shocked at my seemingly out-of-context statement. However, I explained that the outfielder had actually performed a highly sophisticated set of mathematical calculations that, among other things, accurately projected the trajectory of the ball—estimating its hang time as a function of the ball's momentum and the competing forces of gravity, wind resistance, and humidity—ultimately allowing her to set up and solve the equation that mapped the correct velocity and running angle necessary to meet the ball at the precise point at which it could be caught. All of this without the help of a calculator or pencil and paper! I ended by informing the team that each time they stepped up to bat, or made a play on the field, they solved problems far more challenging than anything present in their school textbooks.

For many of these young women, this was the first time anyone had connected mathematics or science to their daily lives. Whereas they had previously been abstractions, my connection made them real, and it attached a future to a seemingly unrelated talent. By the end of the game, team members were approaching me with questions.

"Do you mean when I'm late for school and run to catch the bus, I'm doing a physics experiment?"

"Absolutely!" I replied. "You need only to learn the language that will allow you to communicate this exceptional ability to others—and that language is mathematics."

I was not able to follow the academic progress of these women, but I would venture to guess that they approached mathematics with a new attitude—one that said, "I can do this!"

Attaching a future to talent is of paramount importance for girls, because the culture directs them to define their self-esteem more by popularity than achievement. As a consequence, simply acknowledging mathematics talent may be insufficient, because a girl might deem other people's assessment of her social skills as far more significant than mathematics ability. Thus, girls may elect not to take mathematics and science courses despite having the talent and ability to successfully negotiate through them. In my opinion, girls need a regular diet of future-oriented connections to their daily lives.

An outcome of not properly supporting female achievement is that large numbers of women enter college with little or no sense of purpose and direction. Among college seniors, women are often more unsure of what their next step will be, a trend also evident with women graduate students and postdoctoral research associates. This should come as no great surprise given that the culture does not promote girls to think in terms of having a professional future. Furthermore, women often bring an alternative outlook to the masculine professional arena, one that is not always welcome. When women join a male-dominated profession, they frequently confront complications that originate as an outcome of conflicts between the conventional roles women have occupied in society and the new functions that they are undertaking within the profession (Coe & Dienst, 1990). Duane Schmidt (1987) provided a classic example:

> Dr. Harper made an appointment with a new attorney. Shortly after he was ushered into the attorney's office, a young woman entered. "Honey, would you get me a cup of coffee?" the doctor asked. "Black. No sugar." "Certainly," she said and left the room.
>
> In minutes, she returned with his coffee, then took the seat behind the attorney's desk.
>
> "You're the attorney?" Dr. Harper gasped. "Oh, I'm so sorry. I thought you were the secretary."
>
> "No problem," she said. "But I think you should know my rate is $100 per hour. That cup of coffee just cost you $25." (p. 37)

We can applaud the courage of the attorney in her response to sexist assumptions, but in the long run she may pay a heavy price for her action. Dr. Harper may ignore his own offensive action and instead focus on her assertiveness—viewing it as a negative quality. In retelling his experience in the attorney's office to others, he may describe her as irrational, combative, stubborn, or even a "man-hater." Although all untrue, the news about her will travel fast and she may find herself effectively "walled out" of her profession.

There are costs to violating social norms. There are also rewards. Girls and women need to be informed so that they can make appropriate choices. Value struggles can result in emotional fatigue as a woman strives to satisfy both the demands of a professional role and social expectations that are tied to her gender. Girls and women often see themselves as being trapped into undesirable "all or nothing" options, and the consequences are lasting. In the process, women fear loss of their femininity and their affectional relationship. And it is a fear that is based on reality. For example, women medical students more often lose their affectional partner through separation or divorce during medical school compared to males (Bowers, 1987). Furthermore, the loss of primary partners reported by several contributors to this volume suggests that the trend is not unique to medical school. Hilary Cosell (1985) commented:

I think we're being terribly misled about how much success women as a group have achieved and about how real that success actually is. I think there may be a bitter day of reckoning for many of us that's not too far off. A day where women will say, "I gave up my personal life, I destroyed my marriage, I didn't have children, I gave up this and I gave up that and what was it for? I still haven't been able to achieve the way men do, in the same arena they do, the way I was told I could." Let's face it: women are no longer disenfranchised, but we don't have anything like the power of the white male corporate establishment. I don't know if we'll ever acquire that kind of power, but if we do, it's not going to be anytime in the near future. (p. 135).

As students, women face challenges of isolation, animosity, and harassment while striving to integrate their professional roles with their identities as women. The pressure becomes particularly intense during graduate school, where the competition for limited resources tends to suppress women from articulating their concerns. Consider the response of a medical student when asked about reporting incidents of discrimination and harassment.

[E]veryone here is paranoid about residencies, and getting good letters of recommendations from clinicians, therefore, they seldom create waves . . . in order for prejudices to be weeded out of the system, some waves have to be created . . . but who will do it and risk a good shot at a residency? (Coe & Dienst, 1990, p. 337)

Women are human. The sense of powerlessness to effect change combined with an increasing weariness from fighting a lone crusade for integrity and respect in a system that judges women more by stereotype than ability eventually takes its toll. It is therefore understandable that a significant number of women in the science career track elect to search for a scientific niche where male competition is low, to seek alternative employment opportunities such as science writing and reporting, to seek teaching positions in which tenure is based on one's effectiveness in educating students rather than fund-raising ability, or to exit the field entirely. In short, they settle for something different from their original dream. Women do this not because they view themselves as lacking in talent and ability, but because they want *relief.*

Source: Mahlab, Minna. 1998. "Science and Women: From the Vantage Point of a 'Leak in the Pipeline.'" Pp. 26–34 in Angela M. Pattatucci (ed.), *Women in Science: Meeting Challenges and Transcending Boundaries.* Thousand Oaks, CA: Sage Publications. (Reprinted with permission of Sage Publications.)

Oh, the Places You'll Go . . .

Natalie M. Bachir

I am from a city of 4,000 people in the northern part of Wisconsin. Because only 25% of my graduating class would go on to attend college, and only two would endeavor to leave the state, and only one (me) was able to gain admittance into an Ivy League school, I suppose that I was always considered the smart one. The brain. The hard-working, well-mannered, goal-driven girl. I was never really in competition with any men, or any women, for that matter. I was just me. My being female did not make a difference to my aspirations or my thoughts, my grades or my relations with teachers and classmates. If anything, I was much more uncomfortable with the fact that I was the only first-year student in my chemistry class. I did not even notice the fact that I was the only woman until the day my teacher made a big deal about whether or not he should address us as "you guys." I told him that I really did not mind either way.

Sexual harassment was something that women on television were shouting about. But I was not sure what it was, exactly, that they were fighting against. It was most certainly not anything I had ever encountered in my little protected part of the world.

When I received my college acceptances in the mail, I was both excited and overwhelmed. How would I choose? I narrowed down my choices to three or four. What was it about Dartmouth College that finally made it the right choice for me? My parents liked it because it was a small Ivy League college located in the wilderness of New Hampshire. That assured them of their daughter's safety. I liked it because it was a small Ivy League college located in the wilderness of New Hampshire. That assured me that I would be able to obtain a great education and still have fun in the outdoors, without an incredibly competitive atmosphere. In addition, there was a bonus: Dartmouth's Women In Science Project (WISP). I had remembered my father showing me an article about the internships offered through WISP, and mentioning that one day, maybe I would be fortunate enough to be offered such an internship. That cinched it. In the fall I headed off to Hanover, a town not much larger than my own hometown, to follow my dreams. After all, as the quote from Dr. Seuss, a Dartmouth alum, pointed out in the Dartmouth brochure, "I had brains in my head, and feet in my shoes. I could direct myself in any direction I chose."

I had wanted to be a doctor since the fifth grade. I was so sure of my aspirations that when asked, in college applications, to outline my goals, I always provided the following idealistic answer.

My goal in life is to improve society. Possibly by educating the people of our country, promoting freedom and justice, fighting against discrimination, or striving for international peace. To reach this goal would mean a great deal to me, which is why, after much thought and consideration, I decided to become a physician. As a physician, I feel that I would be closer to humanity than in any other profession. For a physician is more than just a "body repairman," he or she is an advisor, and a confidant. With my constant desire for knowledge, I hope to learn as much as I possibly can in the field of medicine. With this knowledge, I want to further medical progress and accomplish the goal that every physician, I believe, should have: to reduce suffering and prolong life. I not only want to succeed in my career, but also succeed in keeping up that great American tradition of raising a family on the principles of love, trust, and integrity. If I accomplish these goals, I will surely enjoy my life and be happy. The biggest barometer of success to me is that I will be satisfied that I have lived my life for an honorable and worthwhile purpose.

At that point in my life, I had never taken a college-level science course. I had never conducted research in a laboratory with a professor. I knew the compassionate side of myself that was needed in order to be a good doctor, but I was not sure whether I had enough interest in the sciences. The study of English, foreign languages, and art had always come easily to me. Would I be able to compete with students who had come from science magnet or private schools? Would I be able to take on the heavy course work and the laboratories, while still enjoying my favorite extracurricular activities? Or would I be one of the first-year students who, after taking the introductory courses in biology and chemistry, were "weeded out" of the system?

To be honest, I had a rough start. Adjusting to college life in itself was difficult. There were all kinds of people to deal with, decisions to make, and serious study habits to attain. In addition, I was intent on finally becoming an "independent woman." I refused to seek advice from deans, professors, and, most important, my parents. The term did not end with straight A's, as every other term in my life had. In retrospect, it is a good thing it did not. My grades forced me to take the time to think about my reasons for studying at Dartmouth, my reasons for wanting to become a doctor, and how I planned to attain my goals in life.

At first, I was a bit confused. Did I really want to go into medicine? Or had my father's influence, as a general surgeon, been so strong that I had never even bothered to explore any other areas of study? What were the reasons for my not performing up to my usual par in the fall term? Was the tough course load too much for me, or did I just have a difficult time adjusting to college life? Had *I* been "weeded out"? Was it time for me to change course, and think about pursuing another career?

Every year, the Women in Science Project at Dartmouth offers first-year women the opportunity to engage in paid hands-on research internships with science faculty members or researchers in nearby industrial or government laboratories. In order to obtain such a position, one has to go through both an application and interview process. I applied, and was later notified that I had both been accepted into the program and awarded my first choice of research projects. I would be working with Dr. Daubenspeck, a professor in the Physiology Department of the Borwell Research Building of the Dartmouth-Hitchcock Medical Center. I was excited!

Working in a laboratory was an amazing experience. I worked side by side with professors, PhD candidates, and postdoctoral research associates. I learned about muscle control in the respiratory system, the scientific method, and analyzing data. Everyone was willing to help me, teach me, and listen to my ideas. I worked hard, and tried to learn as much as I possibly could from my research. Needless to say, I was enthralled when Dr. Daubenspeck asked me if I would like to continue to do full-time research in the laboratory over the summer.

In the meantime, the Ethics Institute at Dartmouth was planning to offer a multidisciplinary course on the technology of assisted reproduction. An intern for the summer was needed to conduct library research for this course. I applied for the job, went through an interview process, and was selected.

Working in the physiology laboratory during the summer proved to be even more fascinating, as I was able to work longer hours and take part in more aspects of the research. I gave more of my input on the design and execution of the experiments. In addition, I attended engineering and physiology seminars with Dr. Daubenspeck, and enjoyed working in the company of a variety of scholarly and enthusiastic professionals and graduate students.

Conducting research for the Ethics Institute was equally invigorating. When I first began, I could see a potential problem associated with such technologies as in vitro fertilization and surrogate motherhood: women being misinformed, women being manipulated, women being used . . . so I decided to focus on how reproductive technologies affect the health of women. I was somewhat upset by my findings. I had never really cared to be sensitive to women's needs or pay any attention to the way in which women were treated. However, through this research experience, I came to realize women are more likely to be overlooked or mistreated. In the area of assisted reproduction, most researchers and doctors are men, but, of course, the primary patient is the woman who wants to bear a child. As a woman in science, this frustrates me, and drives me to persist in my scientific endeavors.

WISP had a great impact on my first-year experience. Through all of my disappointments, surprises, and realizations, it was quite beneficial to be constantly exposed to "the big picture," and be able to experience and take part in the dynamics of scientific research. The project helped to keep me focused on my ultimate goal at Dartmouth. There was a reason for trudging through all of those hypercompetitive science courses, after all! Throughout my first year, my WISP internship sustained my interest in the sciences, and lifted my spirits.

Over the course of the summer, I thought a lot about the Women in Science Project's impact on my life, and, most important, its potential impact on the lives of many aspiring women scientists at Dartmouth. However, there was one area of weakness that concerned me. Although the Women in Science Project did a fantastic job of supporting women students in science career tracks, it did not really address the issue of our being women at Dartmouth, as well. Balancing both at the same time is no easy task! I had an idea. The gap could be filled by arranging a discussion group, through WISP, to identify problems and challenges women faced at Dartmouth, and share advice on how to cope with those problems. The atmosphere of the discussion group would be friendly and open to all points of view. It would be expansive in that through discussing obstacles women had to face at Dartmouth, we might be able to come to some conclusions on the problems women in science confront in our society as a whole. I immediately arranged to meet with Mary Pavone, director of the Women in Science Project, and outlined what I saw as a need along with my strategy for addressing it. She said she thought my idea was fantastic, and immediately offered me a position as an intern for WISP. My job included leading discussions once a week, researching prominent issues that arose during discussion, and writing articles in the WISP newsletter. I worked toward identifying the obstacles women in science face at Dartmouth, researched their causes, and attempted to arrive at solutions.

Before coming to Dartmouth, I had read and heard that my college of choice had a problem. It was only in 1972 that Dartmouth began admitting women, and it is still labeled as a "male-bonding" school. More than 50% of the male population are affiliated with fraternities, and correspondingly fraternities more than double the number of sororities. I was informed that at Dartmouth, women and men cannot relate. The dating scene is almost nonexistent. The men are overpowering. Dartmouth is male territory. Drinking at fraternity parties is a prevalent part of the Dartmouth social life. The number of women with eating disorders at Dartmouth is appalling. Women are just too worried about how they look and what men think of them. Women cannot relate to one another or form strong friendships. Dartmouth is a conservative "boys' club." In terms of tolerance and political correctness, I was to discover that the college I planned to attend was behind the times. This is what I was told before coming to Dartmouth. It is also what I hear now as I study at Dartmouth. Is it all true?

I can give personal accounts. I can tell you of things I have seen and situations I have encountered. I can say, as I recently admitted to my sophomore dean, that yes, in a way, I was disappointed after I came to Dartmouth. When I had read that at Dartmouth the students played hard, but worked hard, too, I had no idea what all of the playing was about.

I have experienced the fraternity scene. I have watched women playing the fraternity game. What are they doing? This is not the way my mother taught me to behave.

"It's funny," remarked an acquaintance of mine last year, "But you can tell which women will be going to the fraternities tonight."

It was the winter term and we were eating dinner in the dining hall. She was referring to the tight, low-cut shirts that most of the women were wearing. Hers was black. In the fraternities, I saw first-year women more easily able to obtain beer than the first-year men. I saw the men ask the women to get them beer, after beer, after beer, after beer. I have been approached by men who have made idle chit-chat for a few minutes before ever-so-gallantly proposing, "Hey, want to get out of here, and go upstairs?" To this I can do nothing but laugh, and reply, "No, thanks." My rejection never appears to be a blow to their egos. They simply mumble a few more words, and then move on. Such occurrences can make one cynical. After all, these are Dartmouth students, the leaders of tomorrow.

I have watched women smile coyly as they drink themselves into inebriation at fraternity parties. I have seen women treated without any respect at all, in the fraternity basement; on the dance floor, on the way upstairs . . . and yet they go back. Is there no other way to relieve tension and have fun on this campus? What do women feel they are missing in themselves that they participate in activities that compromise rather than build self-esteem?

It may seem that I have diverted from my topic. However, it is my belief that it is impossible to separate what it is like to be a woman majoring in the sciences at Dartmouth from what it is like for a woman to exist in the culture at Dartmouth. From my description, it should come as no surprise to learn that I have encountered sexual harassment. There were four instances last summer alone. One of the most embarrassing moments was when my family was visiting the campus and my younger sister and I were sitting outside peacefully reading. Two men tossed a tennis ball over to us, expecting that we would welcome the opportunity to toss it back in a flirty fashion. When we made no attempt to play along, they began to whistle at us and point at the ball. It was at that point that I gathered my belongings and, with my sister in tow, walked away. Unless we are with a man, it is generally viewed by the male population at Dartmouth that *all* women are approachable in *all* circumstances.

There was also the time that the technician working in the physiology lab where I was conducting research thought himself quite witty. He sarcastically commented to the other men in the room that I was involved in "ChISP," an acronym for the "*Chicks* in Science Project." Despite this being a laboratory providing internships for WISP, headed by a man that is highly supportive of the project's goals, I still was not exempt from taunting and demeaning remarks. Stereotypes about women are deeply ingrained in the social fiber. His statement, which for him probably amounted to no more than incidental, mindless play, upset me. I told him what he had said was not appropriate, and let it go at that, not wanting to pay him any more attention. The culture at Dartmouth has made me cynical.

My research with the Ethics Institute brought me to the office of a fascinating professor. I had met her through a workshop sponsored by the Ethics Institute in the summer.

"I was lucky," Professor Sokol remarked, a smile spreading across her face, "That the high school I attended had a good academic curriculum, as my initial reason for choosing that particular high school was that it had a good dressmaking and design program!"

At this, she could not help but laugh, and I could not help but join her. Dressmaking and design? Professor Sokol? I was speaking to an endocrinologist; a professor of physiology; an ardent feminist: a powerful, independent, assertive woman. Born to immigrants who were quite unfamiliar with the educational system in the United States, Professor Sokol had, as an undergraduate, to make her own decision about majoring in biology and going on to graduate school at Harvard.

"We felt we were somewhat different," she replied.

In 1961, when Professor Sokol came to Dartmouth, there were 2 full-time women professors and 1 part-time woman professor, out of a total of 12 professors, in the Physiology Department. I smiled when she related this to me. I looked at Professor Sokol with admiration. She was a pioneer; a woman of perseverance and courage. Of course, in 1961, Dartmouth was still an all-male school. It was only in 1972 that women were first admitted. "So, what do the numbers look like now?" I asked, a knowing tone in my voice. We would surely laugh together at the fact that "back then," there were so few women involved in science. How fortunate I am to live in an era in which women are encouraged to achieve up to their full potential, even in a typically male-dominated field! Professor Sokol smiled at me; the reason would soon become apparent.

"Hmm, I'll have to count," she said, whirling around at her desk to look at the list of professors in the Physiology Department. She turned and addressed me with the results: "There are 5 women out of a total of 24 professors."

I was shocked. I expected that there would at least have been a slight increase from 1961 to 1994. However, the ratio has remained exactly the same.

I have decided on a double major in chemistry and philosophy. Researching the Chemistry Department at Dartmouth, I found that there are active research programs in subfields of organic, bio-organic, inorganic, physical polymer, and laser chemistry. Although the graduating class of 1995 consisted of an estimated 112 women compared to 152 male science majors, the numbers for the Chemistry Department were very different. In the class of 1995, there were 6 female and 13 male chemistry majors. But that does not upset me. Thus far, my classroom and laboratory experiences in general chemistry have revealed that professors tend to treat students equally and hold the same expectations for females and males. I have found that they are always willing to listen to ideas and encourage women to pursue their interests in chemistry.

In contrast, I have found attitudes in the student population to be much less favorable. For example, last year my laboratory teaching assistant behaved in a condescending manner toward me. When I asked questions, he would automatically assume that they were the typical questions of an uninterested science student who had not taken the time to read the book beforehand. He unnecessarily focused on tangential areas and I had to repeatedly remind him of my initial question. When he found out that I wanted to major in chemistry, he laughed out loud. According to him, I did not "look like a science person." Now what on earth must I possess to look like a stereotypical "science person"? I do not even want to attempt a guess.

Curious about the situation, I went to talk to Dr. Karen Wetterhahn, a professor of chemistry and cofounder of the Women in Science Project at Dartmouth. I hoped that there would be more than a few female role models to whom I could go for advice in my field. I found that the numbers were quite different for female professors in the Chemistry compared to the Physiology Department. Since Professor Wetterhahn had come in 1976, the number of female chemistry professors has doubled. On the surface, this might seem encouraging until one learns that in 1976, Professor Wetterhahn was the *only* woman in the department! Now there are 2, out of a total of 16 professors. This is not acceptable.

My research and classroom experiences at Dartmouth have solidified my goal to become a physician. I would like to practice medicine, to teach, and to conduct research. In one respect, my goals are the same as when I was typing all of those college applications. However, unlike before, my aspirations now come from serious thought and experience. Thus, they are more real to me. No longer an abstract dream, they are ambitions, desires, and attainable goals.

I have discovered that science encompasses everything. It is embedded in our ideas, the knowledge we attempt to acquire, and the interactions we have every day of our lives. To be able to learn and be successful in science, one must be able to communicate effectively. A science major must be able to develop ideas, ask questions, and share thoughts. To be knowledgeable and proficient in an area of the "hard" sciences is a power. With this power, one can enlighten some, anger others, and in the end, it is hoped, benefit society. With this power, one can change the world. Although I am less ignorant about the obstacles I have to face as a woman interested in pursuing a career in the sciences, and although I am more aware of the disappointments and hardships I likely will encounter along the way, I still remain as idealistic as I was before I matriculated to the college. I firmly believe that my college experience will be largely what I make of it. I plan to make mine the best and most productive experience possible, as a woman in science, and as a student ready to face the challenges of changing our society for the better. After all, "I have brains in my head, and feet in my shoes. . . ."

Science has much to offer to women, and women have much to offer to science. It would seem to be a perfect combination. However, women's representation in scientific fields has been historically low and we have a significantly higher percentage of derailment at all points along the career continuum compared to men. It is my sincere belief that one of the best ways to effect a reversal of this trend is to create a pool of women that enter the field informed about the challenges and boundaries to success that they will potentially encounter. The probability for success will be further enhanced if, in addition to being aware of the potential obstacles, women carry with them an assortment of strategies for circumventing these obstacles. This has been the work of this volume. Although there is no such thing as a magic formula, a catalog of steps, which if properly taken will universally guarantee success for all women, the volume does expose the extra stuff and provides a number of suggestions for negotiating through it. For example, a major lesson of this volume is that isolation should be avoided at all costs. It impedes progress and ultimately leads to career derailment. Thus, it would be wise for a budding female scientist or engineer to take the steps necessary to make sure that she does not become isolated. If no organized support system for women exists at your institution, or if the one available is not meeting your needs, it is then incumbent on you to develop your own. Complaining about the lack of adequate support may place you in the company of a group of individuals where you can be collectively miserable, but without action it will get you absolutely nowhere. Find people who will help and support you even if they are in unrelated fields. Take charge of, and responsibility for, your career.

Source: Natalie M. Bachir. 1998. "Oh, the Places You'll Go. . . ." Pp. 271–279 in Angela M. Pattatucci (ed.), *Women in Science: Meeting Career Challenges.* Thousand Oaks, CA: Sage Publications. (Reprinted with permission of Sage Publications.)

Bibliography

Introduction—Marie Curie: An Icon for Women Scientists

Bensaude-Vincent, Bernadette. 1996. "Star Scientists in a Nobelist Family: Irene and Frederic Joliot-Curie." Pp. 57–71 in Helena Pycior, Nancy G. Slack, and Pnina G. Abir-Am (eds.), *Creative Couples in the Sciences*. New Brunswick, NJ: Rutgers University Press.

Bertsch McGrayne, Sharon. 1993. *Nobel Prize Women in Science: Their Lives, Struggles, and Momentous Discoveries*. Secaucus, NJ: Carol Publishing Group.

Boudia, Soraya. 1998. "Marie Curie: Scientific Entrepreneur." *Physics World* 11 (December):35–39.

Coppes-Zantinga, A. R. 1998. "Madame Marie Curie (1867–1934): A Giant Connecting Two Centuries." *American Journal of Roentgenology* 171, 6 (December):1453–1457.

Curie, Eve. 1937. *Madame Curie: A Biography*. Garden City, NY : Doubleday, Doran.

Curie, Marie. 1963. *Pierre Curie*. New York: Dover Publications, Inc.

Elena, Alberto. 1997. "Skirts in the Lab: 'Madame Curie' and the Image of the Woman Scientist in the Feature Film." *Public Understanding of Science* 6:269–278.

Giroud, Francoise. 1986. *Marie Curie, a Life*. New York: Holmes and Meier.

Hallock, Grace T. 1938(?). *Marie Curie*. Health Heroes Series. New York: Metropolitan Life Insurance Company.

LaFollette, Marcel C. 1988. "Eyes on the Stars: Images of Women Scientists in Popular Magazines." *Science, Technology, and Human Values* 13, 3–4 (summer and autumn):262–275.

Langevin-Joliot, H. 1998. "Radium, Marie Curie, and Modern Science." *Radiation Research* 150, 5, suppl. 1 (November):S3–S8.

Lindee, Susan M. 1998. "The Scientific Romance: Purity, Self-Sacrifice, and Passion in Popular Biographies of Marie Curie." Paper read at the American Association for the Advancement of Science in Philadelphia, February.

McKown, Robin. 1971. *Marie Curie.* New York: Putnam.

Noble, Deborah. 1993. "Marie Curie: Half-Life of a Legend." *Analytical Chemistry* 65, 4 (February 15):215A–219A.

Pflaum, Rosalynd. 1989. *Grand Obsession: Madame Curie and Her World.* New York: Doubleday.

Pycior, Helena M. 1987. "Marie Curie's 'Anti-natural Path': Time Only for Science and Family." Pp. 191–214 in Pnina G. Abir-Am and Dorinda Outram (eds.), *Uneasy Careers and Intimate Lives: Women in Science, 1789–1979.* New Brunswick, NJ: Rutgers University Press.

———. 1993. "Reaping the Benefits of Collaboration while Avoiding Its Pitfalls: Marie Curie's Rise to Scientific Prominence." *Social Studies of Science* 23:301–323.

———. 1996. "Pierre Curie and 'his eminent collaborator Mme. Curie': Complementary Partners." Pp. 39–56 in Helena Pycior, Nancy G. Slack, and Pnina G. Abir-Am (eds.), *Creative Couples in the Sciences.* New Brunswick, NJ: Rutgers University Press.

Pycior, Stanley. 1999. "Marie Curie and Einstein: A Professional and Personal Relationship." *Polish Review* 44, 2:131–142.

Quinn, Susan. 1995. *Marie Curie: A Life.* New York: Simon and Schuster.

Rayner-Canham, Marelene, and Geoffrey Rayner-Canham. 1998. *Women in Chemistry: Their Changing Roles from Alchemical Times to the Mid-Twentieth Century.* Washington, DC: American Chemical Society and Chemical Heritage Foundation.

Reid, Robert William. 1974. *Marie Curie.* New York: Saturday Review Press.

Roque, Xavier. 1993. "Marie Curie and the Radium Industry: A Preliminary Sketch." *History and Technology* 13:267–291.

Rossiter, Margaret W. 1982. *Women Scientists in America: Struggles and Strategies to 1940.* Baltimore: Johns Hopkins University Press.

Schiebinger, Londa L. 1988. "Feminine Icons: The Face of Early Modern Science." *Critical Inquiry* 14:661–691.

Symposium Celebrating the Centenary of the Birth of Maria Sklodowska-Curie. Maria Sklodowska-Curie: Centenary Lectures. 1967. Proceedings of a symposium held in Warsaw on October 17–20 and organized in Poland by the

Maria Sklodowska-Curie Centenary Committee in cooperation with the International Atomic Energy Agency and the United Nations Educational, Scientific, and Cultural Organization. Vienna: International Atomic Energy Agency.

Webb, Michael. 1991. *Marie Curie: Discoverer of Radium.* Scientists and Inventors Series. Mississauga, Ontario: Copp Clark Pitman.

Chapter 1—Constructing a New Science: The Masculine Tradition

Battigelli, Anna. 1998. *Margaret Cavendish and the Exiles of the Mind.* Lexington: University Press of Kentucky.

Blaydes, Sophia B. 1988. "Nature Is a Woman: The Duchess of Newcastle and 17th-Century Philosophy." Pp. 51–64 in Donald C. Mell Jr., et al (eds.), *Man, God, and Nature in the Enlightenment.* East Lansing, MI: Colleagues Press.

Cook, Alan. 1997. "Ladies in the Scientific Revolution." *Notes and Records of the Royal Society of London* 51:1–12.

Daston, Lorraine. 1992. "The Naturalized Female Intellect." *Science in Context* 5:209–235.

Davis, Natalie Zemon. 1995. *Women on the Margins: Three Seventeenth-Century Lives.* Cambridge, MA: Harvard University Press.

de Baar, Mirjam, Machteld Lowenstyn, Marit Monteiro, and A. Agnes Sneller. 1996. *Choosing the Better Part: Anna Maria van Schurman (1607–1678).* Dordrecht: Kluwer Academic Publishers.

Douglas, Grant. 1957. *Margaret the First: A Biography of Margaret Cavendish, Duchess of Newcastle, 1623–1673.* Toronto: University of Toronto Press.

Eck, Caroline van. 1996. "The First Dutch Feminist Tract? Anna Maria van Schurman's Discussion of Women's Aptitude for the Study of Arts and Sciences." Pp. 43–53 in Mirjam de Baar, et al. (eds.), *Choosing the Better Part: Anna Maria van Schurman (1607–1678).* Dordrecht : Kluwer Academic Publishers.

Findlen, Paula. 1999. "Masculine Prerogatives: Gender, Space, and Knowledge in the Early Modern Museum." Pp. 29–57 in Peter Galison and Emily Thompson (eds.), *The Architecture of Science.* Cambridge, MA: MIT Press.

———. 2003. "Becoming a Scientist: Gender and Knowledge in Eighteenth-Century Italy." *Science in Context* 16(1/2):59–87.

Harding, Sandra. 1991. *Whose Science? Whose Knowledge? Thinking from Women's Lives.* Ithaca, NY: Cornell University Press.

Harkness, Deborah. 1997. "Managing an Experimental Household: The Dees of Mortlake and the Practice of Natural Philosophy." *Isis* 88, 2 (June):247–262.

Harth, Erica. 1992. *Cartesian Women: Versions and Subversions of Rational Discourse in the Old Regime.* Ithaca and London: Cornell University Press.

Herzenberg, Caroline L. 1990. "Women in Science during Antiquity and the Middle Ages." *Interdisciplinary Science Reviews* 15:294–297.

Hunter, Lynette. 1997a. "Sisters of the Royal Society: The Circle of Katherine Jones, Lady Ranelagh." Pp. 178–197 in Lynette Hunter and Sarah Hutton (eds.), *Women, Science, and Medicine, 1500–1700: Mothers and Sisters of the Royal Society.* Stroud, Gloucestershire: Sutton.

———. 1997b. "Women and Domestic Medicine: Lady Experimenters, 1570–1620." Pp. 89–107 in Lynette Hunter and Sarah Hutton(eds.), *Women, Science, and Medicine, 1500–1700: Mothers and Sisters of the Royal Society.* Stroud, Gloucestershire: Sutton.

Hunter, Lynette, and Sarah Hutton, eds. 1997. *Women, Science, and Medicine, 1500–1700: Mothers and Sisters of the Royal Society.* Stroud, Gloucestershire: Sutton.

Hutton, Sarah. 1997. "The Riddle of the Sphinx: Francis Bacon and the Emblems of Science." Pp. 7–28 in Lynette Hunter and Sarah Hutton (eds.), *Women, Science, and Medicine, 1500–1700: Mothers and Sisters of the Royal Society.* Stroud, Gloucestershire: Sutton.

———, ed. 1992. *Marjorie Hope Nicholson: The Conway Letters.* Oxford: Clarendon Press.

Jones, Kathleen. 1988. *A Glorious Fame: The Life of Margaret Cavendish, Duchess of Newcastle, 1623–1673.* London: Bloomsbury Publishing.

Keller, Evelyn Fox. 1985. "Baconian Science: The Arts of Mastery and Obedience." Pp. 33–42 in Evelyn Fox Keller, *Reflections on Gender and Science.* New Haven: Yale University Press.

———. 1985. *Reflections on Gender and Science.* New Haven: Yale University Press.

Lennon, Thomas M. 1992. "Lady Oracle: Changing Conceptions of Authority and Reason in 17th-Century Philosophy." Pp. 39–61 in Elizabeth D. Harvey and Kathleen Okruhlik (eds.), *Women and Reason.* Ann Arbor: University of Michigan Press.

Merchant, Carolyn. 1980. *The Death of Nature: Women, Ecology, and the Scientific Revolution.* New York: Harper and Row.

Merrens, Rebecca. 1996. "A Nature of 'Infinite Sense and Reason': Margaret Cavendish's Natural Philosophy and the 'Noise' of a Feminized Nature." *Women's Studies* 25:421–438.

Noble, David F. 1992. *A World without Women: The Christian Clerical Culture of Western Science.* New York: Knopf.

Okruhlik, Kathleen. 1992. "Birth of a New Physics or Death of Nature?" Pp. 63–67 in Elizabeth D. Harvey and Kathleen Okruhlik (eds.), *Women and Reason.* Ann Arbor: University of Michigan Press.

Salvaggio, Ruth. 1988. *Enlightened Absence: Neoclassical Configurations of the Feminine.* Urbana: University of Illinois Press.

Sarasohn, Lisa T. 1984. "A Science Turned Upside Down: Feminism and the Natural Philosophy of Margaret Cavendish." *Huntingdon Library Quarterly* 47:289–304.

Schiebinger, Londa L. 1987. "Maria Winkelmann at the Berlin Academy: A Turning Point for Women in Science." *Isis* 78:174–200.

———. 1988. "Feminine Icons: The Face of Early Modern Science." *Critical Inquiry* 14:661–691.

———. 1989. *The Mind Has No Sex? Women in the Origins of Modern Science.* Cambridge, MA: Harvard University Press.

———. 1993. *Nature's Body: Gender in the Making of Modern Science.* Boston: Beacon.

Soble, Alan. 1995. "In Defense of Bacon." *Philosophy of the Social Sciences* 25:192–215.

Tebeaux, Elizabeth, and Mary Lay. 1992. "Images of Women in Technical Books from the English Renaissance." *IEEE Transactions on Professional Communications* 35:196–207.

Chapter 2—Women's Bodies, Women's Minds: The Science of Women

Benjamin, Marina, ed. 1991. *Science and Sensibility: Gender and Scientific Enquiry, 1780–1945.* Oxford: Blackwell.

Birke, Lynda, and Gail Vines. 1987. "Beyond Nature versus Nurture: Process and Biology in the Development of Gender." *Women's Studies International Forum* 10:555–570.

Browne, Janet. 1989. "Botany for Gentlemen: Erasmus Darwin and the Loves of the Plants" *Isis* 80:593–621.

Cadden, Joan. 1993. *Meanings of Sex Difference in the Middle Ages: Medicine, Science, and Culture.* Cambridge: Cambridge University Press.

Daston, Lorraine. 1992. "The Naturalized Female Intellect." *Science in Context* 5:209–235.

Digby, Anne. 1989. "Women's Biological Straitjacket." Pp. 192–220 in Susan Mendus and Jane Rendall (eds.), *Sexuality and Subordination: Interdisciplinary Studies of Gender in the Nineteenth Century.* London: Routledge.

Donnison, Jean. 1977. *Midwives and Medical Men: A History of Inter-Professional Rivalries and Women's Rights.* London: Heinemann.

Douglas, Aileen. 1994. "Popular Science and the Representation of Women: Fontenelle and After." *Eighteenth-Century Life* 18:1–14.

Fausto-Sterling, Anne. 1986. *Myths of Gender: Biological Theories about Women and Men.* New York: Basic Books.

———. 2000. *Sexing the Body: Gender Politics and the Construction of Sexuality.* New York: Basic Books.

Fee, Elizabeth. 1979. "Nineteenth-Century Craniology: The Study of the Female Skull." *Bulletin of the History of Medicine* 53:415–433.

———. 1983. "Woman's Nature and Scientific Objectivity." Pp. 9–28 in Marian Lowe and Ruth Hubbard (eds.), *Woman's Nature: Rationalizations of Inequality.* New York: Pergamon.

Harding, Sandra, and Jean F. O'Barr (eds.). 1987. *Sex and Scientific Inquiry.* Chicago: University of Chicago Press.

Jacobus, Mary, Evelyn Fox Keller, and Sally Shuttleworth. 1990. *Body/politics: Women and the Discourses of Science.* New York : Routledge.

Jordanova, Ludmilla J. 1980. "Natural Facts: A Historical Perspective on Science and Sexuality." Pp. 42–69 in Carol P. MacCormack and Marilyn Strathern (eds.), *Nature, Culture, and Gender.* Cambridge: Cambridge University Press.

———. 1989. *Sexual Visions: Images of Gender in Science and Medicine between the Eighteenth and Twentieth Centuries.* Madison: University of Wisconsin Press.

———. 1999. *Nature Displayed: Gender, Science, and Medicine, 1760–1820.* London: Longman.

Koerner, Lisbet. 1993. "Goethe's Botany: Lessons of a Feminine Science." *Isis* 84:470–495.

Laqueur, Thomas. 1990. *Making Sex: Body and Gender from the Greeks to Freud.* Cambridge, MA: Harvard University Press.

Laslett, Barbara, et al. 1996. *Gender and Scientific Authority.* Chicago and London: University of Chicago Press.

Martin, Emily. 1996. "The Egg and the Sperm: How Science Has Constructed a Romance Based on Stereotypical Male-Female Roles." Pp. 323–339 in Barbara Laslett, et al. (eds.), *Gender and Scientific Authority.* Chicago: University of Chicago Press.

Moscucci, Ornella. 1990. *The Science of Woman: Gynaecology and Gender in England, 1800–1929.* Cambridge: Cambridge University Press.

Mosedale, Susan Sleeth. 1978. "Science Corrupted: Victorian Biologists Consider the 'Woman' Question." *Journal of the History of Biology* 11:1–55.

Newman, Louise Michele, ed. 1984. *Men's Ideas/Women's Realities: Popular Science, 1870–1915.* New York: Pergamon.

Outram, Dorinda. 1989. *The Body and the French Revolution: Sex, Class, and Political Culture.* New Haven: Yale University Press.

Overfield, Kathy. 1991. "Dirty Fingers, Grime, and Slag Heaps: Purity and the Scientific Ethic." Pp. 237–248 in Dale Spender (ed.), *Men's Studies Modified.* New York: Pergamon.

Owen, Alex. 1990. *The Darkened Room: Women, Power, and Spiritualism in Late Victorian England.* Philadelphia: University of Pennsylvania Press.

Russett, Cynthia Eagle. 1989. *Sexual Science: The Victorian Construction of Womanhood.* Cambridge: MA: Harvard University Press.

Schiebinger, Londa L. 1989. *The Mind Has No Sex? Women in the Origins of Modern Science.* Cambridge, MA: Harvard University Press.

———. 1993. *Nature's Body: Gender in the Making of Modern Science.* Boston: Beacon.

Squier, Susan. 1993. "Sexual Biopolitics in 'Man's World': The Writings of Charlotte Haldane." Pp. 137–155 in Angela Ingram and Daphne Patai (eds.), *Rediscovering Forgotten Radicals: British Women Writers, 1889–1939.* Chapel Hill: University of North Carolina Press.

Spanier, Bonnie B. 1995. *Im/partial Science: Gender Ideology in Molecular Biology.* Bloomington: Indiana University Press.

Steinbrügge, Lieselotte. 1995. *The Moral Sex: Woman's Nature in the French Enlightenment.* Translated by Pamela E. Selwyn. New York: Oxford University Press.

Tomaselli, Sylvana. 1991. "Reflections on the History of Science of Women." *History of Science* 29:185–205.

Tuana, Nancy. 1993. *The Less Noble Sex: Scientific, Religious, and Philosophical Conceptions of Woman's Nature.* Bloomington: Indiana University Press.

van den Wijngaard, Marianne. 1997. *Reinventing the Sexes: The Biomedical Construction of Femininity and Masculinity.* Bloomington: Indiana University Press.

Verbrugge, Martha H. 1997. "Recreating the Body: Women's Physical Education and the Science of Sex Differences in America, 1900–1940." *Bulletin of the History of Medicine* 71, 2:273–304.

Wils, Kaat. 1999. "Science, An Ally of Feminism? Isabelle Gatti de Gamond on Women and Science." *Revue Belge de Philogie et d'Histoire* 77, 2:416–439.

Winter, Alison. 1998. *Mesmerized: Powers of Mind in Victorian Britain.* Chicago: University of Chicago Press.

Zahm, Reverend John A. 1991. *Woman in Science, with an Introductory Chapter on Woman's Long Struggle for Things of the Mind.* Notre Dame, IN: University of Notre Dame Press. Originally published in 1913 under the pseudonym H. J. Mozans.

Chapter 3—Women Doing Science: Multiple Avenues

Abir-Am, Pnina G., and Dorinda Outram, eds. 1997. *Uneasy Careers and Intimate Lives: Women in Science, 1787–1979.* New Brunswick, NJ: Rutgers University Press.

Ainley, Marianne Gosztonyi, ed. 1990. *Despite the Odds: Essays on Canadian Women and Science.* Montreal: Vehicule Press.

Alic, Margaret. 1986. *Hypatia's Heritage: A History of Women in Science from Antiquity to the Late Nineteenth Century.* Boston: Beacon Press.

Allen, David E. 1980. "The Women Members of the Botanical Society of London, 1836–56." *British Journal for the History of Science* 13:240–254.

Baym, Nina. 2002. *American Women of Letters and the Nineteenth-Century Sciences: Styles of Affiliation.* New Brunswick, NJ: Rutgers University Press.

Bennett, Jennifer. 1991. *Lilies of the Hearth: The Historical Relationship between Women and Plants*. Camden East, Ontario: Camden House.

Bonta, Marcia Myers. 1991. *Women in the Field: America's Pioneering Women Naturalists*. College Station: Texas A&M University Press.

Bruck, Mary. 2002. *Agnes Mary Clerke and the Rise of Astrophysics*. Cambridge: Cambridge University Press.

Brush, Stephen. 1985. "Women in Physical Science: From Drudges to Discoverers." *The Physics Teacher* (January):11–19.

Buettinger, Craig. 1997. "Women and Antivivisection in Late Nineteenth-Century America." *Journal of Social History* 30, 4:857–872.

Edward, James T., ed. 1971. *Notable American Women, 1607–1950: A Biographical Dictionary*. Cambridge, MA: Belknap Press of Harvard University Press.

Ferguson, Moira. 1998. *Animal Advocacy and Englishwomen, 1780–1900: Patriots, Nation, and Empire*. Ann Arbor: University of Michigan Press.

Findlen, Paula. 2003. "Becoming a Scientist: Gender and Knowledge in Eighteenth-Century Italy." *Science in Context* 16(1/2):59–87.

Gates, Barbara T. 1998. *Kindred Nature: Victorian and Edwardian Women Embrace the Living World*. Chicago: University of Chicago Press.

Gray, Charlotte. 1999. *Sisters in the Wilderness: The Lives of Susanna Moodie and Catherine Parr Traill*. Toronto: Viking.

Johnston, Judith. 1998. "Colonising Botany: Louisa Anne Meredith and *The Romance of Natural History*." *Journal of Victorian Culture* 3:30–44.

Keeney, Elizabeth B. 1992. *The Botanizers: Amateur Scientists in Nineteenth-Century America*. Chapel Hill: University of North Carolina Press.

Kelly, Farley. 1993. *On the Edge of Discovery: Australian Women in Science*. Melbourne: University of Melbourne.

Kolbl-Ebert, Martina. 2002. "British Geology in the Early Nineteenth Century: A Conglomerate with a Female Matrix." *Earth Sciences History* 21:3–25.

Lindee, Susan. 1993. "The American Career of Jane Marcet's *Conversations on Chemistry*, 1806–1853." *Isis* 84:470–495.

Lindsay, Debra. 1998. "Intimate Inmates: Wives, Households, and Science in Nineteenth-Century America." *Isis* 89, 4:631–652.

Maroske, Sara. 1993. "'The Whole Great Continent as a Present': Nineteenth-Century Australian Workers in Science." Pp. 13–34 in Farley Kelly (ed.), *On the*

Edge of Discovery: Australian Women in Science. Melbourne: University of Melbourne.

Mazzotti, Massimo. 2001. "Maria Gaetana Agnesi: Mathematics and the Making of the Catholic Enlightenment." *Isis* 92:657–683.

Myers, Greg. 1989. "Science for Women and Children: The Dialogue of Popular Science in the Nineteenth Century." Pp. 171–200 in John Christie and Sally Shuttleworth, *Nature Transfigured: Science and Literature, 1700–1900.* Manchester: Manchester University Press.

Neeley, Kathryn A. 1992. "Women as Mediatrix: Women as Writers on Science and Technology in the Eighteenth and Nineteenth Centuries." *IEEE Transactions on Professional Communications* 35:208–215.

———. 2001. *Mary Somerville: Science, Illumination, and the Female Mind.* Cambridge: Cambridge University Press.

Ogilvie, Marilyn Bailey. 1986. *Women in Science, Antiquity through the Nineteenth Century: A Biographical Dictionary with Annotated Bibliography.* Cambridge, MA: MIT Press.

Phillips, Patricia. 1990. *The Scientific Lady: A Social History of Woman's Scientific Interests, 1520–1918.* London: Weidenfeld and Nicholson.

Pycior, Helena M., Nancy G. Slack, and Pnina G. Abir-Am, eds. 1995. *Creative Couples in the Sciences.* New Brunswick, NJ: Rutgers University Press.

Rayner-Canham, Marelene F., and Geoffrey Rayner-Canham. 1998. *Women in Chemistry: Their Changing Roles from Alchemical Times to the Mid-twentieth Century.* Washington, DC: American Chemical Society and the Chemical Heritage Foundation.

Reed, Elizabeth Wagner. 1992. *American Women in Science before the Civil War.* Minneapolis: Author.

Richards, Eveleen. 1989. "Huxley and Woman's Place in Science: The 'Woman Question' and the Control of Victorian Anthropology." Pp. 253–284 in James Moore (ed.), *History, Humanity, and Evolution: Essays for John C. Greene.* Cambridge: Cambridge University Press.

Rothman, Patricia. 1988. "Genius, Gender, and Culture: Women Mathematicians of the Nineteenth Century." *Interdisciplinary Science Review* 13:64–72.

Rudolph, Emanuel D. 1990. "Women Who Studied Plants in the Pre-Twentieth Century United States and Canada." *Taxon* 39:151–205.

Sheffield, Suzanne Le-May. 2001. *Revealing New Worlds: Three Victorian Women Naturalists.* London: Routledge.

Shteir, Ann B. 1984. "Linnaeus's Daughters: Women and British Botany." Pp. 67–73 in Barbara J. Harris and JoAnn K. McNamara (eds.), *Women and the Structure of Society: Selected Research from the Fifth Berkshire Conference on the History of Women*. Durham, NC: Duke University Press.

———. 1987. "Botany in the Breakfast Room: Women and Early Nineteenth-Century British Plant Study." Pp. 31–44 in Pnina G. Abir-Am and Dorinda Outram (eds.), *Uneasy Careers and Intimate Lives: Women in Science, 1789–1979*. New Brunswick and London: Rutgers University Press.

———. 1990. "Botanical Dialogues: Maria Jacson and Women's Popular Science Writing in England." *Eighteenth Century Studies* 23:301–317.

———. 1996. *Cultivating Women, Cultivating Science: Flora's Daughters and Botany in England, 1760 to 1860*. Baltimore: Johns Hopkins University Press.

———. 1997a. "Elegant Recreations? Configuring Science Writing for Women." Pp. 236–255 in Bernard Lightman (ed.), *Victorian Science in Context*. Chicago: University of Chicago Press.

———. 1997b. "Gender and 'Modern' Botany in Victorian England." *Osiris* 12:29–38.

Shteir, Ann B., and Barbara T. Gates. 1997. *Natural Eloquence: Women Reinscribe Science*. Madison: University of Wisconsin Press.

Torrens, Hugh. 1995. "Mary Anning (1799–1847) of Lyme: 'The Greatest Fossilist the World Ever Knew.'" *British Journal for the History of Science* 28:257–84.

Torrens Hugh, Elana Benamy, Edward B. Daescheler, Earle E. Spamer, and Arthur E. Bogan. 2000. "Etheldred Benett of Wiltshire, England, the First Lady Geologist—Her Fossil Collection in the Academy of Natural Sciences of Philadelphia, and the Rediscovery of 'Lost' Specimens of Jurassic Trigoniidae (Mollusca: Bivalvia) with Their Soft Anatomy Preserved." *Proceedings of the Academy of Natural Sciences of Philadelphia* 150:59–123.

Walters, Alice N. 1997. "Conversation Pieces: Science and Politeness in Eighteenth-Century England." *History of Science* 35:121–154.

White, Paul. 1996. "Science at Home: The Space between Henrietta Heathron and Thomas Huxley." *History of Science* 34:33–56.

Zahm, John Augustine. 1991. *Woman in Science, with an Introductory Chapter on Woman's Long Struggle for Things of the Mind*. Notre Dame, IN: University of Notre Dame Press. Originally published 1913.

Chapter 4—Women's Education in Science

Albisetti, James C. 1988. *Schooling German Girls and Women: Secondary and Higher Education in the Nineteenth Century.* Princeton: Princeton University Press.

Apple, Rima D. 1997. "Liberal Arts or Vocational Training? Home Economics Education for Girls." Pp. 79–95 in Sarah Stage and Virginia B. Vincenti (eds.), *Rethinking Home Economics: Women and the History of a Profession.* Ithaca and London: Cornell University Press.

Arnold, Lois B. 1984. *Four Lives in Science: Women's Education in the Nineteenth Century.* New York: Schocken.

———. 1999. "Becoming a Geologist: Florence Bascom in Wisconsin, 1874–1887." *Earth Sciences History* 18, 2:159–179.

Attar, Dena. 1990. *Wasting Girls' Time: The History and Politics of Home Economics.* London: Virago.

Ausejo, Elena. 1994. "Women's Participation in Spanish Scientific Institutions (1868–1936)." *Physis* 31:537–551.

Baym, Nina. 2002. *American Women of Letters and the Nineteenth-Century Sciences: Styles of Affiliation.* New Brunswick, NJ: Rutgers University Press.

Bonner, Thomas Neville. 1992. *To the Ends of the Earth: Women's Search for Education in Medicine.* Cambridge, MA: Harvard University Press.

Dyhouse, Carol. 1995. *No Distinction of Sex? Women in British Universities, 1870–1939.* London: University College London.

Eschbach, Elizabeth Seymour. 1993. *The Higher Education of Women in England and America, 1865–1920.* New York: Garland.

Gibson, Mary. 1990. "On the Insensitivity of Women: Science and the Woman Question in Liberal Italy, 1890–1910." *Journal of Women's History* 2:11–41.

Gould, Paula. 1997. "Women and the Culture of University Physics in Late Nineteenth-Century Cambridge." *British Journal for the History of Science* 30:127–149.

Johanson, Christine. 1987. *Women's Struggle for Higher Education in Russia, 1855–1900.* Kingston and Montreal: McGill-Queens Press.

Keeney, Elizabeth B. 1992. *The Botanizers: Amateur Scientists in Nineteenth-Century America.* Chapel Hill: University of North Carolina Press.

Kelley, Mary. 1992. "'Vindicating the Equality of Female Intellect': Women and

Authority in the Early Republic." *Prospects: An Annual of American Cultural Studies*17:1–27.

Koblitz, Ann Hibner. 1987. "Career and Home Life in the 1880s: The Choices of Mathematician Sofia Kovalevskaia." Pp. 172–190 in Pnina G. Abir-Am and Dorinda Outram (eds.), *Uneasy Careers and Intimate Lives: Women in Science, 1789–1979*. New Brunswick and London: Rutgers University Press.

———. 1988. "Science, Women, and the Russian Intelligentsia: The Generation of the 1860s." *Isis* 79:208–226.

———. 1993. *A Convergence of Lives: Sofia Kovalevskaia: Scientist, Writer, Revolutionary*. New Brunswick, NJ: Rutgers University Press.

———. 2000. *Science, Women, and Revolution in Russia*. Amsterdam: Harwood Academic Publishers.

Kohlstedt, Sally Gregory. 1990. "Parlors, Primers, and Public Schooling: Education for Science in Nineteenth-Century America." *Isis* 81:425–445.

Macleod, Roy, and Rusel Moseley. 1979. "Fathers and Daughters: Reflections on Women, Science, and Victorian Cambridge." *History of Education* 8:321–333.

Manthorpe, Catherine. 1986. "Science or Domestic Science? The Struggle to Define an Appropriate Science Education for Girls in Early Twentieth Century England." *History of Education* 15:195–213.

Morantz-Sanchez, Regina Markell. 1985. *Sympathy and Science: Women Physicians in American Medicine*. Oxford: Oxford University Press.

Nerad, Maresi. 1987. "Gender Stratification in Higher Education: The Department of Home Economics and the University of California, Berkeley, 1916–1962." *Women's Studies International Forum* 10:157–164.

Orr, Clarissa Campbell. 1995. "Albertine Necker de Saussure, the Mature Woman Author, and the Scientific Education of Women." in *Women's Writing* 2:141–153.

Parker, Joan E. 2001. "Lydia Becker's 'School for Science': A Challenge to Domesticity." *Women's History Review* 10:269–650.

Petrovich, Vesna Crnjanski. 1999. "Women and the Paris Academy of Sciences." *Eighteenth-Century Studies* 32:383–390.

Richards, Eveleen. 1989. "Huxley and Woman's Place in Science: The 'Woman Question' and the Control of Victorian Anthropology." Pp. 253–284 in James Moore (ed.), *History, Humanity, and Evolution: Essays for John C. Greene*. New York: Cambridge University Press.

Richmond, Marsha L. 1997. "'A Lab of One's Own': The Balfour Biological Laboratory for Women at Cambridge University, 1884–1914." *Isis* 88:422–455.

Rosenberg, Rosalind. 1982. *Beyond Separate Spheres: Intellectual Roots of Modern Feminism.* New Haven: Yale University Press.

Rossiter, Margaret W. 1982. *Women Scientists in America: Struggles and Strategies to 1940.* Baltimore: Johns Hopkins University Press.

———. 1997. "The Men Move In: Home Economics in Higher Education, 1950–1970." Pp. 96–117 in Sarah Stage and Virginia B. Vincenti (eds.), *Rethinking Home Economics: Women and the History of a Profession.* Ithaca and London: Cornell University Press.

Stage, Sarah. 1997. "Ellen Richards and the Social Significance of the Home Economics Movement." Pp. 17–33 in Sarah Stage and Virginia B. Vincenti (eds.), *Rethinking Home Economics: Women and the History of a Profession.* Ithaca and London: Cornell University Press.

Stage, Sarah, and Virginia B. Vincenti, eds. 1997. *Rethinking Home Economics: Women and the History of a Profession.* Ithaca and London: Cornell University Press.

Trecker, Janice Law. 1974. "Sex, Science, and Education." *American Quarterly* 26:352–366.

Warner, Deborah Jean. 1978. "Science Education for Women in Antebellum America." *Isis* 69:58–67.

Wills, Chris. 1999. "All Agog to Teach the Higher Mathematics: University Education and the New Woman." *Women: A Cultural Review* 10:56–66.

Zschoche, Sue. 1989. "Dr. Clarke Revisited: Science, True Womanhood, and Female Collegiate Education." *History of Education Quarterly* 29:545–569.

Chapter 5—Professionalizing Women Scientists

Alarez, Amaya Jane. 1993. "Invisible Workers and Invisible Barriers: Women at the CSIRO in the 1930s and 1940s." Pp. 77–103 in Farley Kelly (ed.), *On the Edge of Discovery: Australian Women in Science.* Melbourne: University of Melbourne.

Appel, Toby A. 1999. "Physiology in American Women's Colleges: The Rise and Decline of a Female Subculture." Pp. 305–335 in Sally Gregory Kohlstedt (ed.), *History of Women in the Sciences: Readings from* Isis. Chicago: University of Chicago Press.

Ausejo, Elena. 1994. "Women's Participation in Spanish Scientific Institutions (1868–1936)." *Physis* 31:537–551.

Babbitt, Kathleen R. 1997. "Legitimizing Nutrition Education: The Impact of the Great Depression." Pp. 145–162 in Sarah Stage and Virginia B. Vincenti (eds.), *Rethinking Home Economics: Women and the History of a Profession*. Ithaca and London: Cornell University Press.

Bartusiak, Marcia. 1993. "The Stuff of Stars: When a Woman Graduate Student Discovered Abundant Hydrogen in Stellar Spectrums, She Was Bullied into Suppressing Her Results." *The Sciences* (September/October):34–39.

Bell, Morag, and Cheryl McEwan. 1996. "The Admission of Women Fellows to the Royal Geographical Society, 1892–1914: The Controversy and the Outcome." *Geographical Journal* 162:295–312.

Blaszczyk, Regina Lee. 1997. "'Where Mrs. Homemaker Is Never Forgotten': Lucy Maltby and Home Economics at Corning Glass Works, 1929–1965." Pp. 163–180 in Sarah Stage and Virginia B. Vincenti (eds.), *Rethinking Home Economics: Women and the History of a Profession*. Ithaca and London: Cornell University Press.

Boney, A. D. 1995. "The Botanical 'Establishment' Closes Ranks: Fifteen Days in January 1921." *Linnean* 11, 3:26–37. [On the controversy over the appointment of Agnes Arber as president of Section K (Botany) for the 1921 meeting of the British Association for the Advancement of Science.]

Bruck, M. T. 1995. "Lady Computers at Greenwich in the Early 1890s." *Quarterly Journal of the Royal Astronomical Society* 36:83–95.

Brush, Stephen G. 1999. "Nettie M. Stevens and the Discovery of Sex Determination by Chromosomes." Pp. 337–346 in Sally Gregory Kohlstedt (ed.), *History of Women in the Sciences: Readings from* Isis. Chicago: University of Chicago Press.

Bud, Robert. 1991. "The Chemical Society—A Glimpse at the Foundations." *Chemistry in Britain* 27, 3 (March):pp.230–232.

Clark, Shirley M., and Mary Corcoran. 1986. "Perspectives on the Professional Socialization of Women Faculty: A Case of Accumulative Disadvantage?" *Journal of Higher Education* 57, 1:20–43.

Creese, Mary. 1991. "British Women of the Nineteenth and Early Twentieth Centuries who Contributed to Research in the Chemical Sciences." *British Journal for the History of Science* 24, part 3 (September):275–305.

Drachman, Virginia G. 1986. "The Limits of Progress: The Professional Lives of Women Doctors, 1881–1926." *Bulletin of the History of Medicine* 60:58–72.

Gallager, Teresa Catherine. 1989. "From Family Helpmeet to Independent Professional: Women in American Pharmacy, 1870–1940." *Pharmacy in History* 31:60–77.

Grossman, Atina. 1993. "German Women Doctors from Berlin to New York: Maternity and Modernity in Weimar and in Exile." *Feminist Studies* 19, 1:65–88.

Herzenberg, Caroline L., and Ruth Hege Howes. 1993. "Women of the Manhattan Project." *Technology Review* 96, 8:34–40.

Howes, Ruth Hege, and Caroline Herzenberg. 1993. "Women in Weapons Development: The Manhattan Project." Pp. 95–110 in Ruth Hege Howes and Michael R. Stevenson (eds.), *Women and the Use of Military Force.* Boulder, CO: Lynne Riener.

———. 1999. *Their Day in the Sun: Women of the Manhattan Project.* Philadelphia: Temple University Press.

Jacob, Margaret C., and Dorothee Sturkenboom. 2003. "A Woman's Scientific Society in the West: The Late Eighteenth-Century Assimilation of Science." *Isis* 94, 2:217–252.

Johnson, Jeffrey A. 1998. "German Women in Chemistry, 1895–1925 (Part I)." *Zeitschrift fur Geschichte der Naturwissenschaft, Technik und Medizin (International Journal of History and Ethics of Natural Sciences, Technology and Medicine)* 6, 1:1–21.

Jones, Greta. 1995. "Women and Eugenics in Britain: The Case of Mary Scharlieb, Elizabeth Sloan Chesser, and Stella Browne." *Annals of Science* 52, 5:481–502.

Kelly, Farley. 1993. *On the Edge of Discovery: Australian Women in Science.* Melbourne: University of Melbourne.

Mack, Pamela E. 1990. "Strategies and Compromises: Women in Astronomy at Harvard College Observatory, 1870–1920." *Journal for the History of Astronomy* 21:65–76.

Mason, Joan. 1991. "A Forty Years' War." *Chemistry in Britain* (March):233–238.

———. 1992. "The Admission of the First Women to the Royal Society of London." *Notes and Records of the Royal Society of London* 46:279–300.

Metropolis, N., and E. C. Nelson. 1982. "Early Computing at Los Alamos." *Annals of the History of Computing* 4, 2:348–357.

Morantz-Sanchez, Regina M. 1987. "The Many Faces of Intimacy: Professional Options and Personal Choices among Nineteenth- and Twentieth-Century Women Physicians." Pp. 45–59 in Pnina G. Abir-Am and Dorinda Outram (eds.),

Uneasy Careers and Intimate Lives: Women in Science, 1789–1979. New Brunswick, NJ: Rutgers University Press.

Nyhart, Lynn K. 1997. "Home Economists in the Hospital, 1900–1930." Pp. 125–144 in Sarah Stage and Virginia B. Vincenti (eds.), *Rethinking Home Economics: Women and the History of a Profession.* Ithaca and London: Cornell University Press.

Rayner-Canham, Marelene F., and Geoffrey W. Rayner-Canham. 1997. *A Devotion to Their Science: Pioneer Women of Radioactivity.* McGill: Queen's University Press.

———. 1998. *Women in Chemistry: Their Changing Roles from Alchemical Times to the Mid-Twentieth Century.* Washington, DC: American Chemical Society and the Chemical Heritage Foundation.

———. 1999. "British Women Chemists and the First World War." *Bulletin for the History of Chemistry* 23:20–27.

Richards, Eveleen. 1989. "Huxley and Woman's Place in Science: The 'Woman Question' and the Control of Victorian Anthropology." Pp. 253–284 in James Moore (ed.), *History, Humanity, and Evolution: Essays for John C. Greene.* Cambridge: Cambridge University Press.

Richmond, Marsha L. 2001. "Women in the Early History of Genetics: William Bateson and the Newnham College Mendelians, 1900–1910." *Isis* 92, 1:55–90.

Rossiter, Margaret W. 1980. "Women's Work in Science, 1880–1910." *Isis* 71:381–398.

———. 1982. *Women Scientists in America: Struggles and Strategies to 1940.* Baltimore: Johns Hopkins University Press.

———. 1993. "The (Matthew) Matilda Effect in Science." *Social Studies of Science* 23:325–341.

———. 1995. *Women Scientists in America: Before Affirmative Action, 1940–1972.* Baltimore: Johns Hopkins University Press.

Royal Society of Canada—Medals and Awards: The Alice Wilson Award. http://www.rsc.ca/english/awards_wilson.html.

Schiebinger, Londa. 2002. "European Women in Science." *Science in Context* 15, 4:473–481.

Shellard, E. J. 1982. "Some Early Women Research Workers in British Pharmacy, 1886–1912." *Pharmaceutical Historian* 12:2–3.

Sime, Ruth Lewin. 1996. *Lise Meitner: A Life in Physics.* Berkeley: University of California Press.

Stage, Sarah, and Virginia B. Vincenti, eds. 1997. *Rethinking Home Economics: Women and the History of a Profession.* Ithaca and London: Cornell University Press.

Terrall, Mary. 1995. "Gendered Spaces, Gendered Audiences: Inside and Outside the Paris Academy of Sciences." *Configurations: A Journal of Literature, Science, and Technology* 3:207–232.

Wilson, Jane S., and Charlotte Serber, eds. 1988. *Standing By and Making Do: Women of Wartime Los Alamos.* Los Alamos, NM: Los Alamos Historical Society.

Witz, Anne. 1999. *Professions and Patriarchy.* New York: Routledge.

Wright, Margaret. 1999. "Marcella O'Grady Boveri (1863–1950): Her Three Careers in Biology." Pp. 347–372 in Sally Gregory Kohlstedt (ed.), *History of Women in the Sciences: Readings from* Isis. Chicago: University of Chicago Press.

Chapter 6—Women's Advancement in Science since World War II

Ambrose, Susan A., Kristin L. Dunkle, Barbara B. Lazarus, Indira Nair, and Deborah A. Harkus. 1997. *Journeys of Women in Science and Engineering: No Universal Constants.* Philadelphia: Temple University Press.

Aqueno, Frank R. 1997. "Exploding the Gene Myth: A Conversation with Ruth Hubbard." http://eserver.org/gender/exploding-the-gene-myth.html.

Bertsch McGrayne, Sharon. 1993. *Nobel Prize Women in Science: Their Lives, Struggles, and Momentous Discoveries.* Secaucus, NJ: Carol Publishing Group.

Bhathal, Ragbir. 1999. *Profiles: Australian Women Scientists.* Canberra: National Library of Australia.

Birke, Lynda. 1986. *Women, Feminism, and Biology: The Feminist Challenge.* New York: Methuen.

Bleier, Ruth. 1984. *Science and Gender: A Critique of Biology and Its Theories on Women.* Elmsford, NY: Pergamon.

———, ed. 1986. *Feminist Approaches to Science.* Elmsford, NY: Pergamon.

Bosch, Mineke. 2002. "Women and Science in the Netherlands: A Dutch Case?" *Science in Context* 15, 4:483–527.

Brown, Andrew. 2000. "The Guardian Profile—Evelyn Fox Keller: Fox among the Lab Rats." *Guardian Unlimited*, November 4, 2000. http://www.guardian.co.uk/ARchive/Article/0,4273,4085786,00.html.

Brush, Stephen. 1985. "Women in Physical Science: From Drudges to Discoverers." *The Physics Teacher* (January):11–19.

Carroll, Berenice A. 1990. "The Politics of 'Originality': Women and the Class System of the Intellect." *Journal of Women's History* 2:136–163.

Cohen, Estelle. 1997. "'What the Women at All Times Would Laugh at': Redefining Equality and Difference, circa 1660–1760." Pp. 121–142 in Sally Gregory Kohlstedt and Helen Longino (eds.), *Women, Gender, and Science: New Directions*. Chicago: University of Chicago Press

Easlea, Brian. 1981. *Science and Sexual Oppression: Patriarchy's Confrontation with Woman and Nature*. London: Weidenfeld and Nicolson.

———. 1983. *Fathering the Unthinkable: Masculinity, Scientists, and the Nuclear Arms Race*. London: Pluto Press.

Etzkowitz, Henry, Carol Kemelgor, and Brian Uzzi. 2000. *Athena Unbound: The Advancement of Women in Science and Technology*. New York: Cambridge University Press.

Evans, Sara M. 2003. *Tidal Wave: How Women Changed America at Century's End*. New York: Free Press.

Garry, Ann, and Marilyn Pearsall. 1989. *Women, Knowledge, and Reality: Explorations in Feminist Philosophy*. Boston: Unwin Hyman.

Gross, Paul R., and Norman Levitt. 1994. *Higher Superstition: The Academic Left and Its Quarrels with Science*. Baltimore: Johns Hopkins University Press.

Gross, Paul R., Norman Levitt, and Martin W. Lewis. 1996. *The Flight from Science and Reason*. New York: New York Academy of Sciences.

Hanson, Sandra L. 1996. *Lost Talent: Women in the Sciences*. Philadelphia: Temple University Press.

Harding, Sandra. 1986. *The Science Question in Feminism*. Ithaca: Cornell University Press.

———. 1991. *Whose Science? Whose Knowledge?: Thinking from Women's Lives*. Ithaca: Cornell University Press.

Harding, Sandra, and Merill B. Hintikka, eds. 1983. *Discovering Reality: Feminist Perspectives on Epistemology, Metaphysics, Methodology, and Philosophy*. Dordrecht: D. Reidel.

Harding, Sandra, and Jean F. O'Barr, eds. 1987. *Sex and Scientific Inquiry*. Chicago: University of Chicago Press.

Hausman, Bernice L. 1999. "Ovaries to Estrogen: Sex Hormones and Chemical

Femininity in the Twentieth Century." *Journal of Medical Humanities* 20, 3 (fall):165–176.

Hubbard, Ruth. "Curriculum Vitae." http://www.cameron.edu/festival/speakers/hubbard-bio.html.

Hubbard, Ruth. 1979. "Have Only Men Evolved?" in Janet A. Kournay, *The Gender of Science*. Upper Saddle River, NJ: Prentice Hall. Originally published in Hubbard, Ruth, Mary Sue Henifin and Barbara Fried. 1982. *Biological Woman: The Convenient Myth. A Collection of Feminist Essays and a Comprehensive Bibliography*. Rochester, VT: Schenkman.

————. 1981. "The Emperor Doesn't Wear Any Clothes: The Impact of Feminism on Biology." Pp. 213–235 in Dale Spender (ed.), *Men's Studies Modified*. New York: Pergamon.

————. 1990. *The Politics of Women's Biology*. New Brunswick, NJ: Rutgers University Press.

Jacobus, Mary, Evelyn Fox Keller, and Sally Shuttleworth. 1989. *Body/Politics: Women and the Discourses of Science*. Boston: Routledge.

Kass-Simon, G., and Patrial Farnes, eds. 1990. *Women of Science: Righting the Record*. Bloomington: Indiana University Press.

Keller, Evelyn Fox. 1982. "Feminism and Science." *Signs* 7:589–602.

————. 1983. *A Feeling for the Organism*. San Francisco: Freeman.

————. 1985. *Reflections on Science and Gender*. New Haven: Yale University Press.

————. 1992. *Secrets of Life, Secrets of Death: Essays on Language, Gender, and Science*. New York: Routledge.

————. 1995. "Gender and Science: Origin, History, and Politics." *Osiris* 10:26–38.

————. 1997. "Developmental Biology as a Feminist Cause? Women, Gender, and Science: New Directions." *Osiris* 12:16–28.

Kelly, Farley. 1993. *On the Edge of Discovery: Australian Women in Science*. Melbourne: University of Melbourne.

Kohlstedt, Sally Gregory. 1995. "Women in the History of Science: An Ambiguous Place." *Osiris* 10:39–58.

Kourany, Janet A. 2002. *The Gender of Science*. Upper Saddle River, NJ: Prentice Hall.

Longino, Helen E. 1988. "Science, Objectivity, and Feminist Values." *Feminist Studies* 14:561–574.

Maddox, Brenda. 2002. *Rosalind Franklin: The Dark Lady of DNA*. New York: Harper Collins.

Merchant, Carolyn. 1982. "Isis Consciousness Raised." *Isis* 73:398–409.

Mies, Maria. 1990. "Women's Studies: Science, Violence, and Responsibility." *Women's Studies International Forum* 13:433–441.

Morse, Mary. 1995. *Women Changing Science: Voices from a Field in Transition*. New York and London: Plenum Press.

Nash, Jessica. 1999. "Freaks of Nature: Images of Barbara McClintock." *Studies in the History and Philosophy of Biology and Biomedical Science* 30:21–43.

Norwood, Vera L. 1987. "The Nature of Knowing: Rachel Carson and the American Environment." *Signs* 12:740–760.

Ogilvie, Marilyn, and Joy Harvey. 2000. *The Biographical Dictionary of Women in Science: Pioneering Lives from Ancient Times to the Mid-Twentieth Century*. Vols. 1 and 2. New York: Routledge.

Olson, Richard. 1990. "Historical Reflections on Feminist Critiques of Science: The Scientific Background to Modern Feminism." *History of Science* 28:125–148.

Paxton, Nancy L. 1991. *George Eliot and Herbert Spencer: Feminism, Evolutionism, and the Reconstruction of Gender*. Princeton: Princeton University Press.

Rasmussen, Carolyn. 1993. "'Science Was so Much More Exciting': Six Women in the Physical Sciences." Pp. 105–131 in Farley Kelly (ed.), *On the Edge of Discovery: Australian Women in Science*. Melbourne: University of Melbourne.

Reed, Evelyn. 1978. *Sexism and Science*. New York: Pathfinder Press.

Rogers, Lesley J. 1981. "Biology: Gender Differentiation and Sexual Variation." Pp. 44–57 in Norma Grieve and Patricia Grimshaw (eds.), *Australian Women: Feminist Perspectives*. Oxford: Oxford University Press.

———. 1988. "Biology, the Popular Weapon: Sex Differences in Cognitive Function." Pp. 43–51 in Barbara Caine, E. A. Grosz, and Marie De Lepervanche (eds.), *Crossing Boundaries*. Sydney: Allen and Unwin.

Rosner, Mary. 1998. "Plotting a Middle Ground: An Account of a Failed Science Story." Pp. 37–51 in John T. Battalio (ed.), *Essays in the Study of Scientific Discourse*. Stamford, CT: Ablex.

Rosser, Sue V. 1982. "Androgyny and Sociobiology." *International Journal of Women's Studies* 5:435–444.

————. 1992. *Biology and Feminism: A Dynamic Interaction.* New York: Twayne.

Rossiter, Margaret W. 1995. *Women Scientists in America: Before Affirmative Action, 1940–1972.* Baltimore: Johns Hopkins University Press.

Sayers, Janet. 1987a. "Feminism and Science—Reason and Passion" *Women's Studies International Forum* 10:171–179.

————. 1987b. "Science, Sex Difference, and Feminism." Pp. 68–91 in Beth B. Hess and Myra Marx Ferree (eds.), *Analyzing Gender: A Handbook of Social Science Research.* Newbury Park, CA: Sage Publications.

Sayre, Anne. 1975. *Rosalind Franklin and DNA.* New York: W.W. Norton.

Schiebinger, Londa L. 1987. "The History and Philosophy of Women in Science." *Signs* 12:305–322.

————. 1999. *Has Feminism Changed Science?* Cambridge, MA: Harvard University Press.

Tuana, Nancy. 1986. "Re-Presenting the World: Feminism and the Natural Sciences." *Frontiers* 8:73–78.

van den Wijngaard, Marianne. 1991. "The Acceptance of Scientific Theories and Images of Masculinity and Femininity, 1959–1985." *Journal of the History of Biology* 24:19–49.

————. 1997. *Reinventing the Sexes: The Biomedical Construction of Femininity and Masculinity.* Bloomington: Indiana University Press.

Watson, James. 1968. *The Double Helix.* New York: Atheneum.

Wertheim, Margaret. 1995. *Pythagoras' Trousers: God, Physics, and the Gender Wars.* New York: Times Books/Random House.

Whitton, Natasha. 1999. "Evelyn Fox Keller: Historical, Psychological, and Philosophical Intersections in the Study of Gender and Science." *Women Writers* (July). http://www.womenwriters.net/archives/whittoned1.htm.

Zuckerman, Harriet, and Jonathan R. Cole. 1975. "Women in American Science." *Minerva* 13:82–102.

Chapter Seven—Creating a Future for Women in Science

Allen, Felicity. 1993. "Girls and Women in Mathematics and Science: Let 'X' Equal a Known Quantity, 'Y' an Unknown Motivation?" Pp. 229–254 in Farley

Kelly (ed.), *On the Edge of Discovery: Australian Women in Science*. Melbourne: University of Melbourne.

American Women in Science. http://www.awis.org.

Bart, Jody, ed. 2000. *Women Succeeding in the Sciences: Theories and Practices across Disciplines*. West Lafayette, IN: Purdue University Press.

Bosch, Mineke. 2002. "Women and Science in the Netherlands: A Dutch Case?" *Science in Context* 15, 4:483–527.

Byrne, Eileen M. 1991. *Women in Science: The Snark Syndrome*. New York: Falmer Press.

Callahan, Joanne. 1998. "Ruth Hubbard: Excerpt from Nomination." *Women in Technology International*. http://www.witi.com/center/witimuseum/womenin-sciencet/1998/061598.shtml.

Costas, Ilse. 2002. "Women in Science in Germany." *Science in Context* 15, 4:557–576.

Eisenhart, Margaret A., and Elizabeth Finkel. 1998. *Women's Science: Learning and Succeeding from the Margins*. Chicago: University of Chicago Press.

ETAN Report on Women and Science: Science Policies in the European Union: Promoting excellence through mainstreaming gender equality, 2000 - http://www.cordis.lu/improving/women/documents.htm.

Etzkowitz, Henry, Carol Kemelgor, and Brian Uzzi. 2000. *Athena Unbound: The Advancement of Women in Science and Technology*. Cambridge: Cambridge University Press.

European Report: National Policies on Women and Science in Europe, 2002. http://www.cordis.lu/improving/women/documents.htm

Fehrs, Mary, and Roman Czujko. 1992. "Women in Physics: Reversing the Exclusion." *Physics Today* 45:33–40.

Haraway, Donna. 1989. *Primate Visions: Gender, Race, and Nature in the World of Modern Science*. New York: Routledge.

———. 1991. *Simians, Cyborgs, and Women: The Reinvention of Nature*. New York: Routledge.

Harding, Sandra. 1989. "Women as Creators of Knowledge." *American Behavioral Scientist* 32:700–707.

Hermann, Claudine, and Francoise Cyrot-Lackmann. 2002. "Women in Science in France." *Science in Context* 15, 4:529–556.

Hrdy, Sarah Blaffer. 1999. *Mother Nature: A History of Mothers, Infants, and Natural Selection.* New York: Pantheon.

Hubbard, Ruth. 1981. "The Emperor Doesn't Wear Any Clothes: The Impact of Feminism on Biology." Pp. 213–235 in Dale Spender (ed.), *Men's Studies Modified.* New York: Pergamon.

Jones, M. Gail, and Jack Wheatley. 1988. "Factors Influencing the Entry of Women into Science and Related Fields." *Science Education* 72:127–142.

Kelly, Farley, ed. 1993. *On the Edge of Discovery: Australian Women in Science.* Melbourne: University of Melbourne.

Lewis, Sue. 1993. "Lessons to Learn: Gender and Science Education." Pp. 255–280 in Farley Kelly (ed.), *On the Edge of Discovery: Australian Women in Science.* Melbourne: University of Melbourne.

Longino, Helen E. 1987. "Can There Be a Feminist Science?" *Hypatia: A Journal of Feminist Philosophy* 2, 3 (fall):51–64.

Love, Rosaleen. 1993. "'Doing the Herky-Jerky': Women in the Public Life of Science." Pp. 179–198 in Farley Kelly (ed.), *On the Edge of Discovery: Australian Women in Science.* Melbourne: University of Melbourne.

Morse, Mary. 1995. *Women Changing Science: Voices from a Field in Transition.* New York and London: Plenum Press.

National Research Council. 1980. *Committee on the Education and Employment of Women in Science and Engineering: Women Scientists in Industry and Government: How Much Progress in the 1970s?: An Interim Report to the Office of Science and Technology Policy from the Committee on the Education and Employment of Women in Science and Engineering, Commission on Human Resources, National Research Council.* Washington, DC: National Academy of Sciences.

———. 1991. *Committee on Women in Science and Engineering: Women in Science and Engineering: Increasing Their Numbers in the 1990s: A Statement on Policy and Strategy.* Washington, DC: National Academy Press.

———. 1994. *Committee on Women in Science and Engineering: Women Scientists and Engineers Employed in Industry. Why So Few?* Washington, DC: National Academy Press.

National Science Foundation. http://www.nsf.gov/.

Natural Sciences and Engineering Research Council of Canada. 1996. *Task Force on Women in Science and Engineering: Report of the Task Force on Women in Science and Engineering to the Natural Sciences and Engineer-*

ing Research Council: Towards Building a New Scientific and Engineering Culture in Canada. Ottawa: Natural Sciences and Engineering Research Council of Canada.

Norwood, Vera L. 1987. "The Nature of Knowing: Rachel Carson and the American Environment." *Signs* 12:740–760.

Parker, Lesley H., Lonie J. Rennie, and Barry J. Fraser. 1996. *Gender, Science, and Mathematics: Shortening the Shadow.* Boston: Kluwer Academic Publishers.

Pattatucci, Angela M., ed. 1998. *Women in Science: Meeting Career Challenges.* London: Sage Publications.

Rose, Hilary. 1983. "Hand, Brain, and Heart: A Feminist Epistemology for the Natural Sciences." *Signs* 9:73–90.

———. 1986. "Beyond Masculinist Realities: A Feminist Epistemology for the Sciences. " Pp. 57–76. in Ruth Bleier (ed.), *Feminist Approaches to Science.* Elmsford, NY: Pergamon.

———. 1988. "Dreaming the Future." *Hypatia* 3:119–137.

———. 1994. *Love, Power, and Knowledge: Towards a Feminist Transformation of the Sciences.* Bloomington: Indiana University Press.

Rosser, Sue, ed. 1995. *Teaching the Majority: Breaking the Gender Barrier in Science, Mathematics, and Engineering.* New York and London: Teachers College Press.

Rosser, Sue V. 1985. "Integrating the Feminist Perspective into Courses in Introductory Biology." Pp. 258–276 in Marilyn R. Schuster and Susan R. Van Dyne (eds.), *Women's Place in the Academy: Transforming the Liberal Arts Curriculum.* Totawa, NJ: Rowman and Allanheld.

———. 1986. *Teaching Science and Health from a Feminist Perspective: A Practical Guide.* New York: Pergamon Press.

———. 1987. "Feminist Scholarship in the Sciences: Where Are We Now and When Can We Expect a Theoretical Breakthrough?" *Hypatia: A Journal of Feminist Philosophy* 2:5–17.

———. 1988. "Good Science: Can It Ever Be Gender Free?" *Women's Studies International Forum* 11:13–19.

———. 1990. *Female-Friendly Science: Applying Women's Studies Methods and Theories to Attract Students.* Elmsford, NY: Pergamon.

———. 1992a. *Biology and Feminism: A Dynamic Interaction.* New York: Twayne Publishers.

———. 1992b. "Are There Feminist Methodologies Appropriate for the Natural Sciences and Do They Make a Difference?" *Women's Studies International Forum* 15:535–550.

Rothschild, Joan. 1988. *Teaching Technology from a Feminist Perspective: A Practical Guide.* New York: Pergamon Press.

Samuels, Linda S. 1999. *Girls Can Succeed in Science!: Antidotes for Science Phobia in Boys and Girls.* Thousand Oaks, CA: Corwin Press.

Schiebinger, Londa. 1999. *Has Feminism Changed Science?* Cambridge, MA: Harvard University Press.

———. 2002. "European Women in Science." *Science in Context* 15, 4:473–481.

Shepherd, Linda Jean. 1993. *Lifting the Veil: The Feminine Face of Science.* Boston: Shambhala.

Shiva, Vandana. 1996. "Science, Nature, and Gender." Pp. 264–285 in Ann Garry and Marilyn Pearsall (eds.), *Women, Knowledge, and Reality.* New York: Routledge.

Stolte-Heiskanen, Veronica, Feride Acar, Nora Ananieva, and Dorothea Gaudart, eds. in collaboration with Ruza Furst-Dilic. 1991. *Women in Science: Token Women or Gender Equality.* Oxford: Berg.

Women in Engineering Programs and Advocates Network. http://www.wepan.org.

Women in Science and Engineering. http://www.engr.washington.edu/wise/.

Women in Science and Technology, Equal Opportunity Act of 1980. 1980. Washington, DC: U.S. Government Printing Office.

Index

About the Author

Suzanne Le-May Sheffield is an assistant professor at Dalhousie University, Halifax, Nova Scotia, and is the author of *Revealing New Worlds: Three Victorian Women Naturalists*.